Crop and Forage Production Using Saline Waters

ABOUT THE CENTRE

The Centre for Science and Technology of Non-Aligned and Other Developing Countries (NAM S&T Centre) is an inter-governmental organisation with a membership of 40 countries spread over Asia, Africa, Middle East and Latin America. Besides this, 11 research and academic institutions of Bolivia, Botswana, Brazil, India, Nigeria and Turkey are presently the members of the S&T-industry Network of the Centre. The Centre was set up in 1989 to promote South-South cooperation through mutually beneficial partnerships among scientists and technologists and scientific organisations from developing countries. The Centre implements a variety of programmes including workshops and meetings, symposiums, training courses and collaborative projects and brings out scientific publications. These activities provide opportunity for scientist-to-scientist contact and interaction; familiarising the participants on the latest developments and techniques in the subject areas, identification of training requirements and expert assistance, identification of technologies for transfer between member countries etc. The Centre has so far published 32 publications and has organised 65 workshops and training programmes. It is operating 2 fellowship schemes, namely, NAM S&T Centre Research Fellowship and Joint NAM S&T Centre - ICCBS Karachi Fellowship.

For further details, please visit www.namstct.org or write to the Director, NAM S&T Centre, Core 6A, 2nd Floor, India Habitat Centre, Lodhi Road, New Delhi-110003, India (Phone: +91-11-24645134/ 24644974; Fax: +91-11-24644973; E-mail: namstct@gmail.com; namstct@bol.net.in.

Foreword

During the years 2005/2006 the Centre for Science and Technology of the Non Aligned and Other Developing Countries (NAM S&T Centre) decided to organize a seminar in the University of Birjand, Centre of South Khorasan province, Iran, with the title "International Workshop on Crop and Forage Production Using Saline Waters in Dry Areas". The Seminar took place during May 7–10, 2006 under the direction of Dr. M. R. Miri, the president of the University. I was invited to that seminar as former President of the International Society for Halophyte Utilization which had just been founded by a group of scientists analyzing the possibilities to irrigate halophytes with unconventional waters on waste land in countries around the Mediterranean Sea.

The group had started their work in 1996 as a Concerted Action Group of the European Union. The results of their work had been published in many books and journal articles and were finally demonstrated in an internet home page under the address USF. Uni-osnabrueck/EXPO2000. The result of this demonstration showed me how superior the information flows through the internet as through the commonly used media. How effective the internet presentation, was shown to me at the presentation of one of the Seminar participants, who followed my first trials in the UAE deserts and planted mangroves in the desert near the Persian Gulf as well, and that obviously with great success.

The attendees of the symposium in Birjand were mainly concerned with the conditions of inland desert as they are present around Birjand and are probably among the worst kinds world wide. Names like "Kavier–E- Lout"and "Dasht–E-Kavir" are notorious for each Iranian scientist as extreme deserts with saline groundwater where nomadic populations try to use the land for animal grazing. The improvement of their lives was obviously the intent of the government when they decided to irrigate this land and try to convert it to a CO_2 fixing region in order to contribute to the climate stabilization.

The irrigation with any water will eventually render the landscape saline and make the use of halophytes mandatory. The question is only to balance the irrigation with the uptake of salt by the halophytes and which species to take among the halophytes and how to use the crop were all questions for research and for the discussion during the seminar. It was very fortunate for Prof. Muhammad Kafi to get Prof. Ajmal Khan from the neighbouring Pakistan to co-organize the Seminar. He had the first great project initiated with a private investor to solve the main problem to make the use of halophytes for agriculture economically feasible. The seminar in Birjand was in many respects a start for new efforts in many countries to search for living and production systems which allow us to populate the deserts and greenify them for the reductions of CO_2 in the air.

All this I had proposed for research and development since many years. The seminar was for me a sign that the young generation had understood the task that many of us will live in the future in saline systems. The little desert village we saw during the excursion appeared to us like a model for the future.

One other activity started to my surprise during the seminar, the yahoo group forum "halopytes" by Dr. Jami al Ahmady, which became an instant success. The forum is well attended by scientists from all countries regardless of their political or religious or social environment. I hope this will develop further and guide us to future improvements of using saline production systems.

An internet forum is good for operational work but for the archives it is needed to present the results in printed form. It is fortunate therefore, that the two well known scientist, Prof. Kafi and Khan have taken the charge to present the results of the symposium in a form that will have a great impact on the scientists in our field as well as for the students in the field. I wish the book a good reception among the academic community. The years from 2000 onward will show us great changes in the attitude towards sustainable saline irrigation. This seminar has contributed much to it.

The number of meetings dealing with saline systems and the attitude towards the salinity in the presentations shows clearly that the deserts of the world are now seen as opportunity rather than waste. The titles and topics of the meetings show that clearly. From Genomics and Proteomics research to new landscaping technologies let us believe that the human population will eventually solve the problem of how to use the deserts in a way that it will allow the human population to live peacefully with nature.

Duerscheid March 2007

Helmut Lieth
Prof. em. University of Osnabrueck
Prof. h.c. IWP Tashkent,
Uzbek Academy of Sciences

Preface

The acute scarcity of water and increasing salinization in the semi-arid and arid regions of the world are posing major challenge for the planners to provide food for the rapidly increasing population. In addition, inappropriate agricultural practices have caused increase in soil salinity which has rendered millions of acres of prime land unfit for conventional agriculture. Underground brackish water and seawater are abundantly available in the semi-arid and arid regions of the world along with the vast stretches of saline land. Saline water and saline land have been considered as waste in the past, however, civilizations are under increasing pressure to utilize these resources for food, feed, fiber, fodder, turf, CO_2 sequestrations, bio-fuels, medicines, oil-seeds, ornamentals etc. The researches conducted during the last few decades indicate that more than 3000 salt tolerant plants are reported in the world and their number has been increasing with each new study reported. It is also heartening to note that most of these plants could be utilized as cash crop using novel ideas and technologies. Among all uses mentioned above few of them could be put to use with little effort and that includes fodder production. In the several countries, grasses and halophytes have been successfully grown under highly saline areas with either brackish or seawater irrigation to provide fodder. However, there is still need to develop technology of saline agriculture suited to the different ecosystems of the world or a major efforts are needed to domesticate them using modern technology.

Salt tolerant plants appear to be a major resource for the survival of civilization particularly in drier regions of the world. Therefore there is a need to understand their ecology, physiology, biochemistry, nutritive value, medicinal usages etc. It is important that we preserve these unusual plants and their habitats, not just for their aesthetic beauty, balancing the ecosystem, combat greenhouse effect but also as a resource for the development of new salt tolerant and halophyte crop of economic importance. Given this scenario and the effort of one of us (Muhammad Kafi) as a crop stress physiologist and in the position of science counselor of I. R. Iran for the Subcontinent resulted in some discussion with Professor Arun P. Kulshreshtha, Director of NAM S&T Centre and they came to decision to hold the "International Workshop on Crop and Forage Production Using Saline Waters in Dry Areas" which was a remarkable effort in this direction. The workshop was co-organized by the Centre for Science and Technology of the Non-Aligned and Other Developing Countries (NAM S&T Centre) and University of Birjand during May 7–10, 2006 at the University of Birjand, South Khorasan, and Iran. The problems and the experience of developing countries could be more applicable for each other, and this the main reason for arranging this type of workshops by NAM S&T Centre.

More than 90 per cent of Iran area is arid or semi-arid with two big desert in the central part in the name of "Kavir-e- Lout" and "Dasht-e-Kavir". Most of the water resources around these two big deserts are saline; so that some of them are high saline that is not suitable to apply for even salt tolerant crops. Khorasan is a symbol of desert areas with high saline water resources, which during last 40 years, Iranian government put tremendous efforts to fix sand dunes and control the progress of desertification, therefore there are some experiences which could be useful for other regions as well, at the same time, research of crop and fodder production could be applicable for other countries.

The Birjand meeting was participated by 55 scientists from 13 countries of the world including Afghanistan, Benin, Cuba, Germany, India, Indonesia, Kuwait, Myanmar, Pakistan, Sri Lanka, Qatar, Uganda and a significant number of the participants from various organizations of Iran. The papers presented in this meeting could be grouped into three categories: 1. Country papers, 2. Halophyte utilization, 3. Salt tolerance and ecosystem. Country papers of Benin, Cuba, India, Iran, Indonesia, Kuwait, Myanmar, Pakistan, Qatar, Sudan, Sri Lanka, Turkey and Uganda is included in this presentation. The authors presented an overview of the salinity problems in their respective countries and the measure and remedies being utilized to overcome this menace. It seems that salinity problems are more severe in the arid regions of the world including India, Pakistan, Iran, Kuwait, Qatar and Uganda whereas more moist regions like Indonesia, Sri Lanka, Myanmar and Cuba the extent of damage caused by salinity is minimal. Halophyte utilization was the second major group where about eight papers presented. Lieth, Tavakoli *et al.*, Ansari *et al.*, and Dehghani *et al.*, presented a global view of the halophyte utilization and possibility and the potential of utilization halophytes as cash-crop. Rastegari *et al.*, have demonstrated that how nuclear techniques could be best utilized to increase crop production in saline soils. Jami-Al-Ahmadi and Kafi has successfully argued the case of *Kochia scoparia* to be utilized as fodder plants in the most arid and saline regions of the world. Similarly Hamid and Ashraf very eloquently presented the case of *Panicum antidotale* a potential fodder grass for the saline soils of Pakistan. Riasi and Mesgaran have evaluated the chemical composition and digestibility parameters of the local halophyte. In order for us to successfully utilize the halophytes as animal fodder nutritive value and digestibility of all halophytes of the regions must be determined to help selecting potential candidates for the fodder. A third group of papers included the discussion mechanism and extent of salt tolerance in halophytes. It is imperative that the mechanism of salt tolerance of halophytes studied in detail so that we could use them in the appropriate conditions and add specific amendments to make them successful in the varying saline arid habitats. Askari and Kafi dealt with the molecular arguments of salt tolerance while Schleif presented water uptake mechanisms during salt stress. Presentations of Saberi *et al.*, and Ranjbar *et al.*, explained salt tolerance of local wheat varieties and different stages of the life cycle. Karimi *et al.*, showed that how nitrogen and sulfur fertilization could improve halophyte biomass production under saline conditions. Professor Lieth presented paper deal with the energy balance of ecosystem and he forcefully argued that the balance of nature primarily depend on this energy exchange during transpiration.

We wish to acknowledge the support provided by Professor Kulshreshtha, Director of NAM S&T Centre, and Governor of South Khorasan Province. Dr. M. R. Aghaebrahimi the ex president and Dr. M. R. Miri, the present president of the University, must be commended for the excellent arrangement made before and during the workshop. We appreciated the support and co-operation of Birjand University administration team under the leadership of Dr. Gh. R. Zamani and Dr. S. H. Hashemi in providing services in connection with the symposium. We especially want to express gratitude to Dr. Gh. R. Zamani, M. Jami-Al-Ahmadi, Dr. Farhang Far, Dr Behdani, Mrs. Ghollasi and Dr. Riasi who was involved with the planning and coordinating of the symposium. Thanks are also due to Dr. Raziuddin Ansari for critically reading the manuscripts and for useful suggestions to improve the quality. We would also like to thank Mr. Zaheer Ahmed and Mr. Irfan-ud-din for their invaluable contribution in editing and making it camera ready for the publication.

M. Kafi

April 20th, 2007 ***Mohammad Kafi***

M. Ajmal Khan

M. Ajmal Khan

Introduction

As we all know, the saline water contains dissolved substances that are essentially the salts in the form of chlorides, sulphates, carbonates and bicarbonates of calcium, magnesium, sodium, and potassium. Salinity can improve the soil structure but at the same time, it can also have adverse effect on the plant growth and crop yields. *'Sodicity'* is a term related to the salinity that specifically refers to the amount of sodium present in irrigation water. Irrigating with highly saline and sodic water can unfavourably impact the soil structure and cause problems in irrigation, depending on the type and amount of salts present, soil type being irrigated, specific plant species and growth stage, and the amount of water able to pass through the root zone. However, the salts of magnesium and calcium being smaller tend to cluster closer to clay particles and have a positive attitude towards the soil keeping it flocculated with fine particles binding together into aggregates.

Salinity becomes a problem when enough salts accumulate in the root zone, thereby hindering the plant roots from withdrawing water from surrounding soil and lowering the amount of water available to the plant, regardless of the amount of water actually in the root zone. In fact, the presence of salt in the water causes the plants to exert more energy in extracting water from the soil and as a consequence leaving the plant with less energy available for growth. Generally, the most likely effect of salinity on plants is stunted growth. Leaves may be darker green, thicker and more succulent than non-stressed plants. Visual symptoms (leaf burn, necrosis and defoliation) sometimes occur particularly in woody species. At high levels, salinity can cause physical damage and mortality. While most crops are relatively tolerant to salinity during germination, young developing seedlings are particularly vulnerable to salinity damage during emergence and early development. The delay of germination may prove fatal if the salt-stressed seedlings encounter additional stresses, such as water stress, extreme temperature fluctuations and/or surface crusting. However, the plants generally become increasingly tolerant of salinity in later growth stages, once they are developed.

So how do we manage the saline soil? Some of the available options are: (a) to grow field crops such as sugar beet, safflower, wheat, barley and cotton that are moderate to highly tolerant of salinity; (b) to use sloping beds where seeds are safely planted on the slope below the zone of high salt accumulation; (c) to employ a variety of physical, chemical and biological soil management measures such as tillage, deep ploughing, sanding the surface, chemical amendments, using organic and green manure, mulching etc.; and (d) to use appropriate water delivery and distribution systems, particularly avoiding over-irrigation. With a view to upgrade the knowledge and skills by making the relevant information easily accessible to researchers, agriculturists, extension workers, planners and policy makers to utilise it profitably in the diagnosis, improvement and exploitation of saline water irrigated agriculture, the Centre for Science and Technology of the Non-Aligned and Other Developing Countries (NAM S&T Centre) organised an international workshop on 'Crop and Forage Production Using Saline Waters in Dry Areas' during 7-10 May, 2006 at Birjand in the Islamic Republic of Iran jointly with the Birjand University. As the national capacity and strengths of the NAM and other developing countries differ from one country to another, the exchanges between the participants from various countries during the event well complemented each other.

Fifty two scientists and senior experts from 11 countries, which included Afghanistan [Mrs. Maliheh Ghafoori of the Ministry of Agriculture], Cuba [Dr. Aurelio Álvarez Menéndez, Director, Pastures and Forages Research Institute, Ministry of Agriculture], Germany [Prof. Helmut Lieth, International Ecological Projects, Institute of Environmental Systems Research, University of Osnabruck], India [Prof. Rakesh Kumar Trivedy, Head, Dept. of Environmental Science, University of Pune and Dr. Balram Sharma, Ex-Head, Division of Genetics, Indian Agricultural Research Institute, Delhi], Indonesia [Prof. Benyamin Lakitan, Minister's Advisor for Food Affairs, Sriwijya University], Kuwait [Mr. Faisal Yousif Ali-Sadeeki, Director, Research and Nurseries Administration and Mr. Faisal H. Aseery, Experiment and Laboratory Division, Abdaly Agriculture Administration], Myanmar [Ms. Lae Yin Win, Assistant Lecturer, Biotechnological Department, Government Technological College, Yangon], Pakistan [Prof. Mohammad Ajmal Khan, Department of Botany, University of Karachi], Qatar [Prof. Yassin Ibrahim, Department of Biological Sciences, College of Arts and Sciences, Doha and Mr. A. Salam Al-Harbi, Head, Research Station for Crops and Wild Plants, Dept. of Agriculture and Water Research, Doha], Sri Lanka [Dr. Amarasinghe Arachchige Yasarathna Amarasinghe, Dean, Faculty of Agricultural Science, Sabaragamuwa University] and the host country Iran (38 participants) attended the workshop.

Prof. H. Lieth, Prof. Ajmal Khan, Dr. Balram Sharma, Dr. Hossain Tavakoli (Head, Agriculture and Natural Resources Research Centre, Khorasan) and Dr. S. A. M. Cheraghi (Director, National Salinity Research Centre) were the principal key speakers during the workshop. Most of the foreign participants presented their respective country status reports; the report on Iran was presented by Prof. Muhammad Kafi; and Dr. Seyed Jalal Rastegari (Head of Salinity Research at the Nuclear Research Centre for Agriculture and Medicine of the Atomic Research Organisation), Dr. M.S. Saberi (Agriculture and Natural Resources Research Centre, Khorasan), Dr. Ahmad Riasi, Dr. Karimi and Dr. Rangbar (all from the Iranian research centres and universities) were the Iranian participants, who presented scientific papers on the subject.

The workshop was organised in seven technical sessions that were chaired by, respectively, Dr. Balram Sharma / Dr. Hossain Tavakoli; Prof. Ajmal Khan / Prof. Muhammad Kafi; Prof. Yassin Ibrahim / Dr. Gholam Reza Zamani (Birjand University); Prof. Benyamin Lakitan / Dr. Homayoun Farhangfar (Veterinary Department, Birjand University); Prof. Dr. Helmut Lieth / Dr. Majid Jami Al-ahmadi (Agronomy and Plant Breeding Department, Birjand University); Prof. Arun Kulshreshtha / Dr. Mohammad Ali Behdani (Agronomy Department, Birjand University); and Dr. A. A. Y. Amarasinghe / Dr. Ebrahim Rowghani (Head, Animal Sciences Department, Shiraz University).

The concluding session was devoted to consolidating on various issues concerning the topic of the event and was held in the form of a panel discussion with Prof. H. Lieth, Prof. Ajmal Khan, Dr. Balram Sharma, Prof. B. Lakitan, Dr. A. Álvarez Menéndez, Prof. M. Kafi and Dr. A. A. Y. Amarasinghe being the panellists.

After the meeting a halophyte group was created at the University of Birjand (http://tech.groups.yahoo.com/group/halophytes/) by Dr. Majid Jami Al-Ahmadi to coordinate the research activity of halophyte biologists from both developing and developed countries. This group was also adopted by the International Society for Halophyte Utilization (ISHU) for the patronage and support. About 50 scientists all over the world have become its member and the membership is growing rapidly. The contribution of the NAM S&T Centre and Birjand University is already making the difference and it is hoped that this coordination would lead to enhanced collaboration among member countries.

A high spot of the workshop was the Birjand Declaration, a copy of which is placed at the following page.

I express sincere gratitude to Prof. Mohammed Aghaebrahimi, former President of Birjand University, who laid the foundation of this event, and to Dr. Mohamad Reza Miri, Asst. Professor and Acting President of Birjand University for successfully organising the Workshop jointly with the NAM S&T Centre. I have special words of appreciation to the colleagues at the Birjand University, especially Dr. Majid Jami Al-Ahmadi, Mrs. Sholeh Ghollassi, Dr. Gholam Reza Zamani who had been the real force behind the successful management of this event at the Iranian end. I am grateful to Prof. Muhammad Kafi, who was heading the Education and Science Wing in the Embassy of the Islamic Republic of Iran in New Delhi at the time of the Workshop and is presently teaching at Ferdowsi University of Iran and Prof. Dr. M. Ajmal Khan from the University of Karachi, Pakistan, who are the eminent experts in the related fields, for their contribution in jointly editing this. I acknowledge with thanks the kind words from Prof. Helmut Leith, University of Osnabrueck, Germany about this publication and suggesting the future course of action. Last, but not the least, I would like to appreciate the valuable services provided by the entire team of the NAM S&T Centre, particularly by Mr. M. Bandyopadhyay, Mr. Gaurav Gaur and Mr. Pankaj Buttan, who coordinated the work on the organization of the Workshop and subsequently compiled the material for bringing-out the publication.

I am confident that the material presented in this volume will serve as a valuable reference material for the scientists, professionals and policy makers engaged in the schemes and programmes for the development of the agriculture sector in the dry and saline regions.

Arun P. Kulshreshtha

Director, NAM S&T Centre

Birjand Declaration

The rapid population growth in the less developed countries of arid and semiarid regions and concomitant decline in productivity of agricultural lands due to shortage of good-quality irrigation water and increasing soil salinity are exerting enormous pressure on the dwindling supplies for human consumption. Equally or even more affected in some cases are other resources like fodder for animals and fuel wood for the rural poor.

False irrigation practices have caused water logging to prime agricultural lands and extensive utilization of underground resources in dry areas has caused the substantial depletion of underground water resources. This has created serious scarcity of fodder and water for other agricultural usages. It is becoming increasingly difficult to grow any conventional crop in areas where they were planted before and therefore it is necessary to look for alternate solutions.

To share ideas and experiences of scientist from various countries an "International Workshop on Crop and Forage Production Using Saline Waters in Dry Areas" was organized on May 7-10, 2006 at the University of Birjand, Iran with the collaboration of the Centre for Science and Technology of the Non-Aligned and Other Developing Countries. More than 50 scientists from 11 countries have participated in this workshop and they agree on the following:

1. Salinization of soil and water resources and management is of great importance. Therefore governments should consider it as a serious challenge.
2. Comprehensive plans for sustainable use of saline waters and halophytes are needed for each region.

3. The conventional water resources and crops do not meet the requirements of human society in dry and saline areas. The new source of water (seawater and/or brackish) and new crop (salt tolerant plants) should be considered for research.
4. This is a global problem therefore there is a need o create national authorities and international funds and collaboration for the management of saline lands.
5. Holding workshops and conferences for exchange of experiences and ideas is highly recommended.
6. The Iran already has a strong human base of the research on saline agriculture. However, coordination among various government agencies and academia is not sufficient. Therefore an aggressive plan may be made to increase collaboration and planning for saline agriculture.
7. International collaboration may be encouraged and financed at the government's levels to make valuable contributions both locally and globally.
8. It is also suggested that the NAM S&T Centre may play a central role in organizing bio-saline agriculture development in the member countries.
9. It is suggested that the different scientific disciplines should cooperate in the solution of sustainable and economically feasible management of saline systems.

Contents

1

Sustainable Utilization of Halophytes and the Significance of that Concept for Future Generations

Helmut Lieth

Environmental Systems Research, University of Osnabrueck, Germany
E-mail: helmum.lieth@t-online.de, helmut.leith@usf.uni-osnabrveck.de

ABSTRACT

The University of Birjand is the centre for the project to implement the use of halophytes for the greenification of the semi-deserts in South Khorasan/Iran. Irrigation water, conducted to the area from a distance is used together with saline groundwater. The present paper provides an overview of the use of halophytes in the context of the local conditions. Some general background remarks of the energy flow in ecosystems are followed by the evaluation of the use of halophytes for the CO_2-fixation from the air. The sustainability of the project is discussed with regards to the soil condition, the ground water pollution and the changes of local environmental conditions. Some possible problems for the sustainability of the production system in inland deserts are discussed; some additional research topics are suggested.

Introduction

Background Information: The Vision of An Ecologist

The problems of population growth, degradation of the environment and the need for the increase of biological production are widely discussed. Our own contribution to this topic was the global pattern of ecosystems production.

The plant production pattern on earth is shown in Figure 1.1. It presents the biomass production in terms of g/m^2/annum. From this unit we can calculate the amount of CO_2 taken from the air as well as the amount of energy generated in terms of Joule for the same surface and time units.

In previous papers we have compared the conduction of solar energy on the surface of earth via the various channels like water evaporation, biomass production, air conductivity reflection and others (Lieth, 2005). This comparison has shown that the evaporation of water is the prime conductor of energy from earth to outer space. Biomass production can be seen as temporary sink for energy. This creates with little

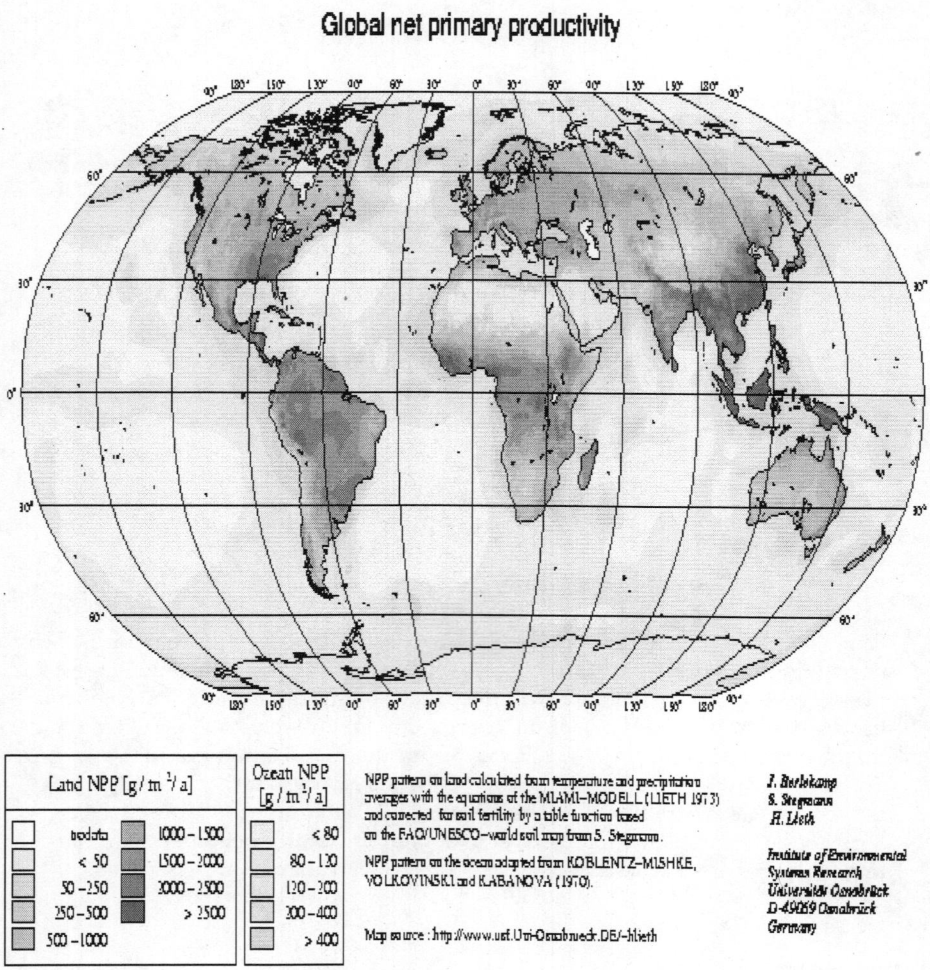

Figure 1.1: This Map is Based on a Model which Predicts the Net Primary Productivity from the Environmental Parameters, Temperature, Precipitation, Length of Growing Season, Soil Fertility and Other Parameters. It shows clearly that the productive power of plants is controlled by environmental forces.

energy input surface structures which enhance the best water evaporation when it is available in unlimited amounts. Under conditions of water shortage develops the ecosystem other structures which may enhance the energy dissipation. From these facts we can postulate, that the biological evolution of ecosystems aims at the maximum possibility of energy flow. The "reward" for the ecosystem is the fact, that maximisation of energy flow achieves the best possible homeostasis of the ecosystem.

The map demonstrates the regions with limitations for the productive powers on land as well as in the ocean. Central Iran is shown as an area with low plant production. In such regions, present patterns of drought alternating with salinities of different severity. We do have a large number of plants available which can tolerate different levels of salinity in soil water which makes it possible to improve the bio-productivity in this region (Lieth and Menzel, 2002).

The Paradigm of Soil Salinity in Agriculture

In common agriculture salinity in soil water is considered as one of the most serious problems for crop production. The reason for this is the fact that agriculture developed the majority of crops from plants of regions with enough rainfall. This stimulated the natural evolution of plants to develop species adapted to low mineral content in soil water.

From these plants the human population selected those which provided food for them and fodder for their animals. For these plants salinity is a major problem, just like excessive drought. From this point of view salinity is destructive for agriculture, a paradigm which is upheld by the human population to this day. We undertake analyses since 30 years to modify that. We think that modern science has means to generate production systems on land irrigating it with waters of ocean level salinity.

The New Paradigm: Cash Crop Halophytes

Within a large number of species in the entire plant kingdom which tolerate salinity we find those which grow in soil and waters with salinities around 3–5 per cent. A large number of species tolerate lower soil salinities at different levels and a few species which tolerate extreme levels of salinity. Many of these species even require salinity at different concentrations and for these species the lack of NaCl in the water appears to be stress in the same way as it is the excess level of salinity. This needs to be tested for all species, because we find for several freshwater species some salinity tolerance although they were classified as glycophytes (Lieth, 1964).

We need to develop different concepts for plant production under such salinities. Our textbooks for agricultural production do not provide presently such information. This is needed because a large number of potential cash crops are among these species. We need to develop for these species special production systems in order to cultivate them in desert regions which are otherwise unproductive.

Definition of a Cash Crop Halophyte

In agriculture we understand as cash crop a plant species which can be sold on the market after its cultivation. When glycophytes are used as cash crops we must have enough rain to provide water in the substrate or we must irrigate salt free water to the production area.

When fresh water is not available one can use saline water for irrigation if the cultivated plants are salinity tolerant. These plants are called halophytes. The salinity tolerance can reach different levels. It is interesting to see that rather elevated salinities are tolerated by certain species if sufficient water is available. We distinguish three main groups of halophytes tolerating low levels up to 0.8 per cent salinity, medium up to 3 per cent and levels above 3 per cent. More details are compiled in Lieth and Mochtchenko (2002).

Halophytic cash crops may have applications which may differ slightly from freshwater cash crops. Table 1.1 shows the different types of uses with which one may be able to achieve financial gain from halophytic crops, up to the point that we can call these plants cash crops. Under certain environmental conditions may we use halophytes as catch crops, when these plants are used to reduce the salinity in the soil water to the point that a common glycophytic crop can be grown after it.

Table 1.1: Possibilities for Utilisation of Halophytes Already Existing and Possibilities under Investigation

1.	**Food**	Starch, Protein, Fat, Vitamins
2.	**Feed**	Starch, Protein, Minerals
3.	**Wood**	Firewood, Building, Furniture, Crates
4.	**Chemicals**	Industrial Chemicals, Pharmaceuticals, Plastics
5.	**Landscaping**	Roadsides, Housing areas, Turf
6.	**Ornamental**	Potting plants, Gardening
7.	**CO_2^- sequestration**	Desert Greenification, Aforestation
8.	**Tertiary treatment**	Water, Soil
9.	**Industrial raw material**	Fiber, Biomass, Ash
10.	**Unconventional irrigation water**	Seawater, Brackishwater, Wastewater
11.	**Environmental protection**	Coastlines, Dune fixation, Erosion prevention
12.	**Wildlife support**	Species diversity, Maintenance

The entries may somewhat stretch the use of the term cash crops, but the new developments of the global population is expected to require plant production systems for new purposes. These new systems will be undertaken against cash payment; hence it appears logical to call the halophytic plants to be used as cash crops. This means that agriculture will be expanded beyond food production to provide industrial raw material in unprecedented quantities.

Various Types of Salinity

Salinity is a common term for the concentration of salts dissolved in water. For the purpose of plant production we must distinguish between salinity (NaCl), sodicity (Na_2SO_4) and alkalinity ($NaHCO_3$). The definition of these terms is used in soil science since a long time by the USDA. The explanation created some confusion among agronomists. In certain conditions, when they tried to use waste water on upland soil for irrigation. While the original distinction remains still valid for the impact on plants on normal soil, it requires careful analysis under conditions where plant production with seawater generates a system comparable to salt marshes. We suggest to consult recent publications about this topic by Van de Graff and Patterson, 2001 in the internet. This paper is mostly concerned predominantly with NaCl salinity in the water.

Additional minerals and compounds in the water or soil can cause problems of various kinds which cannot be covered in this paper. Some aspects are covered in publications by Lieth and Markert (1990) and Markert (1996).

The utilization of halophytes in one of the ways shown in Table 1.1 requires investigations in different sections of science ranging from meteorology and soil science to special management procedures, sales and marketing. Because of the global importance of the new production systems for mankind we need the cooperation from all sections of sciences in order to gain full control over the utilization of desert land for bio-production. The logical sequence of investigations required is shown in Figure 1.2.

A research project following the steps of this flow chart was implemented by the EU concerted action group "Sustainable Utilization of Halophytes" (2000). See publication in www.usf.uni-osnabrueck.de/~h.lieth and Hamdy *et al.* (1997), Hamdy *et al.* (1999), Lieth *et al.* (1999).

We covered the ecological problems with halophytes in several other papers already, see summary in Lieth and Mochtchenko (2002). We want to discuss in this paper some different possibilities for the use of halophytes in wet and dry systems which are important in the climate of Central Iran. Shown in Figure 1.3 is the climate diagram taken from Lieth *et al.* (1999), Climate Diagram World Atlas on CD.

Wet saline systems are naturally occurring in and around the oceans of the world. They form algal food chains for fish and marine animals in the ocean. In coastal areas on land occur salt marshes in regions with lower temperatures and mangroves in warmer climates.

Dry saline systems are found around inland salt lakes and in agricultural production systems using freshwater irrigation. The latter are becoming increasingly more important because the steady increase in soil salinity in time makes them unsuitable for standard cash crops.

In both types of halophytes, hydro- or xerophytes are species available which can be classified as salt includers and excluders. Both types can be useful, but the management of these species must be quite different. More research is required in this respect.

They can reduce soil salinity if they accumulate sufficient salt in their above ground biomass which can be removed from the field. They can be eaten by man or domestic animals if their biomass or some parts of it have a low enough concentration of salinity. They can also be harvested as industrial raw material.

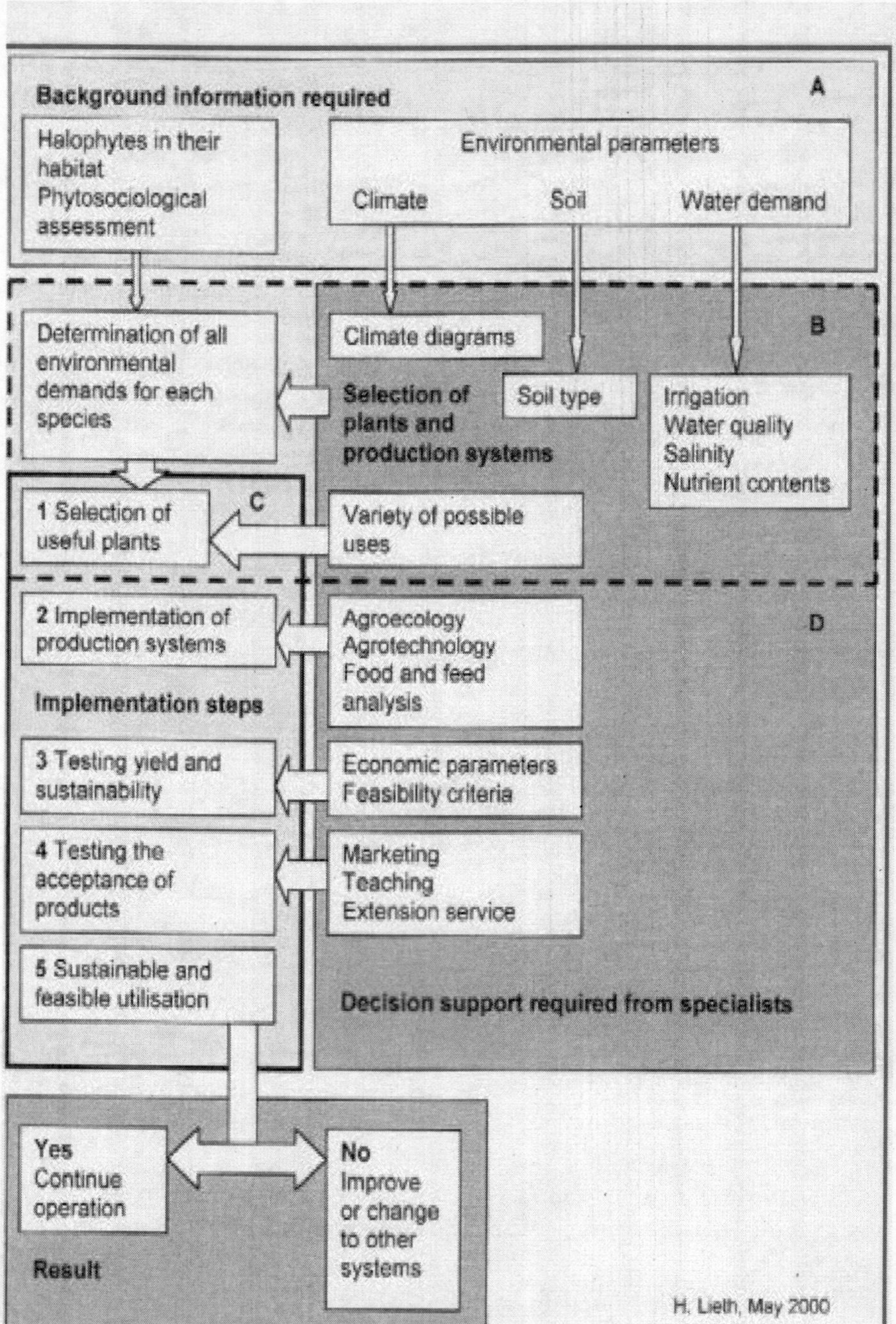

Figure 1.2: Flow Chart showing the Sequence of the Work Steps Required to Reach a Sustainable Saline Production System

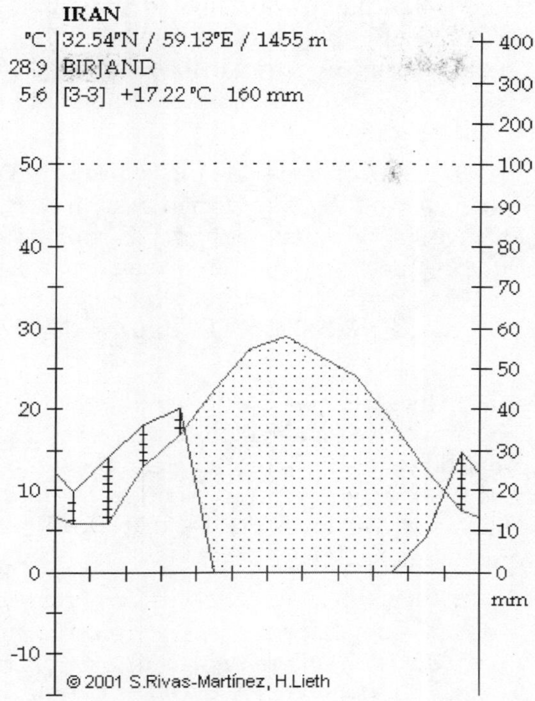

Figure 1.3: Climate Diagram Taken from Lieth *et al.* (1999)

Research Needed for Halophyte Utilization

General Requirements for Growing Halophytes

According to local conditions sustainable utilization of halophytes can be achieved in several ways:

1. We have enough saline water available for irrigation and the possibility to drain the surplus (percolation and surface flow) into the ocean or a salt lake.
2. We have enough rainfall to keep the soil water salinity close above the tolerance level for common crops and halophyte species, which can absorb and store so much salt into their above ground parts, that one can remove the salt with the biomass during harvest.
3. We have locally heavy torrential storms, which bring so much water that the salinity accumulated in the top soil can be washed away by the rivers either towards the ocean or towards a salt lake.

The combined modelling of ecologists with bio meteorologists can support the agronomists and foresters which type of irrigation and which species of halophytes one can select in order to generate a sustainable production system.

The EU concerted action group "Sustainable Utilisation of Halophytes" has investigated various options for the use of halophytes in the geographical regions from Western Europe and Northwest Africa in the west, to Pakistan, India and China in the east. Many halophytic species were tested especially in the Mediterranean region, either alone or in alteration with glycophytic crops.

Some of the crops we saw during the field trip at the Birjand meeting appeared to have similar environmental conditions as we had them in North Africa. That led us to the conclusion, that we should compare shortly the Birjand project concept with the way of thinking, the EU concerted action group had developed. Much of our work was published in the books by Lieth at al. see list in Lieth and Mocthchenko (2003) and the internet- homepage: www.USF.Uni-Osnabrueck. de/~h.lieth, established for the EXPO 2000.

We should start with the statement that the environmental conditions should be known and that they are decisive for the selection of species, plants and animals, but that economic, organisational and social parameters are as important for the proper functioning of rural production systems as the constraints from the natural environment. The flow chart we developed for that is demonstrated in Figure 1.2 (taken from Lieth and Lohmann, 2000).

For some of the blocks shown in Figure 1.2 we can find relevant information in the literature. For soil and climate parameters we recommend the use of locally available meteorological data and soils maps if available. For planning purposes we recommend the relevant publications of FAO, UNESCO and UNEP. For some of the climate data we provided the Climate Diagram World Atlas on CD. Lieth *et al*. (1999). As an example we show in Figure 1.3 the climate diagram of Birjand.

In order to get a slate of suitable species we need to discuss prior to the initiation of new projects with local phytosociologists, which halophyte promise results. They can advise us on the natural salinity tolerance of local plant communities. From these we select the species which may offer either maximum productivity or special chemical compounds which can provide monetary income in a future agro ecosystem, irrigated with saline or other unusual type of irrigation water.

The irrigation system needs special attention. For any kind of long time sustainable system, we need to calculate the evaporation, percolation in the soil column and drainage into the groundwater and from there to the lower situated landscape.

Research Requirements for New Crops

Due to the importance of the need to increase biological production worldwide we should enforce research at all levels. Key problems are the need for more food and fodder and the reduction of climate relevant trace gases in the atmosphere.

As I have demonstrated in Table 1.1 there exist many possibilities to use halophytes for both key problems available. The present interest in biology seems to shift to molecular genetics for the attempt to enhance the salinity tolerance of common freshwater cash crops. While this is certainly one type of research which must be undertaken, there are many other physiological problems to be solved. We do need more insights in the mechanism of salinity tolerance in plants, we do need more information about the impact of saline irrigation on upland soil, and we do need more information about the chemical composition of the biomass we consume and / or utilise for other purposes.

Research is also required for the irrigation of halophytes in coastal deserts. In these environments a large potential for global increase of biomass production exist which can consume large amounts of CO_2 from the air. This is feasible only if we can use the biomass for energy or as industrial raw material. Otherwise its planting is economically impracticable.

As it has been shown in many field trials the type of irrigation and soil surface manipulation can make great differences for the distribution of salts in the top soil. Most trials seek to minimise the amount of water applied. In case of irrigation with brackish water or seawater this may be different. Depending on the species and type of product which a farmer wants to yield, the soil may be kept soaking wet or kept flowing in ditches in which the halophytes grow in the same way as we know it from natural salt marshes. The limiting factor for such production systems may be the energy required for pumping enough water to keep the water temperature and its salinity at proper levels. New crops may require new technologies for growing, storage and marketing. In many cases we may expect that people reject the inclusion of new species into their diet. Considerable effort must be taken to persuade people to use these plants. This requires new efforts and new methods of advertising the purchase of halophytes for food or feed.

Only if we can persuade a sizeable number of farmers to plant the new crops, can we be sure that it will be accepted by the local population.

Remarks and Proposals Concerning the Sustainability of the Birjand Project

General Aspects

To my knowledge the Birjand project is one of the few attempts to greenify inland desert areas. The main water resources are groundwater which is saline already and drinking freshwater is provided by a 20 km pipeline. The major goal seems to be the greenification for the purpose of CO_2-reduction in the atmosphere. This goal appears to be combined with the attempt to provide the rural population with a new task and income, the planting of new forests with trees and shrubs which can store carbon compounds tolerate higher salinity.

The project looks at first sight enormously interesting. The combination of CO_2 harvesting with halophytes for climate stabilisation and the improvement of indigenous rural desert population is a great goal. Similar projects exist to my knowledge only in Pakistan, NE India and China. We recommend that the present team gets contacts with the teams from the other countries. They should also present their project on a homepage to invite comments from other concerned scientists, like the homepage from Hodges, Sato and Lieth.

The project has started with freshwater species which is tolerable in order to teach the rural population how to create a nursery, get experience with modern irrigation technique and to learn how to manage such projects independently. From the viewpoint of salinity tolerance this project is not sustainable for a long time, because the repeating irrigation of the same fields without proper drainage will result in a slow but steady increase in soil salinisation.

I could not discover a plan how the biomass production could be maintained in the long run without a system of using stepwise irrigation water of higher salinity until the salt level reaches the point where it must be collected in an evaporation pond. In such a pond can salt be recovered from the drainage water and stored or used at other plans.

During recent years several experiments were started in this direction and the Birjand project has great possibilities to elaborate such a system with the goal to create a long time sustainable production system. In such a system should more attention be paid to the inclusion of solar energy devices, the creation of desalination units for the production of drinking water for plants and animals and for the initiation of fish and seafood farming in water circulation tanks.

There is much need for large scale production systems. In desert areas in order to achieve both, reduction of climate relevant trace gases in the air and to produce food and drinking water for men and animals.

This will help to improve the life of the rural people in the deserts of Iran.

Observation from the Site Visits

During the meeting we were taken to a very interesting field trip. From this field trip I made notes about the following items:

1. The present water supply for the region.
2. The traditional village with its bathhouse and external water supply.
3. The forestation project nursery.
4. The use of halophytes for greenification on a field.

Based on my experiences from other regions of the world I would make some suggestions:

To Point 1

The water supply for the region we visited, came from the wells of different depth and through a pipeline from a freshwater well about 20 km away.

Both water supplies are not sustainable over a longer time period. From the information of the engineers of the project we heard that the groundwater contained some salinity already. Irrigation with such waters will result in the increase of soil salinity. If no surplus water can drain to a lower basin from where it could be used once more for irrigation of halophytes. After this it could be used to produce drinking water in evaporation basins using solar energy. This technology is available and should be tried in this region.

The same is true to a lower degree with the imported water. The salinity increase on irrigated fields will take place; it will only take a few years longer.

To Point 2

The impact of the availability of fresh water for the village population was demonstrated. The houses still made of abode in the traditional style with dome shaped roofs. In contrast to the village the bathhouse with solar heating device in a more modern technical and practical style is built on the opposite site of the road, but close to the old well. One can only hope that modern science can improve the life of the villagers in a sustainable manner.

To Point 3

The nursery of the project together with some examples of growing trees and shrubs consist mostly of species which can tolerate slightly saline water. This project should be enlarged including salinity tolerant species which have been tested in the respective projects in Pakistan, China, the USA and Central Asia. Many species are compiled in the lists by Menzel and Lieth (2002, 2003).

But we need to warn the local colleagues. All their effort will be in vain if they are unable to create a sustainable level of salinity which in the inland desert condition requires the development of an evaporation pond where the salt can be accumulated. From here it may be harvested and removed for industrial utilisation or be dumped into safe places.

To Point 4

The use of halophytes for the greenification of inland deserts was shown at one place. Under the present conditions it seemed to work, although the productivity per unit area appeared somewhat low. Since I do not know the plan for the experiment I cannot make any recommendation for further actions. We might consider more discussions about this in the future.

As a last comment we may suggest experiments with rain making in certain inland conditions, especially on flat plains close to some mountains it might be possible to generate artificial rain. Depending on the soil salinity it may be possible to use under such conditions cash crops together with halophytes.

References

Berlekamp, J., S. Stegmann and H. Lieth, 2000. Global net primary productivity. www.usf.uni-osnabrueck.de/~h.lieth

Hamdy, A., M. Caldarulo and M. Todorovic. (eds.), 1997. Water management salinity and pollution control towards sustainable irrigation in the Mediterranean region, 127 p. Valenzano / Bari, Italy, CIHEAM / MAI-B.

Hamdy, A., H. Lieth, M. Todorovic and M. Mochtchenko. (eds.), 1999. Halophyte uses in different climates II, 144p., Leiden / NL, Backhuys Publishers.

Lieth, H. 1964. Beziehungen zwischen Bodensalinität und dem Vorkommen von Pflanzenarten in Salzmarschen der Gaspé-Halbinsel, p. 1–18, in Kreeb, K. (ed.), Beiträge zur Phytologie (Festschrift Walter), Stuttgart, Ulmer.

Lieth, H. 2005. External and internal controls of ecosystems. Ecosystems as accelerators of the energy flow from the earth surface, to be published by A. Sommerfeld-Gesellschaft, Leipzig.

Lieth, H. and B. Markert. (eds) 1990. Element concentration cadasters in ecosystems, 448 p., Weinheim, VCH publishers.

Lieth, H., J. Berlekamp, S. Fuest and S. Riediger. 1999. Climate Diagram World Atlas on CD. Leiden/NL, Backhuys Publishers.

Lieth, H., M. Mochtchenko, M. Lohmann, H.W. Koyro and A. Hamdy. (eds.), 1999. Halophyte uses in different climates I, 258 p., Leiden / NL, Backhuys Publishers.

Lieth, H., M. Lohmann, with contributions by M. Gueth and U. Menzel. 2000. Cashcrop halophytes for future halophyte growers, 31 p. and 5 appendages. Osnabrueck, Inst. für Umweltsystemforschung.

Lieth, H. and U. Menzel. 2002. List of Halophytes in use. Appendix in Lieth, H., Mochtchenko, M. (eds.): Halophyte Uses in different climates IV. Leiden / NL, Backhuys Publishers.

Lieth, H. and M. Mochtchenko. 2002. Halophyte Uses in different climates IV, 34 p. Append. Leiden, NL, Backhuys Publishers.

Lieth, H., T. Tadano and A. Khan. (eds.), 2002. Halophyte Utilisation and regional sustainable development of Agriculture (in Chinese), 324 p., ISBN 7-5029-3422-7/S.0402, Shijiazhuang, Inst. of agricultural modernisation.

Lieth, H. and M. Mochtchenko. (eds.) 2003. Cashcrop Halophytes, recent studies. 10 years after the Al Ain meeting. Vol. 38, Tasks for Vegetation Studies. Dordrecht, Springer.

Markert, B. 1996. Instrumental Element and Multi-Element Analysis of Plant Samples, 216 p., translated by Brooks Haderlie,.John Wiley and sons, Chichester, New York.

Taha, F.K., S. Ismail and A. Jardat. (eds.), 2004. Prospects of Saline Agriculture in the Arabian Peninsula, 576p., Amherst MA, Amherst Scientific Publishers.

Van de Graff, R. and R.A. Patterson. 2001. Explaining the mysteries of salinity, sodicity, SAR and ESP in onsite-practice. www./unfaxlabs.com.au, Armidale / NSW, Australia

Wild, A. 1995. Soils and the Environment. German edition translated by Anette Pause, 328 p., Heidelberg, Berlin Oxford, Spectrum Academic Publisher.

2

Potentials of Using Saline Soils and Waters for Forage Production in Dry Regions

H. Tavakoli*, I. Filehkesh, V. Kashki and J. Bashtini

Khorasan Agricultural and Natural Resources Research Centre, Mashad, Iran

**E-mail: tavakoli_res@yahoo.com*

ABSTRACT

Increasing human population and change in his food diets increased the demands for animal products. To meet this demand, more forage and fodder production is needed for feeding livestock. Because non-saline soils or fresh waters are mainly used for more economic and cash crops, there is limitation for development of forage crops in these areas. Therefore, saline soils and waters are potential sources and can be devoted for forage production. Preconditions for success in this approach are: recognizing and selecting suitable plant materials, implementing appropriate salinity management and proper use of forage products. In this respects for example, many native plants which grow in saline soils can be selected and used for plantation development and also releasing salt tolerance species. Establishment of plants in saline soils and utilization of saline waters is very important for expectable stands and/ or sustainable forage production. Preparing suitable seedbed is essential for plant establishment. Seedling transplant gave better results than direct seeding for many species. Irrigation systems that reduce salt accumulation in the root zone are needed. Floodwaters can be accounted as water source for reducing soil salinity and develop plantation of some forage species in saline soils.

Special management is needed for effective use of forage and improving nutritive value of feedstuff. Scheduling the time of using available forage by animals and feed animal as mixed rations have shown relatively good results in improvement of animal nutrition. In this paper the research results and experiences obtained mainly from dry region of Khorasan province have been discussed in details.

Introduction

Rapidly growing demand for livestock products increased due to increasing human population changing his diet regimes. To provide more animal products, more forage and fodder are also needed. Because most of productive soils and fresh underground waters are devoted to conventional crops, expansion of forage cropping into marginal areas such as saline soils or by using more saline waters are viable options in dry regions. In Iran about 25 million hectares of the lands affected by salinity and saline waters are the main sources of available waters in these regions. Therefore these lands and waters can be devoted for halophyte forage production.

Some authors, working on salinity related subjects, such as Ayers and Westcot (1976), Rhoades (1991) and Rhoades *et al*. (1992), Le Houerou (1993) and Squires and Ayoub (1994) believe that using saline soils and waters for crop production requires special care and management, and they have described some principles in this subject. They emphasised the proper method of planting, irrigation practices to control salt deposition in the root zone of plants, reclamation of saline soils and in general adequate understanding about the effect of salts on waters, soils and plants. In this paper some practical experiences and research results which were obtained mainly from dry region of Khorasan province are presented.

Characteristics of Khorasan Province

Khorasan province located in north-east of Islamic Republic of Iran covers about 300,000 square kilometers of the country. Khorasan climatically classified as dry region with mean annual rainfall of 80 to 350 mm in different locations. The extreme maximum and minimum temperature reach +49 and to -20 degrees celcius. The potential evapotranspiration rate varies between 1000 to 3000 mm per year. Available sources of water are mainly underground water and rainfall. Shortage of fresh water and low quality waters are common almost all over the province. In many parts of Khorasan saline soils forced farmers to cultivate special salt tolerant crops such as barley, cotton and sugar beet. There are many large salt desert basins in which floodwaters are drained. Halophytes are the dominant group of plants in all saline areas of Khorasan.

Management of Saline Soils and Saline Waters for Forage Production

Plant selection: Among different methods of saline remediation, cultivation of salt tolerant plants might be the most applicable and economical. Therefore identification of capable native plants with economic benefit for societies that are living in these areas is of importance. Fortunately there is enough genetic variation for salt tolerance that exists at the interspecific (Roger and Noble, 1991), intraspecific (inter-cultivar) and intracultivar levels (Noble and Shannon, 1987). For example, a vast genetic diversity existed between sorghum varieties in the laboratory and field experiment selection in terms of salt tolerance (Ajirlou *et al.*, 2003).

In salt affected regions of Khorasan, some forage crops such as barley, sugar-beet, forage-beet are cultivated by farmers and *Atriplex spp.* are planted for range and desert reclamation. Some other species such as some sorghum cultivars, different types of millets, *Atriplex spp.*, *Kochia spp.* and switchgrass (*Panicum antidotale*) have shown good adaptation and performance in the experiments and pilot projects in this part of Iran. Table 2.1 shows the soil extract electrical conductivity, sodium content, and dominant halophyte species growing in main deserts of Khorasan.

Table 2.1: Identification of Some Salt Tolerant Plant Growing in Natural Soil Habitats in Khorasan

		EC (dSm^{-1})	Na^{+}
Kale shoor Sabzevar			
	Seidletzia rosmarinus–Reaumuria	101.4	880.4
	Halocnemum strobilaceum	341.1	4000.0
	Aeluropus littoralis	197.5	2173.9
	Alhagi persarum	88.9	15826.1
	Seidletzia rosmarinus	107.1	1086.9
	Seidletzia rosmarinus–Reaumuria–Halocnemum strobilaceum	14.8	65.2
Kavir namak			
	Halocnemum strobilaceum	704.2	2775.2
	Reaumuria fruticosa	82.7	1124.3
	Salsola arbuscula	54.9	906.9
	Salsola arbuscula–Artemisia sieberi	117.5	1776.1
	Salsola arbuscula–Seidletzia rosmarinus	72.4	1298.3
	Suaeda fruticosa	665.1	2514.8
Khaf and Dagh petergan			
	Aeleropus litoralis	51.9	815.2
	Salsola arbuscula	71.3	782.6
	Artemisia sieberi-Zygophyllum atriplecoides	16.1	76.1

Natural vegetations of vast deserts of Khorasan are good resources for plant selection for further utilisation. The natural genetic resources of salt desert habitats of Khorasan province was studied by Saghafi *et al.* (2004). In this study different species from high salty areas were recorded. Table 2.1 shows some of those species which grow in salt desert basins. Criteria for forage plant selection on salt-affected lands described by Ayoub and Malcome (1993).

Ecological Requirement for Halophyte Utilisation

Halophyte plant materials selected for cultivation in saline soils or irrigated by saline waters should have the ability to establish and persist under this condition. Understanding some ecological characteristics including germination ability, seedling vigor, regeneration strategy and tolerance to drought, temperature extremes, salinity, competition and herbivory are necessary for extending the vegetation of these plants and maintain the established and existing vegetation. In this case assessing the salt tolerance of plants in different stage of growth, the interactive effect of these plants on environment of cultivated areas, and their nutritive value are very important in terms of management of vegetation in salt affected lands. This information can be collected in laboratory, greenhouse, and field experiments by recording the characteristics of plants in their habitats. The practical implication of some studies in this respects are explained as follow:

Genetic variation observed amongst *Haloxylon spp.* in terms of drought and salinity resistance. For example *Haloxylon persicum* grows over sand dune with low salinity, while *H. aphyllum* is a relatively high salt tolerant plant and grows well in saline soils with heavy texture and high water table.

Atriplex veruciferum (*Halimione veruciferum*) with relatively good palatability grows in soils with high level of salinity (EC > 41 dS m^{-1}). *Puccinellia distans* established with a stands in soils with EC = 36 -54 dS m^{-1} in three locations in Ghorgan region (Hosseini, 1997).

In a study, the relationship between salinity and distribution of *Hammada - salicornica* was considered. Results indicated that this plant resists well in soils with high gypsum and lime, but it is replaced by high contents of sodium (Na) in Kavir Tabas (Table 2.2). The Jemz habitat is located in surround of Kavir Tabas.

Table 2.2: Soil Characteristics in Different Natural Habitats of *Hammada salicornica*

	Soil Parameters			
	EC	*pH*	*Na (meq/lit)*	
Ravar (Kerman)	16.62	8.64	104.3	Loamy sand
Halvan (Tabas)	41.56	8.39	143.4	Sandy loam
Khor and Byabanak (Esfahan)	5.54	8.34	19.5	Loamy sand
Jemz Tabas (1)	69.27	8.72	376.08	Sandy loam
Jemz Tabas (2)	16.6	7.93	434.78	Sandy loam
Khosf–Deyhok	6.7	7.65	19.35	Loamy sand
Khor va Biabanak	4.75	8.41	22.82	Sandy loam
Kavir Tabas	191.49	8.22	15826.08	Loam
Naybandan (1)	4.5	7.4		Loamy sand
Naybandan (2)	1.12	7.75		Loam
Bandar Abbas	3.17	8		Loamy sand

Management of Salt Affected Lands for Forage Production

Bed preparation and planting methods: In arid conditions with salt affected lands, the placement of seed and seedlings is quite critical. It is often a compromise between maximising moisture availability and reducing the toxic effects of soluble salts. So, seeds or plantlet should be placed in a location that reduces the negative effect of salinity on plant establishment and maintenance of planted species.

Plants are generally most sensitive to salinity during seedling stage (Leman and Robinson, 1979; Rhoades, 1990), So proper seedbed preparation is important for satisfactory germination and seedling establishment of halophytes to have good stands and then produce satisfactory forage yields in saline soils or under irrigation by saline waters.

Experiences in dry regions indicated that for establishment of most species under saline and drought conditions, growing seed in nursery and transplanting seedling to salt lands gives better results than direct seeding, however, preparing nurseries, planting and caring seedling until establishment is costly and expensive, but better stand will compensate the extra cost. Presently in many part of Iran including Khorasan, *Atriplex sp.* and *Haloxylon sp.* are being transplanted. We found that establishment of *Hammada salicornica.* and *Salsola rishteri* need transplanting. By transplanting method, plants escape stresses in early stage in nursery and also dryness of later season because it finishes a part of its growth in the nursery.

The distribution of salts in the soil is also influenced by seedbed shape. Salts tend to accumulate to excess levels in top of the furrows under furrow irrigation (Bernstein and Fireman, 1975). Seed placement and surface irrigation strategies that can be used to optimise plant establishment under saline conditions are described by Kruse *et al.* (1990). In Khorasan province seeds are placed in the edge of fault mound of double furrow by farmers. To prevent accumulation of salt around the root zone, planting inside the row can be practiced. In single furrow it is recommended that the seeds are put in the slope of mound rows just below the zone of salt accumulation. In an experiment carried out by Hosseini (1997), placement of seed or seedling on the top and centre of an "M" shapped mound gave good results in terms of plant establishment in all three plantation sites in Ghorgan.

The time of planting is also important. Farmers are growing barley, wheat and canola in autumn on the top of raised furrow. Because of low evaporation during the autumn, winter and early spring and also precipitation season, accumulated salts is low and washed by rainfall. For planting cotton and barley in flood irrigation, the land irrigated first and then seed cultivation is done. In flood irrigation system, the accumulated surface salts leached to the lower layers. Leveling the lands for uniform water distribution is very important.

We replaced the saline soils of planting zone with sand in Sabzevar. This method made it possible to establish *Atriplex canescence* in very high salty soil (EC = 47 dS m^{-1}). However, changing soil of seedbed would not be economic for forage production in the vast areas, but it is feasible for aesthetic improvement, wildlife habitat, soil conservation or a combination of these.

Deep ploughing is needed for breaking hard layer in some saline soils. This layer is often located under 25 cm of soil surface. This is a common practice in desert reclamation for plantation of the lands.

Irrigation Methods

Irrigation of saline soils or with saline waters is different from non saline soils and fresh waters. Any irrigation practices should control salinity and provide sustainable use of lands. Sprinkler irrigation in saline soils can leach the salts, but may cause leaves burning. So, changing methods of irrigation to for example furrow irrigation may prevent this damage. Furrow irrigation is common in dry region because by wetting the seedbed, it prevents soil crusting. Salt accumulation in the raised bed is the problem of this method that can be solved by time of planting and form of furrow as mentioned earlier. In a field with available saline water in Kahnoj of Ferdows, plantlets of *Atriplex* transplanted inside the deep and wide furrow resulted a good vegetation stand and produced forage between 1000 to 3000 kg/ha in different years, with one to two times irrigation per year by saline water. In 2006 available *Atriplex* forage prevented immigration of herder to other regions while in the similar region herders had to immigrate to other regions.

Basin flood irrigation is common in places with small pieces of lands or in farms with many owners. In this method level of soil salinity is controlled better because of leaching the salts, but crusting and aeration problems may occur. Leveling the soil, increasing soil organic matter is required for amendment of these soils under this irrigation method.

In general, increasing irrigation frequency and higher quantity of irrigation water if available in each irrigation period is effective in solving the salinity problems. By basin irrigation, farmers use to irrigate these lands in winter time in which the water is more than plant requirement. Using floodwater for this purpose is also helpful.

Atriplex canescence were transplanted to a saline soils with nearly no vegetation in Borabad of Sabzevar region and individual transplants irrigated with fresh water to about 40 liter per plant for two to three times in the first and the second years. No success was obtained by this method of planting and irrigation. In another experiment floodwater was used as a source of available water in arid region for establishment of *Atriplex canescence*. The electrical conductivity of floodwater was between 8.5–18 dS m^{-1}. *Atriplex* transplanted into wide and deep furrow and irrigated with floodwater two times per year for two years. More than 80 per cent of plants well established. The plantation areas is about 3000 ha and the mean forage production of 6 years has been measured about 600 kg/ha. Presently, vegetation of this area is visited by many scientists and administrators as a pilot project.

It was concluded that the main reason for the failure of previous plant establishment was shortage of water for draining soil salinity. Floodwater in large volume has reduced the soil salinity and provided suitable condition for growth of *Atriplex*. The effect of floodwater in changing soil salinity was evaluated by analysis of the soil samples in the first year of transplanting and after the sixth year. Results are presented in Table 2.3. Although, soil salinity was reduced by floodwater irrigation during the first year however it gradually increased after six years. So, to sustain forage production in such as this site, irrigation with flood water should continue and repeated every two to three years as soon as flood water become available. Also, to use flood water for forage production, the selected plants should have tolerance to sedimentation and aeration.

Table 2.3: Soil Chemical Characteristics Under Floodwater Irrigation (F) and Without Irrigation (W) in the First and After the Sixth Year in Sabzevar Region

Soil Characteristics	*First Year (0–20)*		*Sixth Year (0–20)*	
	W	*F*	*W*	*F*
pH	8.18	7.52	8.18	8.64
EC (dSm^{-1})	97.8	30	97.8	177.2
TDS (g/lit)	62.6	19.2	62.60	113
Na^{+} (meq/lit)	560.2	250	560.2	860
Mg^{+2} (meq/lit)	2.3	3.5	2.3	4.75
Ca^{+2} (meq/lit)	14.4	19.5	14.4	32.3
K^{+} (meq/lit)	1.43	1.1	1.43	1.01
Cations (meq/lit)	578.3	277.1	578.3	878
SO_4^{-2} (meq/lit)	335	376	335	614
HCO^{-3} (meq/lit)	1.12	0.1	1.12	2.28
CO_3^{-2} (meq/lit)	0.22	0.1	0.22	0.02
Cl^{-} (meq/lit)	242	132.1	242	261.6
Ex.Na^{+} (meq/lit)	16.27	17	18	17.3
SAR	200	73.7	200	178
ESP (per cent)	70.9	50.8	70.9	71
KAR	0.31	0.32	0.31	0.4
$CaSO_4$ (per cent)	4	2.1	4	3.35
$CaSO_3$ (per cent)	23.4	5.5	23.4	20.16

Soil Fertility

Improving soil fertility in saline soils is essential for using these soils for forage production. Generally soil organic matter in the soils of dry regions is often low. So, based on our experience we recommend converting most of residual material such as straw by ploughing and adding some amount of nitrogen to the soil. In arid region of Iran most of residual material is often grazed by livestock. Organic matter reduces soil pH, and as a consequence improves water retention and act as buffer for seedling establishment, so, it is much beneficial to add plant residues for plant production in saline soils. In recent years, the cotton residues are chopped and added to the soil.

The experimental results obtained from different research station in this province have shown good benefit for improving of soil fertility by this way. We are recommending using it for improvement of soil organic matter in the rainfed lands. Species such as canola, barley, wheat are suitable for cultivation as green manures.

In salt affected soils there is competition for absorption suitable and harmful ions by plants and it causes negative effect and toxicity in plants. One of the important elements for plants is nitrogen. Practically, experts recommend use of 25 per cent more nitrogen in saline soils relative to non- saline soils. Care must be made if other stresses such as desiccation occur, plants may damage severely in soils with high content of nitrogen. Ammonium nitrate and then ammonium sulfate are better sources of nitrogen fertilizer than urea fertilizer in saline soils. It is recommended to use fertilizer in two to three times instead of one time in saline conditions (Keshavars, 2001). Keshavars and Malakuti (2005) recommended using zinc fertilizers due to increasing tolerance of plants to salinity by decreasing the absorbtion of Na by plants.

Utilisation of Forage Plants

Generally forages which grow in saline condition have lower quality and nutritive value than conventional forage species or forages produce in non saline conditions. Some of halophytes contain high fiber, some low in nitrogen and some accumulate high amount of salt in their cells, they also may contain moderate levels of secondary metabolites that may inhibit digestion rate or reduce palatability (Ayoub and Malcolm, 1993).

To improve feed efficiency and utilisation of forage species provided from saline condition, determination of energy content, protein, minerals and other components (such as oxalate in chenopod family), determination of animal preference of forages is necessary. This kind of data determine the feasibility of halophyte production as forages and fodder and guides the manager to how they should be fed to the animals.

In this case, data of Table 2.4 show differences between chemical composition and voluntary intake of five common annual halophyte species of natural saltlands. In another experiment seven species of perennial *Atriplex* show different chemical composition (Table 2.5) and digestibility (Table 2.6) when fed to the sheep (Table 2.5). The ash contents of both annual and perennial species (Tables 2.4–2.6) are high indicating a low feed value.

Table 2.4: Chemical Composition and Voluntary Intake of Five Annual Dominant Species of Halophytes in Desert Area of Khorasan Province

Species	*DM*	*CP*	*Fat*	*CF*	*Ash*	*VI*
Ghamanthus gamocarpus	92	6.2	5.8	8.6	40.6	267bc
Petrosimonia glauca	89.8	11.6	6.7	10.3	31.0	268bc
Salsola crassa	86.3	8.9	6.8	16.9	38.0	342ab
Halocharis sulphurea	87.7	8.04	4.21	20.5	39.5	362a
Halotis occulta	93.6	9.1	6.7	13.1	40.9	215c

Table 2.5: Chemical Composition of Seven Atriplex Species

	A. canescence	*A. lentiformis*	*A. halimus*	*A. leucoclada*	*A. nummularia*	*A. undolata*	*A. deserticola*
DM	90	93	93	94	89	87	88
OM	85	72	79	67	83	79	76
CP	12	12	13	9	18	12	13
NDF(1)	48	50	52	37	41	42	54
ADF(2)	46	39	49	36	40	35	37
ASH	15	28	21	33	17	21	24
Ca	2	2.4	1.9	2.2	1.9	1.8	2.2
P	0.15	0.12	0.1	1.5	1.5	0.1	1.5

Table 2.6: Forage Digestibility of Different Atriplex Species

Atriplex spp	*DM*	*OM*	*OM*	*NDF*	*ADF*	*DV*
A. deserticola	68.22ab	74.65b	79.23ab	63.01bc	50.69b	57ab
A. undolata	73.55a	81.51a	82.75a	76.47a	58.50ab	65a
A. nummularia	66.46bc	79.37ab	75.29b	64.64b	79.84a	53ab
A. leucoclada	45.31e	47.05d	61.74c	62.25bc	33.08c	54c
A. halimus	73.27a	81.65a	66.57c	54.96c	59.34ab	64a
A. lentiformis	60.04d	64.19c	65.26c	67.57b	48.83b	46bc
A. canescence	62.45dc	66.76c	81.18ab	68.79ab	64.75ab	57ab
SE	1.6878	1.8495	1.8935	2.5913	5.1311	-

For improvement the feed value of feedstuff, *Atriplex canescence* and *Panicum antidotale* were fed in different proportion to the sheep. Both species are accounted as salt and drought tolerant. Chemical composition of these species has been shown in Table 2.7. The results (Table 2.8) show that digestibility of dry matter (DM), organic matter (OM), crude protein (CP), natural detergent fiber (NDF) and acid detergent fiber (ADF) of feedstuffs increased when fed as mixed ration with proportion of 75 per cent Atriplex and 25 per cent Panicum and or even 50:50 of both. So this findings show the possibility of improvement in animal nutrition by proper feeding of salt tolerant plants. It may be possible to cultivate both of these plants in one rangeland of farm and grazed by animals simultaneously.

Table 2.7: Plant Chemical Composition of Atriplex and Panicum

	DM (%)	*E kcal/agr*	*CP (%)*	*NDF (%)*	*ADF (%)*
Atriplex canesence	89	4106	13.72	32.75	16.43
Panicum antidotale	91.5	4416	12.2	62.03	29.17

Table 2.8: Digestibility of *Atriplex canesence* and *Panicum antidotale* by *in vivo* Methods in Baluchi Sheep

A. ca / P.an	D. Value (%)	DE (%)	DADF (%)	DNDF (%)	DCP (%)	DOM(%)	DDM (%)
0:100	39.51a	29.73d	43.76bc	45.37a	63.70b	40.25c	34.39d
25:75	41.14a	39.59c	43.33bc	38.04a	70.76b	46.59b	45.46c
50:50	41.47a	46.92b	48.25b	38.57a	84.97a	50.61ab	52.54b
75:25	42.07a	53.97a	57.21a	38.57a	86.63a	54.47a	57.99a
100:0	34.31b	52.29a	40.39c	32.96a	91.79a	49.10b	56.43a
SE	1.575	1.279	2.265	4.387	2.884	1.487	1.139
P	0.03459	0.0001	0.0011	0.9058	0.0001	0.0001	0.0001

In another experiment which is in progress, *Alhagi spp.* collected and chopped to certain size was fed to goat and sheep as mixed ration in different proportion with lucerne. Results indicated that Alhagi can be a promising fodder species in saltlands and it is useful for feeding animals mixed with higher quality forage species.

The time of using forage species is also important. In a field experiment, *Atriplex canescence* was grazed by livestock in late spring–early summer and late autumn-early winter. Late spring–early summer grazing provide higher animal feed requirement of animals than late autumn - early winter grazing due to availability of annual and some other green and palatable herbs in the field (Table 2.9). Barley cultivated in saline soils, has high palatability and nutritive value in vegetative stage and before maturity as green forage which could be considered as complementary of halophyte grazing.

Table 2.9: The Mean Animal Liveweight (kg) on Atriplex Vegetation in Different Times

	Start of Grazing	First Month	Second Month
		Spring and Summer Grazing	
Sheep	39.52	41.88	42.18
Goat	32.41	33.6	35.23
		Autumn and Winter Grazing	
Sheep	38.21	36.82	35.42
Goat	35.56	35.26	34.70

Conclusion

Salinity and aridity are the main limiting factors of agriculture development in dry region. If there is no more lands or fresh water for forage production, but still there are potentials to produce forage in salt affected lands and by using saline waters and salt tolerant species. In dry region, valuable genetic resources are available for recognizing, selecting and application for forage production in saline soils or irrigated by saline waters. Producing forage and its proper utilisation requires the adaptation of appropriate management strategies, systems and practices. Because of lack of information on forage production in saline conditions, local and international investigations, education and training are needed.

References

Ajirlou, A.F. E. Majidi Heravan and Y. Nakano 2003. Evaluation of forage sorghum varieties for salt tolerance. Seventh International Conference on Development of Dry Lands. Tehran, Iran. p107.

Ayers, R.S. and D.W. Westcot. 1976. Water quality for agriculture. F.A.O. Rome, Paper No 29.

Ayoub, A. T. and C.V. Malcolm. 1993. UNEP Environmental Management Guidelines for Halophytes for livestock, rehabilitation of degraded land and sequestering atmospheric carbon. United Nations Environment Programme. Nairobi, Kenya. 59 p.

Bashtini, J. and H. Tavakoli. 2002. Determination nutritive value of five dominant species of halophyte plants in desert area of Khorasan province. Pajohesh-va-Sazandegi. No 55. p. 1-5.

Bernstein, L. and M. Fireman. 1957. Laboratory studies on salt distribution in furrow-irrigated soil with special reference to the pre-emergence period. Soil Science. 83:249-263.

Hosseini, A. 1997. Autecology of *Puccinellia distance* in Ghorgahan region. Pajohesh-va-Sazandegi. No 36. p. 21-27.

Kashki, V. 2000. Determination of nutritive value of some *Atriplex species* by in-vivo method in sheep. M. S. Thesis. Ferdowsi University of Mashhad. 87p.

Keshavarz, P. 2001. Effect of sources of nitrogen fertilizer on growth, Cl and Na content of wheat in saline condition. Journal of Soil and Water. 15: 232-242.

Keshavarz, P. and M. J. Malakuti. 2005. Effect of zinc and salinity on growth, chemical composition and membrane tissue of wheat. Journal of Soil and Water. 19: 115-123.

Le Houerou, H.N. 1993. Salt tolerant plants for the arid regions of the Mediterranean isoclimatic zone. The towards the Rational Use of High Salinity Tolerant Plants, edited by H. Leith and A. Al Masoom. pp403-422. Dordrencht: Kluwer Academic Publishers.

Leman, W.F. and F.E. Robinson. 1979. Progress in developing salt tolerance in alfalfa. Proceeding of 9th California Alfalfa Symposium. p. 73-75.

Noble, C.L. and M.C. Shannon. 1987. Strategies for irrigating white clover with saline water on the heavy soils of northern Victoria. Proceeding of the 4th Australian Agronomy Conference. Melbourne. p. 308.

Rogers, M.E. and C.L. Noble. 1991. The effect of NaCl on the establishment and growth of balansa clover (*Trifolium michelianum* Savi var. *balansae* Boiss). Australian Journal of Agricultural Research. 42: 847-857.

Rhoades, J.D., A. Kandieh and A.M. Mashali. 1992. The use of saline waters for crop production. FAO Irrigation and Drainage Paper. No 48. Scientific Publishers, Jodhpur. Italy. 133p.

Rhoades, J.D. 1990. Soil salinity–causes and controls. In Techniques for desert reclamation. Edited by A. S. Guoudie. P109-132.

Saghafi, F., G.A. Shad, H. Anghoshtari, A. Safarnejad and H. Zarifketabi. 2004. Identification of saline region and halophyte plant vegetation. A research Report. Khorasan Agricultural and Natural Resouces Research Centre. 54p.

Squires, V.R. and A.T. Ayoub. 1994. Halophytes as a resource for Livestouk and for Rehabilition of Degraded Lands. Dordrecht: Kluwer Acadamic publishers.

Tavakoli, H., R. Kakhki and A.R. Beheshti. 1999. The effect of salt stress and temperature on germination of different cultivars of alfalfa. Scientific Report. Agriculture and Natural Resources Research Centre of Khorasan, Agricultural and Education Organization. Iran. 22p.

3

Gainful Utilization of Salt Affected Lands: Prospects and Precautions

Raziuddin Ansair, M. Ajmal Khan* and Bilquees Gul

Institute of Sustainable Halophytes Utilization,
University of Karachi, Karachi – 75270, Pakistan
**E-mail: ajmalk@super.net.pk, khan74@yahoo.com*

ABSTRACT

Soil salinity which is rampant and unevenly distributed in arid and semi-arid areas of the world, retards plant growth through osmotic and specific ion effects. In order to gainfully utilize these soils, we need either to improve the soil or enhance salt tolerance limit of existing plants through genetic manipulation. Both are theoretically possible but the former *i.e.* soil improvement is more feasible and has a history of success while the later requires sophisticated instrumentation and better comprehension of the underlying processes involved in achieving the desired objectives. Taking into consideration the inherent ability of certain plants to tolerate salinity, there is yet another easier and more practical approach of screening and selection of suitable species/varieties from the already available germplasm and uses them for cultivating saline lands as such. The paper discusses the possibility of using woody species in saline lands, proposes some methods for better growth under such conditions and warns of probable consequences in following this approach.

Introduction

Salinity, generally synonymous with sodium, is the presence of excess salts in the root zone in concentrations that retard plant growth. This reduction in growth depends primarily on the inherent salt tolerance of the plant which is exacerbated by environmental factors like vapour pressure deficit, radiation, temperature and a host of other interacting influences both within and outside the plant.

The estimates of salt affected lands vary widely *i.e.* between 340 and 950 x 10^6 ha because the data are imprecise and different agencies have used different criteria of classification (Flowers *et al.*, 1986). According to conservative estimates however, about 7 per cent of total land area of the world is saline to varying degrees and is unevenly spread over the continents, generally in arid and semi-arid regions (Szabolcs, 1987; UNEP, 1992). Inter-regional variation exists both among the countries and in the regions within a country. In Pakistan for instance, where substantial salinization is reported in all the provinces (Table 3.1, Anonymous, 2004), the southern part is comparatively more saline both quantitatively as well as qualitatively.

Table 3.1: Extent of Salinity (Area in Million ha) in Pakistan

Province	*Total area*	*Cultivated*	*Salt affected*
Punjab	20.63	12.27	2.67
Sindh	14.09	5.65	2.11
Balochistan	34.72	2.11	1.35
NWFP	10.17	1.84	0.05
Total	**79.61**	**21.87**	**6.18**

Note: About 40000 ha become saline each year due to secondary salinization

Annual loss to the economy is in the region of approx US$ 3 billion

The salinity may be due to saline parent material or it may have developed by the intervention of mankind. Irrigated agriculture is often practiced without taking into consideration of the requirement of the crop or applying enough quantities to meet the leaching requirement. Over or under-irrigation both create problem, the former leads to water logging and later leaves salts in the root zone. Under arid climates where saline soils are abundant, high evapouration from the soil surface continuously brings up more water from root zone and below through capillary rise and results in salt accumulation on the surface. Keeping in view the total quantity of water applied during a given season, it happens even when good quality irrigation water (this contains 600-1000 ppm salts) is used, highlighting the necessity to adopt proper management practices to avoid or minimize this development.

Options Available to Combat the Hazard

Under such conditions, according to the old adage, either improves the soil or changes the existing crop and use a tolerant species/variety suited to the prevailing conditions. A third option, yet to materialize fully, suggests bringing changes in the genetic makeup of a sensitive plant to make it grow under harsher conditions. Considerable progress has been and is being made on all the three fronts. Lowering underground water table, using suitable agronomic practices and adding ameliorants to soil help in improving the situation. On the other hand, plant breeders and physiologists have realized the importance of joining forces to identify the causative factors and genes involved in conferring tolerance or sensitivity in order to attempt gene manipulation.

The main hindrance in transferring desirable character(s) from a salt tolerant species to improve the less tolerant one lies in the fact that unlike biotic stresses, insect or disease for instance; salinity tolerance is a multigenic character (Chaubey and Senadhira, 1994) and there are, as yet, few readily recognizable genes to try and select (Winicov, 1994). It may also be worth mentioning here that tinkering with a number of genes at a time may create a situation where back crossing fails to dilute the alien undesirable genes effect and while attempting to introduce one desirable character; it also opens the possibility of knocking out other good characters or consecutively introducing undesirable traits in the progeny (Flowers and Yeo, 1995). Although this is a viable option but requires a cautious approach.

The Way Forward

The above leads us to the comparatively easier and more practical option of identifying species/cultivars from the available germplasm which are tolerant of salinity and can be directly planted in the affected areas. A large number of studies have been conducted in the past and new papers continue to appear on the subject but the need continues to expand this activity further depending on particular requirements. We discuss here the results from a project conducted in collaboration with the Australian CSIRO, which aimed at exploring the potential of some woody species from Australia and Pakistan for utilization of saline areas. The experiments were conducted simultaneously in Pakistan and Australia but the paper draws heavily from the work in Pakistan. The experiments were conducted at Nuclear Institute for Agriculture experimental farm located at 25°, 24′north, 31′ east in southern Sindh, Pakistan. The soil was medium in texture, showing high heterogeneity in salinity ranging from 7.3 to 26.5 dS m^{-1} in 0-60 cm soil layer recorded with EM-38, and with a shallow ground water table.

Based on growth performance, observations recorded after three years of growth, the species under test showed diverse response to salinity levels and accordingly separated into three categories (Table 3.2). *Acacia species*, both from Pakistan (there was only one- *Acacia nilotica*) and others from Australia proved to be performing very well in their respective category. *Acacia nilotica*, although it was not very salt tolerant, but survived occasional water logging and once established, performed very well subsequently. Because of its socio-economic importance (almost all parts from root to fruit are used in one way or another, Ansari *et al.*, 1998) it may be an important component of the system in moderately saline conditions prone to occasional, but not sustained, inundation. *Prosopis juliflora* and *Tamarix* species, also from Pakistan, performed very well and survived salinity above 16 mS cm^{-1}.

The acacias from Western Australia were generally tolerant of salinity but *Acacia ampliceps* was the best, followed by other species *i.e. A. machonochieana, A. stenophylla* and *salicinia*. Provenance variation within species also existed. Australian acacias however, performed poorly under salinity plus prolonged water logging condition. None of the Eucalyptus survived above 10 mS cm^{-1}.

Table 3.2: Classification of Some Woody Species Based on their Salt Tolerance

Salinity (Electrical Conductivity of Soil Saturation Extract in dSm^{-1})		
4–8	*8–16*	*>16*
*Acacia auriculiformis**	*Acacia salicinia*	*Acacia ampliceps*
*Acacia nilotica**	*Casuarina glauca**	*Acacia machonochieana**
Acacia saligna	*Casuarina obesa*	*Acacia stenophylla**
Acacia tortilis	*Conocarpus lancifolius*	*Melaleuca halmaturorum*
Albizzia lebbeck	*Eucalyptus camaldulensis**	*Prosopis juliflora**
Azadirachta indica	*Eucalyptus occidentalis*	*Tamarix aphylla**
*Casuarina cunninghamiana**	*Eucalyptus rudis*	*Tamarix articulata**
Casuarina equisetifolia	*Eucalyptus tereticornis*	*Tamarix gallica**
*Eucalyptus citriodora**	*Melaleuca leucodendra**	
*Eucalyptus grandis**	*Zizyphus jujube*	
Eucalyptus robusta		
Leucaena leucocephala		
Melaleuca arcane		
*Melaleuca bracteata**		
*Parkinsonia aculeate**		

Notes:

1. Provenance variation within species is often very important for response
2. Species marked with an [*] may tolerate occasional water logging
3. 1 dSm^{-1} = 1 mScm^{-1} = 10 mM or 0.06 per cent NaCl approximately

The following cultural practices were recommended for improving growth of woody species planted in saline/sodic fields.

1. Excess water should be removed from soil surface as well as from the root zone, either through vertical drainage (tube well) and/or horizontally (ditches/trenches, tile drain). In case of presence of hard pan near root zone, mechanical ripping or digging of auger holes in the field may be helpful.
2. In areas affected by water logging, planting on mounds is advised to give seedlings elevated position above the water. The combination of these two stresses is any way hazardous in most cases.
3. Application of mulches to reduce evapouration and accumulation of salts at or near the soil surface helps in moisture conservation and sustained growth.
4. Addition of organic and inorganic fertilizers *e.g.* farm yard manure, NPK, Zinc is helpful. Inoculation with suitable rhizobia/frankia often improves growth of N-fixers. Gypsum may be needed against the threat of sodicity.

Using heat pulse technique, studies on water consumption by three year old trees of *Acacia nilotica* and subsequent effect on under ground water table showed a gradual water draw down under the plantation during summer, the water started rising with cooler temperature and returned to initial level at the end of winter (Figure 3.1, Khanzada *et al.*, 1998). The area surrounding the experimental field remained fallow during summer but was planted with irrigated wheat in winter. The wheat planting starts in middle October, which coincides with the onset of rising water table under the plantation. It however, demonstrates the potential use of trees to act as biological pump and lower the under ground water table.

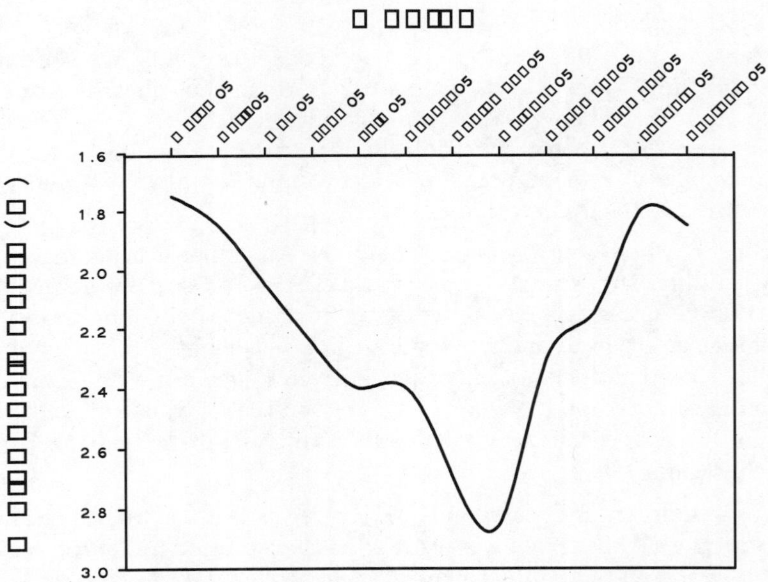

Figure 3.1: Fluctuation in Underground Water Table of a Three Year Old Plantation of *Acacia nilotica* Trees

Precautions and Pitfalls

The seed of Australian species was brought from areas which were carefully matched with the environmental conditions of the experimental sites in Pakistan. However, introducing alien material to a new habitat requires great precautions in view of the ecological consequences, which could be drastic. The Australian species of this project have, to the best of our knowledge, not spread widely in Pakistan, probably because of lack of proper back up and extension. However, an earlier introduction, also from Australia-*Eucalyptus camaldulensis*, needs mention here. It was one of the test species of the present study as well but was overshadowed by other species, particularly acacias.

The forest departments of Pakistan spent lots of their resources to popularize *Eucalyptus camaldulensis*, which undoubtedly has fast growth rate and the potential of tolerating moderately saline conditions. It was distributed extensively and people planted it indiscriminately in saline and sweet soils alike, without realizing that it was a heavy consumer of water. Another problem, although debatable, related to allelopathic effects. The leaves contain volatile chemicals, the stem and branches have limited utility as fuel wood and there is a saying that even the birds do not nest on this tree. It may be a habitat for koalas, but there are no koalas outside Australia. *Eucalyptus* growers in Pakistan realized primarily ill effects on the adjoining crops, caused most probably by water depletion, and found it hard to sell their trees when they reached marketable stage.

This does not imply that *Eucalyptus* is entirely useless. It is a source of chemicals and the wood pulp is used in paper manufacturing. The relevant industries in Pakistan should have been geared up in advance to accept the responsibility of proper utilization of the raw material in time to start processing it when the crop became ready to harvest. Without this arrangement the *Eucalyptus* growers were left with a dilemma of what to do with these trees.

The same may hold true to some extent for the Australian acacias, mainly due to their limited utility. They may be acceptable as fuel wood and are definitely good colonizers of high saline areas. Another useful attribute is being non-thorny, which makes them easier to handle and graze, although no systematic animal feeding trials appear to have been conducted. These trees generally have multi-stems from the base (generally undesirable) and the wood is not as good as that from *Acacia nilotica*. The large within-provenance variation offers opportunity of screening and selection for desirable characters.

Lowering underground water table by trees, also observed in the present study, is possible but should be viewed with caution as also reported by Heuperman, 1992; Thorburn, 1996 and more recently by Barrett-Lennard, 2002. The evapo-transpiration increases the hydraulic gradient towards the root zone of the plantation and continuous water consumption by the vegetation results in gradual buildup of salts left behind in the root zone. Ultimately, a stage is reached when the ability of roots to function properly is impaired encouraging rise in water table. This may be a long term scenario (10-20 years) but one that one should be ready to expect and combat with.

Acknowledgements

The seeds of Australian species were supplied by the Australian Tree Seed Centre, Canberra, Australia, while the Australian Centre for International Agricultural Research (ACIAR) provided partial financial support.

References

Anonymous. 2004. Agricultural Statistics of Pakistan. Ministry of Food and Agriculture, Government of Pakistan, Islamabad.

Ansari, R., N.E. Marcar, M.A. Khan, M.U. Sirazi, A.N. Khanzada and D. Crawford. 1998. Acacias for saltland of Pakistan. In 'Recent developments in acacia planting', J.W. Turnbull, H.E. Crompton and K. Pinyopusarek (eds.), pp. 60-65, ACIAR proceedings no. 82, Canberra.

Barrett-Lennard, *E.G.* 2002. Restoration of saline land through revegetation. Agricultural Water Management 53: 213-226.

Chaubey, C.N. and D. Senadhira. 1994. Conventional plant breeding for tolerance to problem soils. In 'Soil mineral stresses: Approaches to crop improvement', A.R. Yeo and T.J. Flowers (eds.), pp. 11-36, Springer-Verlag: Berlin.

Flowers, T.J. and A.R. Yeo. 1995. Breeding for salinity resistance in crop plants: Where next? Australian Journal of Plant Physiology 22: 875-884.

Flowers, T.J., M.A. Hajibagheri and Clipson, N.J.W. 1986. Halophytes. The Quarterly Review of Biology 61: 313-337.

Heuperman, A. 1992. Trees in irrigation areas: the bio-pumping concept. Trees Natural Resource 34: 20-25.

Khanzada, A.N., J.D. Morris, R. Ansari, P.G. Slavich and J.J. Collopy. 1998. Groundwater uptake and sustainability of Acacia and Prosopis plantations in Southern Pakistan. Agricultural Water Management 36: 121-139.

Szabolcs, I. 1987. The global problems of salt affected soils. Acta Agronomica Hungarica 36: 159-172.

Thorburn, P. 1996. Can shallow water tables be controlled by the revegetation of saline lands. Australian Journal of Soil Water Conservation 9: 45-50.

UNEP. 1992. World atlas of desertification. Edward Arnold: London.

Winicov, I. 1994. Gene expression in relation to salt tolerance. In 'Stress-induced gene expression in plants', (ed.), S. Basra, pp. 61-85, Harwood Academic Publishers: Switzerland.

4

Molecular View on Determinants and Effectors of Halophytic Salt Tolerance: A Case Study on *Suaeda aegyptiaca*

H. Askari[1] and M. Kafi[2*]

[1]*Agricultural Biotechnology Research Institute of Iran, Karaj*
[2]*Department of Agronomy and Plant Breeding, Faculty of Agriculture, Ferdowsi University, Mashhad, Iran*
**E-mail: mkafi2003@yahoo.com, m.kavi@ferdowsi.um.ac.in*

ABSTRACT

Evolutionary processes under saline conditions shape some plants which are called halophytes. The plants have evolved some beneficial physiological characters to manage the salinity of the growth medium. It seems that increase in Na^+ content of leaf tissues as well as proper spatial and temporal accumulation of osmolytes, remodelling of cytoskeleton and remediation of toxic compounds which are coordinated through an effective signalling network can be considered as remarkable attributes of halophytes. Halophytic plants have successfully passed some molecular and cellular developments to govern regulation of antioxidative defence systems and cellular tolerance against reactive oxygen species. Employment of high throughput techniques is an efficient approach to discover effectors and determinants of halophytic salt responsiveness. The evidences from cluster analysis of the salt responsive proteins revealed that coordination of the functional circumstances appears to play a crucial role in differential salt tolerance of halophytes and glycophytes. It may create the need for wide application of the techniques in conjunction with bioinformatics algorithms to better understand the biological systems involved in plant salt tolerance. The presented chapter made an effort to concentrate on dimensions of salt responsiveness which have paid less attention to them.

Introduction

Salinity is one of the major limiting factors for economic crop production worldwide. Evolutionary process through natural selection in salt affected areas forms some phylogenetically diverse plant species which are able to complete their life cycle at salinity levels even up to those found in sea water. This can provide a genetic basis for salt tolerance which is established among different plant species.

Salt tolerance in plants is a complex property involving different mechanisms influenced by coordinated and different expression of a network of genes. The greatest challenges faced by salt-affected plants are a consequence of both a water deficit resulting in osmotic stress and the effects of excess sodium ions on vital biochemical processes. At cellular level, sodium ions can enter into the cells through different plasma membrane pores. Plant cells of salt tolerant species should be able to use the toxic ions for osmotic adjustment and make efforts to keep the ions away from the cytosol. There are three ways to determine Na^+ accumulation in the cytosol; amount of Na^+ entry, compartmentalization of Na^+ in vacuole, and outward Na^+ extrusion (Blumwald *et al.*, 2000; Zhu, 2001). In this way, vacuolar compartmentalization can be considered as an effective strategy to give salt tolerance to plants (Flowers *et al.*, 1977). Salt stress signal transduction that may be composed of homeostasis signalling, detoxification, and growth regulation pathways has already been discussed (for review, see Zhu, 2002). One of the major components of those homeostasis signalling is Na^+/H^+ antiporters (Zhu, 2002), ubiquitous membrane proteins that catalyze the exchange of Na^+ for H^+ across membranes. The presence of Na^+/H^+ antiporters has been detected on both plasma membrane and tonoplast in different plant species (Blumwald *et al.*, 2000). Proton pumping H^+-ATPase supplies the driving force for a wide array of H^+-dependent co-transporters such as Na^+/H^+. Simultaneously, function of H^+-ATPase as a detector and as an effector in response to salt (Reinhold *et al.*, 1984) and necessary coordination of PM H^+-ATPase and V-ATPase pumps in response to environmental signals (Sze *et al.*, 1999) along efficient activity of Na^+/H^+ may be required for successful sequestration of sodium ions away from cytosol.

Some organic compounds which are named compatible solutes (Johnson *et al.*, 1968) can provide directly and indirectly beneficial effects on stability of macromolecules and osmotic adjustment (Yeo, 1998). Metabolic engineering that tend to install osmoprotectant biosynthesis to improve plant cellular response under stress conditions may be a reliable approach to achieve tolerant crop plants (for review, see Rathinasabapathi, 2000). But energy cost of production of organic solutes is one of the important issues (Munns, 2002) that can create some challenges for metabolic engineering particularly in agricultural crops. Different ATP requirements for Na^+ accumulation or osmolyte synthesis were 3.5, 34, 41, 50, and 52 moles for Na^+, mannitol, proline, glycinebetaine, and sucrose, respectively (Raven, 1985). Since,

Na^+ inhibits activity of many enzymes in halophytes and glycophytes cells (Flowers *et al.*, 1977), accumulation of osmolytes in cytosol to protect the cytoplasm from detrimental effects and to prevent dehydration of the cytosol is a crucial part of salt tolerance in both halophytes and glycophytes.

There are many potential sources of reactive oxygen species (ROS) production in the plant cells (Bolwell *et al.*, 2002) such as chloroplasts, mitochondria, plasma membrane, apoplast, and peroxisomes (Del Rio *et al.*, 2002). Mitochondria as a major source of ROS production in eukaryotic cells (Maxwell *et al.*, 1999) especially in mammalian cells do not play an important role in ROS production in plant cells (Laloi *et al.*, 2004). Antioxidative defense mechanism determines lifetime of ROS (for review, see Noctor and Foyer, 1998). It seems that under stress conditions, ROS concentrations can be increased to toxic levels (Foyer *et al.*, 1994; Asada *et al.*, 1998). Salt stress can accelerate production of ROS in plant cells (Borsani *et al.*, 2001) and balance between formation and removal of ROS as a determinant factor in the severity of oxidative damage (Alscher *et al.*, 2002).

It seems that regulation and coordination between the mechanisms by which plants can tolerate different salt levels are crucial for successful management of the osmotic stress. Unlike crop plants, halophytes have evolved advanced characteristics to manage and adapt to saline environments. Research efforts on improving salt-tolerance of crop plants and evaluation of salt-adaptive properties of plants revealed that halophytes might be attractive sources of proper adaptive determinants for crop breeding (Inan *et al.*, 2004). There is no doubt that using models facilitate identification of complex molecular mechanisms in salt tolerance. Models such as yeast for sodium transport in plants (Blumwald *et al.*, 2000), *Mesembryanthemum crystallinum*, and *Dunaliella salina* for their ability to tolerate severe osmotic and ionic stresses (Cushman and Bohnert, 2000), and salt cress (*Thellungiella halophila*) that can reproduce under extreme salinity and for having high sequence identity with Arabidopsis (Inan *et al.*, 2004) and finally *Suaeda aegyptiaca*, a halophytic plant which is introduced as a source of genes for high salt tolerance and its classified salt responsive proteins (Askari *et al.*, 2006) have been recognized.

Prerequisites of salt tolerance in halophytes and its differences with those of glycophytes tend to use halophytic plants to decode genetic information of salt tolerance. Based on that, this paper will review some recent findings on molecular mechanisms of salt tolerance of halophytic plants and physiological consideration of halophytism to face saline environments. In the current chapter, our discussion will be on some other dimensions of salt tolerance which have paid less attention to them, until now.

Preservation of Photosynthetic Apparatus of Halophytes from Salt Injuries

Any effort to increase or maintain rate of dry matter accumulation in plants affected by salinity will be partly based on photosynthetic yield. In saline conditions, plants have to manage different processes like guard cell response, balance between water demand and supply, short and long distance transport of toxic ions, exclusion and accumulation of ions out of cytoplasmic space, efficient accumulation of compatible solutes and supply of energetic cost of all physiological and biochemical functions involved in salt management to overcome negative component of salinity. The above mentioned functions make a need to use photosynthetic apparatus to produce sufficient assimilates for different plant parts. Salinity up to 300 mM NaCl increased *Suaeda aegyptiaca* growth rate (Askari *et al.*, 2006). As shown in Figure 4.1, the salt levels for optimal growth were estimated to be between 150 to 300 mM NaCl. There are many reports to confirm increase of dry matter production of halophytes in response to salinity (Ayala *et al.*, 1996; Khan *et al.*, 2000; Inan *et al.*, 2004). It appears that halophytes have come a long way to protect their photosynthetic apparatus from detrimental effects of salinity. The evidence resulted from growth analysis of *Suaeda salsa* revealed that salt treatment induced further CO_2 assimilation and dry matter accumulation in the plant (Lu *et al.*, 2003).

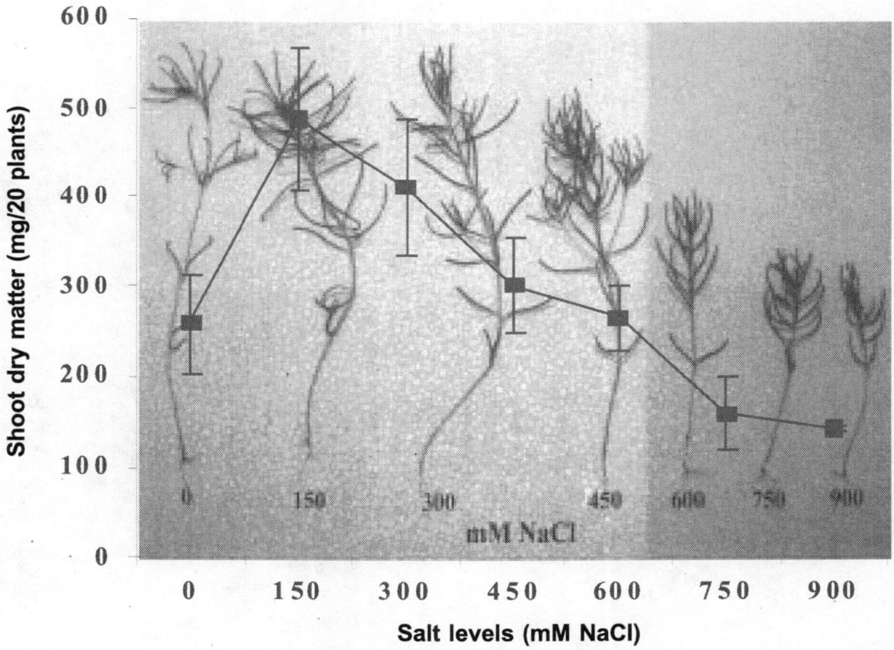

Figure 4.1: Increase in Salt Levels Allowed Further Dry Matter Production in Shoot Tissues of *Suaeda aegyptiaca*. The positive trend continued until NaCl concentration became 150 mM. We did not find significant difference between 150 and 300 mM NaCl.

Photosystem II as a major protein complex of the photosynthetic apparatus in oxygen-evolving species is responsible for photosynthetic oxygen evolution and consists of two main functional parts; the photochemical reaction centre and the water oxidizing complex (Klimov *et al.*, 2003). D2 protein is one of five main intrinsic chloroplast-encoded subunits (P5, P6, D1, D2, and cytochrome b_{559}) of the PS II core (de Vitry *et al.*, 1989). The later study concluded that in the green alga *Chlamydomonas reinhardtii*, the D2 protein is necessary for synthesis of other large subunits like CP5 and D1. Komenda *et al.* (2004) demonstrated that D2 accumulation is a regulatory step for D1 accumulation and subsequent assembly of the PS II reaction centre complex in Synechocystis PCC 6803 (a strain of cyanobacteria that is frequently used as an excellent model to study PS II assembly). On the other hand, expression and accumulation of cytochrome b_{559} constitutes an initial step for synthesis of the D2 protein (Komenda *et al.*, 2004).

Two dimensional electrophoresis-based proteome analysis of *Suaeda aegyptiaca* showed that salt levels (15, 300, 450 and 600 mM NaCl) apart from their effects on plant growth increased expression level of D2 protein (Askari *et al.*, 2006). It seems that the D2 protein plays a key role in increase and maintenance of growth rate under salt treatments (Figure 4.2).

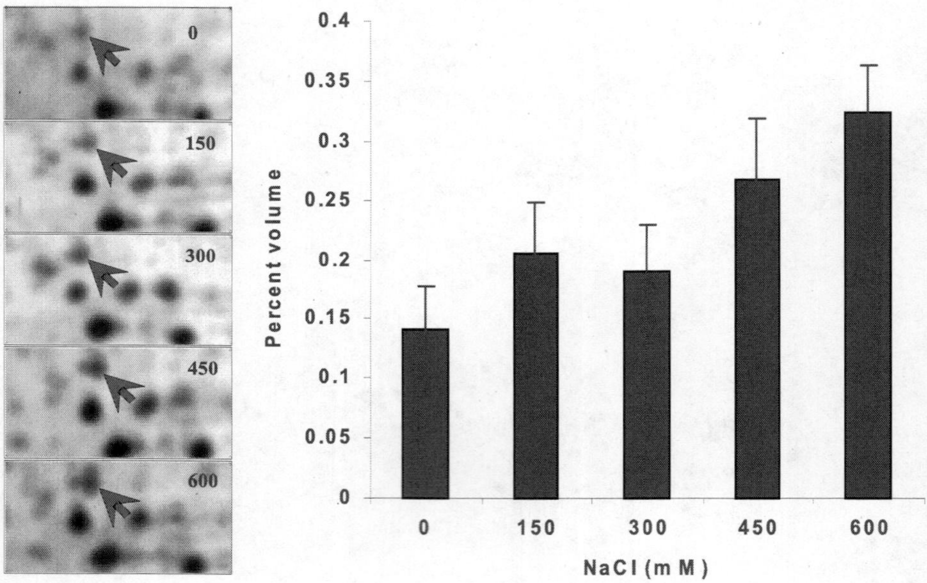

Figure 4.2: Effect of Different Salt Treatments (0, 150, 300, 450, 600 mM NaCl) on Regulation of D2 Protein Expression. The present data showed 2 times over expression of the protein in 600 mM NaCl in comparison with normal condition (0 mM NaCl)

Protection of plant macromolecules and organelles against osmotic pressure of salt ions is an important approach to maintain plant growth ability. As Lu *et al.* (2003) showed salinity can stimulate osmolyte accumulation and subsequently increase thermostability of photosystem II and O_2-evolving centre in salt-adapted plants. Proline accumulated in saltcress leaves is a major compatible solute to counterbalance ion accumulation under saline condition (Inan *et al.*, 2004).

Glycinebetaine is an osmoprotectant that effectively stabilizes enzymes critical to physiological functions (Papageorgiou and Murata, 1995). It was shown that the level of glycinebetaine in halophyte plants rises under salt stress (Khan *et al.*, 2000). Glycinebetaine is synthesized in higher plants by two-step oxidation of choline, and the first step is catalyzed by choline mono-oxygenase (CMO). Gene expression data demonstrated that two isoforms of choline monooxygenase were independent on salt levels up-regulated (Askari *et al.*, 2006). The isoforms with the same molecular mass (41 KDa) were differentiated from each other on the basis of their isoelectric points (6.12 and 6.01).

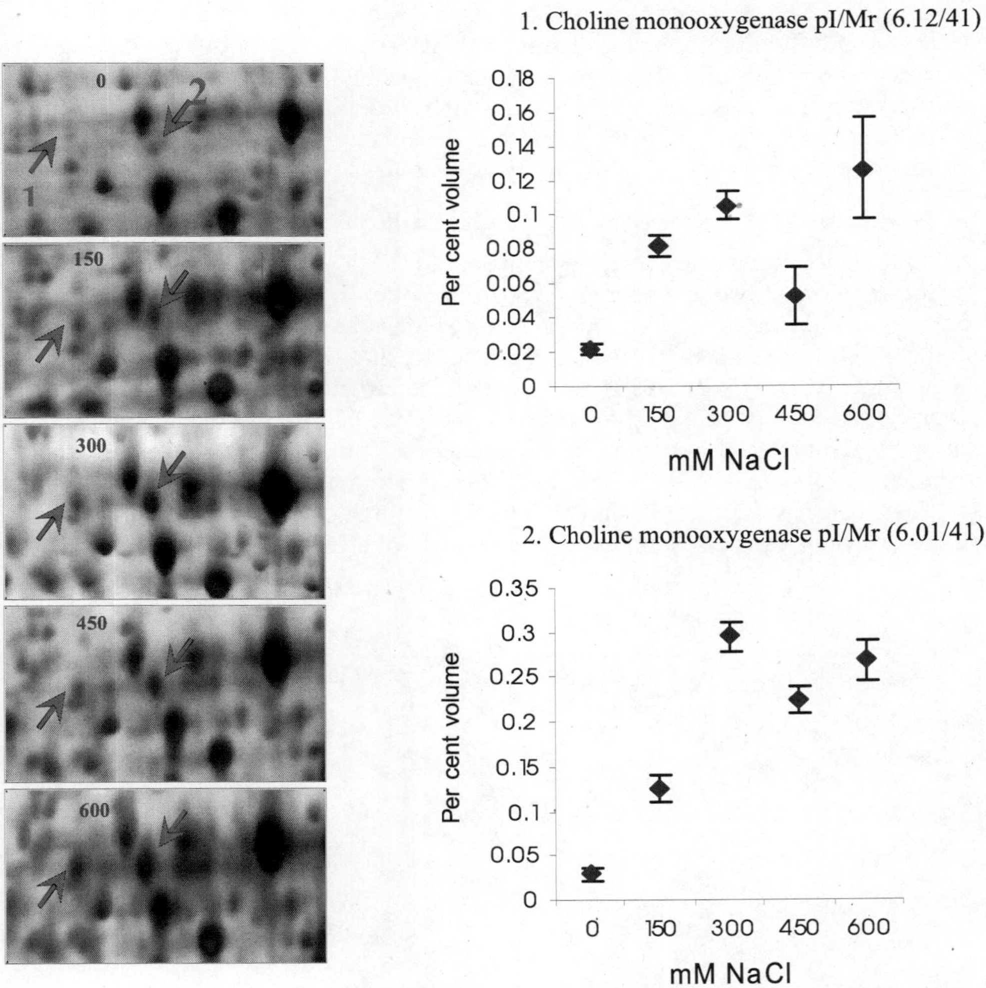

Figure 4.3: Proteome Pattern of Total Soluble Protein Extracted from *S. aegyptiaca* Leaves Showed Two Distinct Isoforms of Choline Monooxygenase with Different pI (6.12 and 6.01). Increase in salinity (150 and 300 mM NaCl) severely induced up-regulation of the isoforms

Expression level of choline monooxygenase isoforms was 5 times greater than those in control condition (0 mM NaCl). Thus, it appears that the two isoforms actively involved in osmotic defence of the plant cells. On the other hand, companion increase in dry matter production and regulation of choline mono-oxygenase quantity may confirm the results reported by Lu *et al.* (2003).

Halophytes follow a growth curve which consists of three distinguishable parts (increase in dry matter production, steady state of growth and decrease in growth rate) when affected by different salt levels, while glycophytes do not have the first part of the halophytic response. In conclusion, it may be considered that osmotic adjustment of halophytic plant cells is not linked only to the salt tolerance stages (Stages 2 and 3) but also to stimulation of their dry matter accumulation (Stage 1). Accumulation of the highest amount of Na ions in leaf tissues through application of the first step of salinity (150 mM NaCl) (Askari *et al.*, 2006) and the stimulating effect of that on dry matter production create a need to up-regulate genes involved in production of compatible solutes like choline mono-oxygenase.

Production and Remediation of Cyanide

Ethylene regulates many diverse metabolic and developmental processes like acting as a stress hormone during biotic and abiotic stresses in plants (Bleecker and Kende, 2000). Ethylene biosynthesis is considered as a major source of cyanide production in plant cells (Walsh *et al.*, 2000; Kozliak *et al.*, 1995). As depicted in Figure 4.4, ACC oxidase, oxidizes ACC to generate ethylene, carbon dioxide and cyanide which must be detoxified to prevent toxicity of its accumulated compounds during high rates of ethylene production. Cyanide is a potent cytotoxic agent especially as an inhibitor of metalloenzymes, which must be immediately detoxified (Meyer *et al.*, 2003). Thus, an increase in cellular demand for ethylene production may indirectly increase cyanide production and a need to its detoxification.

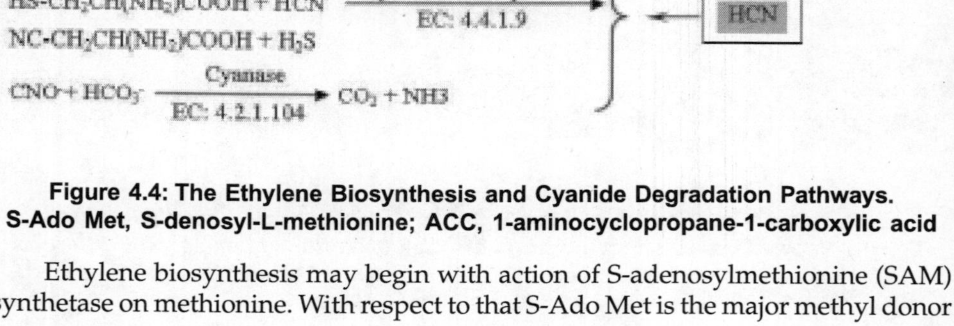

Figure 4.4: The Ethylene Biosynthesis and Cyanide Degradation Pathways. S-Ado Met, S-denosyl-L-methionine; ACC, 1-aminocyclopropane-1-carboxylic acid

Ethylene biosynthesis may begin with action of S-adenosylmethionine (SAM) synthetase on methionine. With respect to that S-Ado Met is the major methyl donor in plants and is used as a substrate for many biological reactions (Ravanel *et al.*, 1998), in fact the first step of ethylene production will be conversion of S-Ado Met to ACC. SAM sythetase is a salt response related protein in *S. aegyptiaca* (Askari *et al.*, 2006). But there is no information by which we can understand how much the enzyme has been employed in ethylene production (Figure 4.5).

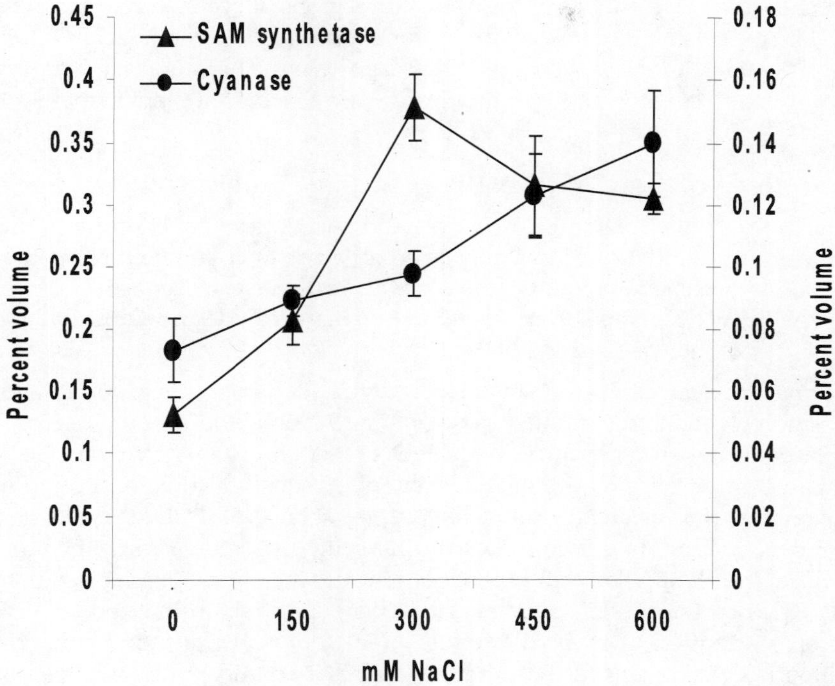

Figure 4.5: Expression Pattern of the Two Proteins are Involved in Ethylene Biosynthesis. The data shows that SAM synthetase than cyanase severely up-regulated and take the highest expression amount in 300 mM NaCl while cyanase expression pattern tend to go on a linear trend.

Cyanase is an enzyme found in plants and bacteria that catalyzes the reaction of cyanate with bicarbonate to produce ammonia and carbon dioxide (Ebbas *et al.*, 2003). Cyanase and cyanate metabolism have mainly been studied in bacteria such as *Escherichia coli*. Cyanate can be fairly toxic to the bacteria, probably due to its reactivity with nucleophilic groups in proteins. However, cyanate can also serve as the sole source of nitrogen for growth of *E. coli* (Kende, 1993). The up-regulation of a gene coding for cyanase indicates cyanate may be formed during exposure of *S. aegyptiaca* plants to high salt levels (Askari *et al.*, 2006) (Figure 4.5). The amino acid sequence of the short peptide fragments obtained from mass spectrometry analysis (LC/MS/MS) showed that the salt responsive protein is a cyanate lyase (or cyanase) (Figure 4.6). The role of the plant cyanase is likely to detoxify the cyanate, but it is also possible that the cyanase-based catabolism of cyanate provides the plant with nitrogen and carbon dioxide.

In theory, SAM synthetase and cyanase can be organized in a biosystem in response to the salinity for regulation of ethylene production. To further study on the relationship between cyanase and ethylene biosynthesis we employed genetic information of two model plants *Arabidopsis thaliana* and *Oryza sativa*. Analysis of cis-acting regulatory elements of the upstream regions of cyanase genes using Plant CARE (Lescot *et al.*, 2002) revealed an ethylene-responsive element (ATTTAAAA) at 2125 and 2169 in *A. thaliana* and *O. sativa*, respectively. This suggests that ethylene may control both cyanide and cyanase production as a part of plant response to salt stress.

Remodelling of Cell Wall: A Basis for Salt Tolerance

The entry of Na^+ ions into cell lumen and subsequent partitioning to the safe cellular compartment like vacuole may be inhibited by an assumed threshold of Na^+ accumulation in the intercellular compartment. In the case, it seems that succulence (it is defined as water content to dry weight ratio) may provide an adaptive advantage to prevention of increasing trend of Na concentration.

Formation of succulent cells in the halophytic plants appears to be explained by cytoskeleton remodelling resulted from polymerization, and de-polymerization of actin filaments which can induces plant cells to respond to salinity. Actin is a major constituent protein of microfilaments in the plant cells which is implicated in cell expansion and development (Ramachandran *et al.*, 2000; Meagher *et al.*, 1999). A number of actin binding proteins like profilins is involved in rearrangement of actin filaments in the plant cells. Profilins are ubiquitous and low-molecular-weight (from 12 to 15 KDa) proteins, which bind actin monomer and cause either polymerization or depolymerization of actin filaments. Also, reported that actin cytoskeleton modelling is determinant for both pollen germination and pollen tube elongation (review by Pierson and Cresti, 1992).

Amino acid sequence of peptide fragments using mass spectrometry showed that salt treatments on *S. aegyptiaca* up-regulate a protein spot which based on results from BLAST search at the National Centre for Biotechnology Information (NCBI), was profilin. The other protein spot named proline-rich protein that was positively regulated by salinity. Based on the concluding remarks of Gibbon *et al.* (1998) pollen profilin function depends on interaction with proline-rich motifs. On the other hand, it was shown that the expression level of profilin is important for cell elongation and cell shape in Arabidopsis (Ramachandran *et al.*, 2000). Cluster analysis of 102 salt responsive proteins using Fuzzy adaptive resonance theory (Fuzzy ART) showed that the two proteins (Profilin and Proline-rich protein) in response to salinity have been co-regulated (Askari *et al.*, 2006).

Increase in water content and diameter of plant leaves and the accompanied increase in expression level of the proteins seems to be involved in creation of succulent leaves. It means up-regulation of profilin may be associated with cytoskeleton remodelling that might be required for *S. aegyptiaca*, a succulent halophyte, to adjust

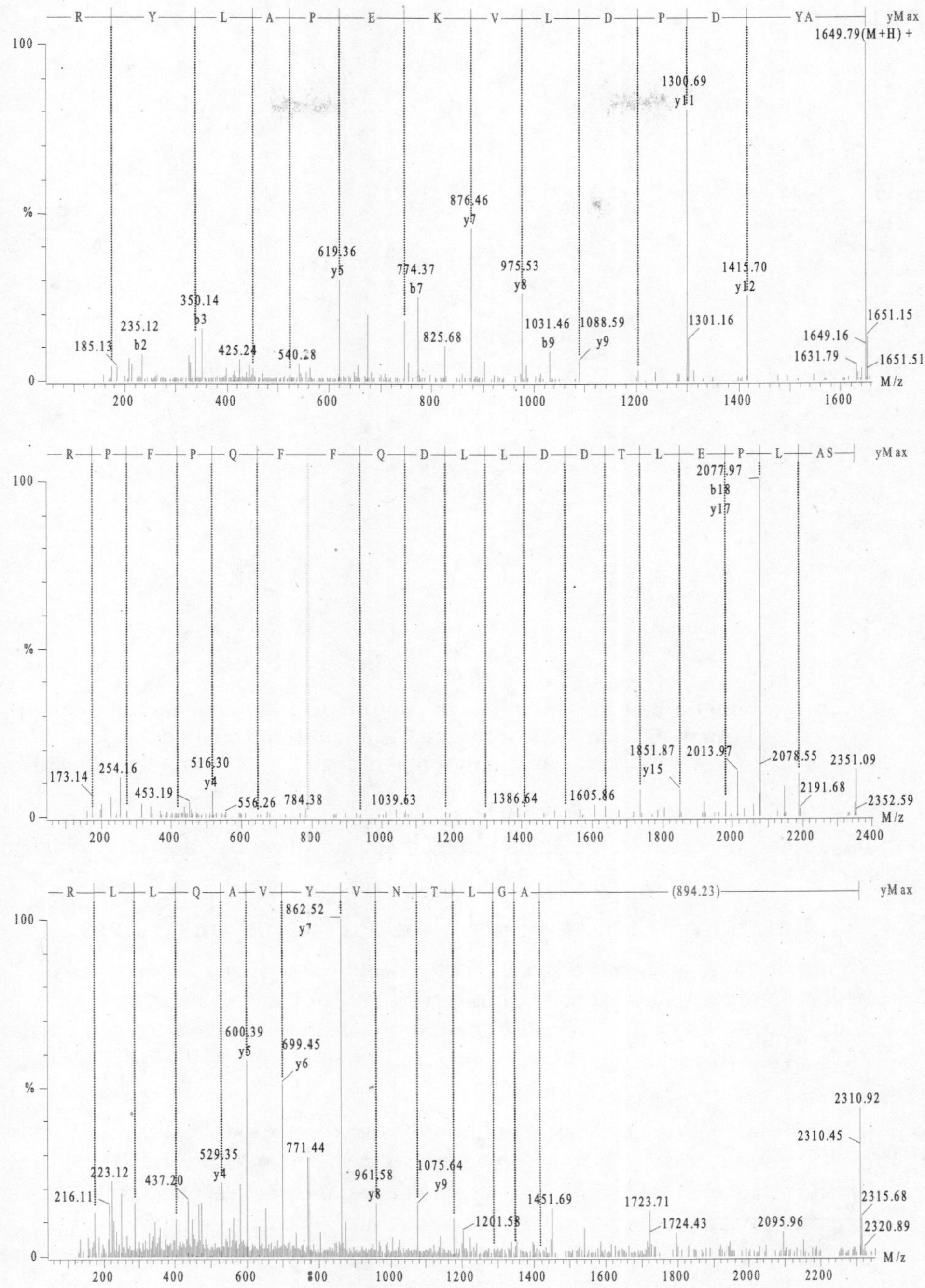

Figure 4.6: LC/MS/MS Spectra of Three Peptides Driven from Cyanase Protein. Vertical axis is intensity (per cent) and horizontal axis shows mass (m/z)

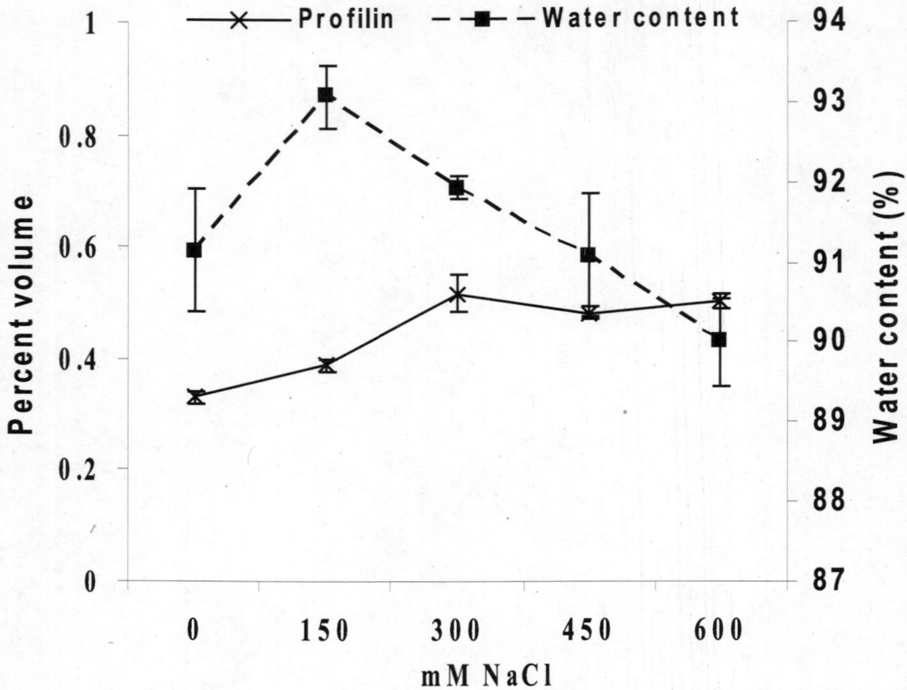

Figure 4.7: With Further Expression Level of Profilin, Leaf Water Content was Increased. After 150 mM NaCl, water content starts to decline while leaf diameter follows a hyperbolic curve under different salt levels (data not shown).

cellular behaviour to substantial quantities of salt and to minimize its toxicity in plant cells.

Salt Acclimation, Behind Acceleration of Halophytic Growth

Salinity induces oxidative stress in plants. Reactive oxygen intermediates (ROI) are unavoidable consequence of aerobic metabolism. The degree of oxidative stress is determined by the balance between the generation rate of the reactive oxygen species and the rate at which the species removed and their detrimental effects repaired (Figure 4.9).

It seems that there is a dynamic equilibration between oxidative defence systems (radical scavengers, and repair and turnover processes) and ROS generation in plants. The reported data shows that 150 mM NaCl have had the greatest effects on increase in dry matter production, sodium accumulation, and water content of *S. aegyptiaca*. During the functions, nine protein involved in plant oxidative defence were coordinately up-regulated (Table 4.1). As shown, all the proteins (with exception of protein No. 1) obtain at least 25 per cent of the highest expression level at the treatment 150 mM NaCl.

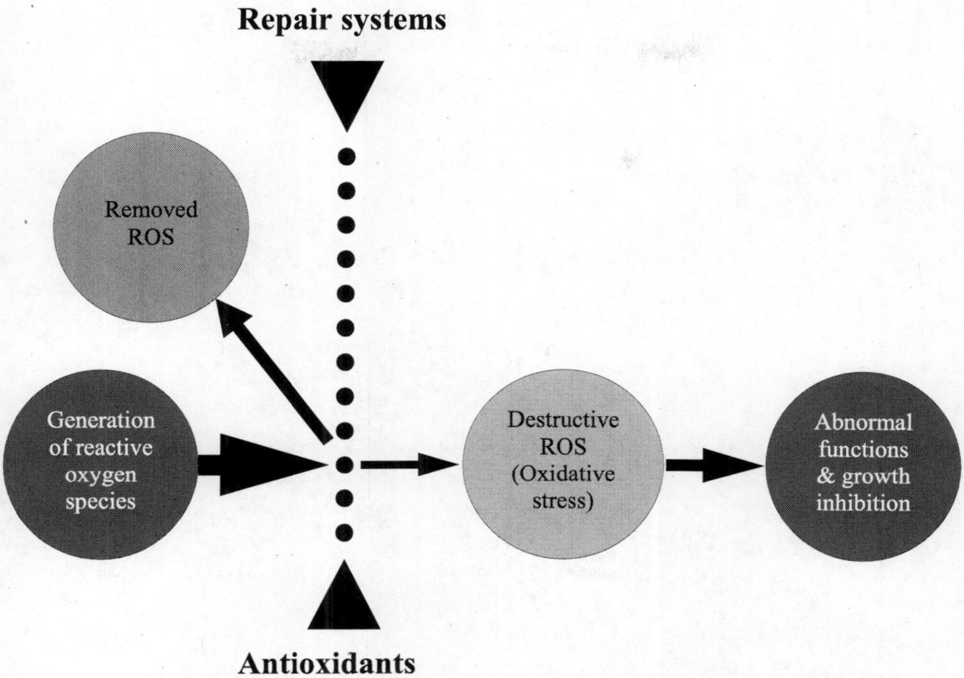

Figure 4.8: An Overall View of Interaction of the Oxidative Defence Systems with the Processes Responsible for ROS Generation

Table 4.1: Comparative Expression Level of Salt Responsive Proteins that are Involved in Plant Oxidative Defence

Sl.No.	*Responsive Proteins*	*Salt Treatments (mM NaCl)*				
		0	*150*	*300*	*450*	*600*
1.	Cytosolic Cu/Zn superoxide dismutase	0	8.04	65.85	78.59	100
2.	Glutathione peroxidase	0	25.37	37.01	62.13	100
3.	Superoxide dismutase	0	49.04	88.79	102.6	100
4.	Peroxiredoxin like protein	0	31.48	68.76	78.45	100
5.	Proteasome subunit beta type 1	0	26.59	55.66	52.10	100
6.	Dehydroascorbate reductase	0	35.8	14.54	25.29	100
7.	Proteasome subunit alpha type 6	0	55.3	90.17	91.44	100
8.	Quinone oxidoreductase	0	25.73	92.89	66.04	100
9.	Stromal ascorbate peroxidase	0	35.23	39.38	74.95	100

On the other hand, application of adaptive-quality based clustering algorithm (De Smet *et al.*, 2002) on the salt responsive proteins showed that all identified proteins which are involved in oxidative defence systems were grouped in the same cluster. It means that the proteins functionally or physically interact together.

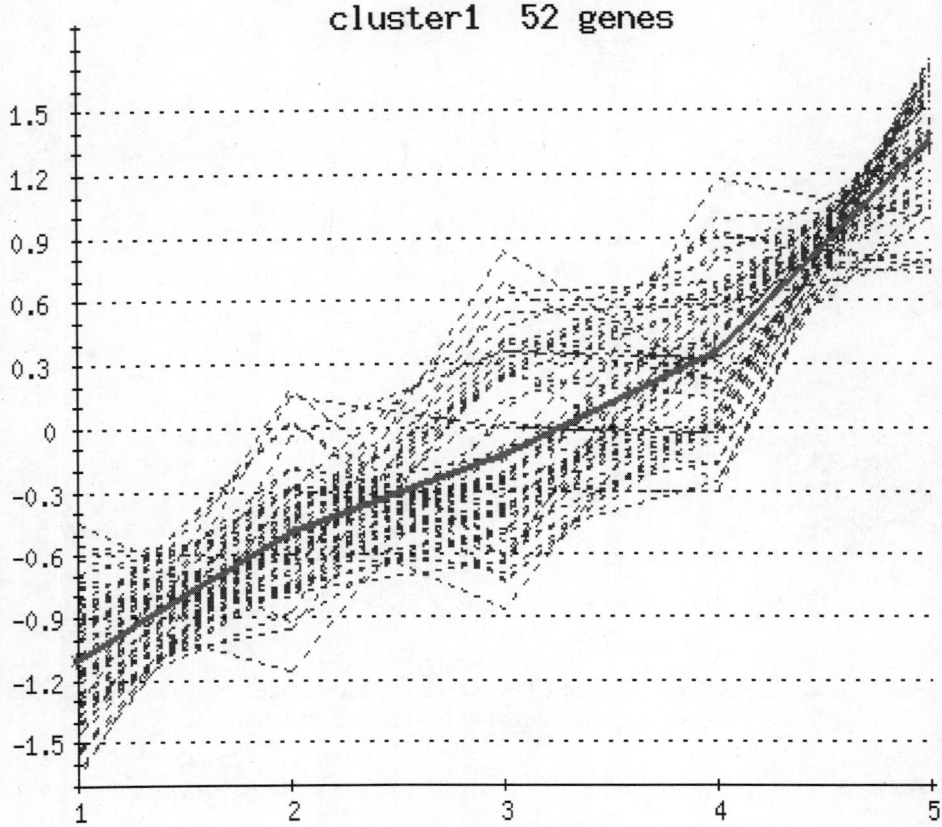

Figure 4.9: Cluster Analysis of 102 Salt Responsive Protein Spot Using Adaptive-Quality Based Clustering Method (De Smet *et al.*, 2002) Classified 52 Protein Spots in the Present Cluster. The plot is generated using the normalized input data which are used during the processing. Vertical axis is normalized data and horizontal axis shows 5 arrays (salt treatments, 0, 150, 300, 450, and 600 mM NaCl). The present dataset was clustered with significant level 90 per cent.

One of the oxidative defence-related proteins was glutathione peroxidase which is also named phospholipid hydroperoxide. According to the studies of Avsian-Kretchmer *et al.* (1999 and 2004), the *Citrus sinensis* gene, gpx1, encoding phospholipids hydroperoxide isoform contains a cis-acting element responsive to ROS. They concluded that this gene might be responsible for salt stress signals mediated by oxidative signals. Mittler *et al.* (2004) mentioned that reactive oxygen species act as signalling molecules or destructive by-products.

If the findings can be extended, one can conclude that 150 mM NaCl without creating salt stress is able to activate oxidative defence system in halophytes and thus oxidative signals are able to be launched before salt stress signals. *S. aegyptiaca*, as a halophyte, may use ROS as signalling molecule for its acclimation to high salt level through induction of oxidative defence enzymes.

As shown in Table 4.1, increase in salt levels generally caused expression of the proteins involved in oxidative defence to be increased. In other words, the same proteins whose expression were primarily up-regulated by 150 mM NaCl are at least partially involved in detoxification of ROS in the higher salt levels (because of their continuous up-regulation). Therefore, the companionship of the up-regulated proteins and increased dry matter accumulation can be speculated that ROS my act as a signalling molecules in this stage.

References

Alscher, R.G., N. Erturk and L.S. Heath. 2002. Role of superoxide dismutases (SODs) in controlling oxidative stress in plants. Journal of Experimental Botany 53: 1331-1341.

Asada, K, T. Endo, J. Mano and C. Miyake. 1998. Molecular mechanism for relaxation of and protection from light stress. In K. Sato, N. Murata, eds, Stress Responses of Photosynthetic Organisms. Elsevier Science Publishing, Amsterdam, pp 37–52.

Askari, H., J. Edqvist, M. Hajheidari, M. Kafi and G. H. Salekdeh. 2006. Effects of salinity levels on proteome of *Suaeda aegyptiaca* leaves. Proteomics 6: 2542- 2554.

Avsian-Kretchmer, O., Y. Gueta-Dahan, S. Lev-Yadun, R. Gollop and G. Ben-Hayyim. 2004. The salt-stress signal transduction pathway that activates the gpx1 promoter is mediated by intracellular H_2O_2, different from the pathway induced by extracellular H_2O_2. Plant Physiology 135: 1685–1696.

Ayala, F., J.W. O'Leary and K.S. Schumaker. 1996. Increased vacuolar and plasma membrane H^+-ATPase activities in *Salicornia bigelovii* Torr. in response to NaCl. Journal of Experimental Botany 47: 25–32.

Bleecker, A.B. and H. Kende. 2000. Ethylene: a gaseous signal molecule in plants. Ann. Rev. Cell Dev. Biol. 16: 1–18

Blumwald, E., G.S., Aharon and M.P. Apse. 2000. Sodium transport in plant cells. Biochimica et Biophysica Acta 1465: 140-151

Bolwell, G.P., L.V. Bindschedler, K.A. Blee, V.S. Butt, D.R. Davies, S.L., Gardner, C. Gerrish. and F. Minibayeva. 2002. The apoplastic oxidative burst in response to abiotic stress in plants: A three-component system. Journal of Experimental Botany 53: 1367-1376.

Borsani, O., V. Valpuesta and M.A. Botella. 2001. Evidence for a role of salicylic acid in the oxidative damage generated by NaCl and osmotic stress in *Arabidopsis* seedlings. Plant Physiology 126: 1024–1030.

Cushman, J.C. and H.J. Bohnert. 2000. Genomic approaches to plant stress tolerance. Current Opinion in Plant Biology 3: 117–124.

De Smet, F., J. Mathys, K. Marchal, G. Thijs, B. De Moor and Y. Moreau. 2002. Adaptive quality-based clustering of gene expression profiles. Bioinformatics 18: 735-746.

De Vitry, C., J. Olive, D. Drapier, M. Recouvreur and F.A. Wollman. 1989. Post-translational events leading to the assembly of photosystem II protein complex: a study using photosynthesis mutants from Chlamydomonas reinhardtii. Journal of Cell Biology 109: 991–1006.

Del Rio, L.A., F.J. Corpas, L.M. Sandalio, J.M. Palma, M. Gomez and J.B. Barroso. 2002. Reactive oxygen species, antioxidant systems and nitric oxide in peroxisomes. Journal of Experimental Botany 53: 1255-1272.

Ebbs, S.D., J. Bushey, S. Poston, D. Kosma, M. Samiotakis and D. Dzombak. 2003. Transport and metabolism of free cyanide and iron cyanide complexes by willow. Plant Cell Environment 26: 1467-1478.

Flowers, T.J., P.F. Troke and A.R. Yeo. 1977. The mechanism of salt tolerance in halophytes. Ann. Rev. Plant Physiology 28: 89-121.

Foyer, C.H., P. Descourvie´res and K.J. Kunert. 1994. Protection against oxygen radicals: an important defence mechanism studied in transgenic plants. Plant Cell Environment 17: 507–523.

Gibbon, B.C., L.E. Zonia, D.R. Kovar, P.J. Hussey and C.J. Staiger. 1998. Pollen profilin function depends on interaction with proline-rich motifs. Plant Cell 10: 981–993.

Inan, G., Q. Zhang, P. Li, Z. Wang, Z. Cao, H. Zhang, C. Zhang, T.M. Quist, S.M. Goodwin, J. Zhu, H. Shi, B. Damsz, T. Charbaji, Q. Gong, S. Ma, M. Fredricksen, D. W. Galbraith, M.A. Jenks, D. Rhodes, P.M. Hasegawa, H.J. Bohnert, R.J. Joly, R.A. Bressan and J.K. Zhu. 2004. Salt Cress. A halophyte and cryophyte Arabidopsis relative model system and its applicability to molecular genetic analyses of growth and development of extremophiles. Plant Physiology 135: 1718–1737.

Johnson, M.K., E.J. Johnson, R.D. MacElroy, H.I. Speer and B.S. Bruff. 1968. Effects of salts on the halophilic alga *Dunaliella viridis*. Journal of Bacteriology 95: 1461-1468.

Kende, H. 1993. Ethylene biosynthesis. Annu. Rev. Plant Physiol. Plant Mol. Biol. 44: 283–307.

Khan, M.A., I.A. Ungar and A.M. Showalter. 2000. The effect of salinity on the growth, water status, and ion content of a leaf succulent perennial halophyte, *Suaeda fruticosa* (L.) Forssk. Journal of Arid Environments 45: 73–84.

Klimov, V.V., S.I. Allakhverdiev, Y. Nishiyama, A.A. Khorobrykh and N. Murata. 2003. Stabilization of the oxygen-evolving complex of photosystem II by bicarbonate and glycinebetaine in thylakoid and subthylakoid preparations. Functional Plant Biology 30: 797-803.

Komenda, J., V. Reisinger, B.C. Müller, M. Dobáková, B. Granvogl and L.A. Eichacker. 2004. Accumulation of the D2 Protein is a Key Regulatory Step for Assembly of the Photosystem II Reaction Centre Complex in Synechocystis PCC 6803. Biol. Chem. 279: 48620-48629.

Kozliak, E. I., J.A., Fuchs, M.B. Guilloton and P.M. Anderson. 1995. Role of bicarbonate/ CO_2 in the inhibition of Escherichia coli growth by cyanate Journal of Bacteriology 177: 3213–3219.

Laloi, C., K. Apel1 and A. Danon. 2004. Reactive oxygen signaling: the latest news Current Opinion in Plant Biology 7: 323–328.

Lescot, M., P. Déhais, Y. Moreau, B. De Moor, P. Rouzé and S. Rombauts. 2002. Plant CARE: a database of plant cis-acting regulatory elements and a portal to tools for in silico analysis of promoter sequences. Nucleic Acids Research 30: 325-327.

Lu, C., N. Qiu, B. Wang and J. Zhang. 2003. Salinity treatment shows no effects on photosystem II photochemistry, but increases the resistance of photosystem II to heat stress in halophyte *Suaeda salsa*. Journal of Experimental Botany 54: 851-860.

Maxwell, D.P., Y. Wang and L. McIntosh. 1999. The alternative oxidase lowers mitochondrial reactive oxygen production in plant cells. Proc. Natl. Acad. Sic. USA 96: 8271-8276.

Meagher, R.B., E.C. Mckinney and A.V. Vitale. 1999. The evolution of new structures: clues from plant cytoskeletal genes. Trends in Genetics 15: 278-284.

Meyer, T., M. Burow, M. Bauer and J. Papenbrock. 2003. Arabidopsis sulfurtransferases: investigation of their function during senescence and in cyanide detoxification. Planta 217: 1–10

Mittler, R., S. Vanderauwera, M. Gollery and F. Van Breusegem. 2004. Reactive oxygen gene network of plants. Trends Plant Science 9: 490–498.

Munns, R. 2002. Comparative physiology of salt and water stress. Plant Cell Environment 25:239-250.

Noctor, G. and C. Foyer. 1998. Ascorbate and glutathione: keeping active oxygen under control. Annu. Rev. Plant Physiol Plant Mol. Biol. 49: 249-279.

Papageorgiou, G.C. and N. Murata. 1995. The unusually strong stabilizing effects of glycine betaine on the structure and function of the oxygen-evolving photosystem II complex. Photosynth. Res. 44: 243–252.

Pierson, E.S. and M. Cresti. 1992. Cytoskeleton and cytoplasmic organization of pollen and pollen tubes. International Review of Cytology 140: 73–125.

Ramachandran, S., H.E.M. Christensen, Y. Ishimaru, C. Dong, W. Chao-Ming, A.L. Cleary and N. Chua. 2000. Profilin plays a role in cell elongation, cell shape maintenance, and flowering in *Arabidopsis*. Plant Physiology 124: 1637–1647.

Rathinasabapathi, B. 2000. Metabolic engineering for stress tolerance: installing osmoprotectant synthesis pathways. Annals of Botany 86: 709–716.

Ravanel, S., B. Gakie're, D. Job and R. Douce. 1998. The specific features of methionine biosynthesis and metabolism in plants. Proc. Natl. Acad. Sci. USA 95: 7805–7812.

Raven, J.A. 1985. Regulation of pH and generation of osmolarity in vascular plants: a cost-benefit analysis in relation to efficiency of use of energy, nitrogen and water. New Phytologist 101: 25-77.

Reinhold, L., A. Seiden and M. Voolkita. 1984. Is modulation of the rate of proton pumping a key event in osmoregulation? Plant Physiology 75: 846–849.

Sze, H., X. Li and M.G. Palmgren. 1999. Energization of Plant Cell Membranes by H-Pumping ATPases: Regulation and Biosynthesis. Plant Cell 11: 677–689.

Walsh, M. A., Z. Otwinowski, A. Perrakis, P.M. Anderson and A. Joachimiak. 2000. Structure of cyanase reveals that a novel dimeric and decameric arrangement of subunit is required for formation of the enzyme active site. Structure 8: 505–514.

Yeo, A.R. 1998. Molecular biology of salt tolerance in the context of whole-plant physiology. Journal of Experimental Botany 49: 915–929.

Zhu, J.K. 2001. Plant salt tolerance. Trends Plant Science 6: 66–71

Zhu, J.K. 2002. Salt and drought stress signal transduction in plants. Annu. Rev. Plant Biol. 53: 247–273.

5

Nuclear Techniques Aided Studies for Sustainable Biomass Production in Salt Affected Soils Using Haloculture Method

J. Rastegari* and M. Farhangi-Sabet
Nuclear Research Centre for Agriculture and Medicine, Karaj, Iran
**E-mail: jrastegari@yahoo.com*

ABSTRACT

More than 25 Mha of salt affected land exists in Iran. Considering that about 90 per cent of the country's area is arid or smi-arid and the fresh water resources are limited population growth of the country and the consequent increase in food demands makes utilization of saline land and water resources for increased food production and socio economic improvement of the salt-affected areas, an important task. Biosaline agriculture (Haloculture) can offer an economically beneficial and practical alternative towards achieving these goals. Haloculture is economical production of highly salt tolerant plants and halophytes in highly saline lands with saline irrigation water. The main objective of this study is to evaluate the adaptability of several local and imported salt tolerant plants with harsh environmental conditions of Iran. Other objectives of this project include utilization of Isotope techniques for irrigation scheduling of plants, and hydrological evaluation of underground water resources. Chah Afzal area in Yazd province, a dry and saline area, was selected as experimental site. Cultivation of salt tolerant species and irrigation with saline water had profound effect on reduction of soil pH, EC, and SAR. Results of neutron meter study revealed that *Eucalyptus* and *Sesbania* consumed 2.5 times more water than Pistachio, wild olive, *Atriplex*, *Haloxylon*, *Tamarix*, *Acacia* and *Kochia*, and 2 times more than pomegranate. Plantation of species with low water requirement for arid region of Chah Afzal is more desirable. In general, the results of this project revealed that haloculture can be utilized effectively in Iran to produce different salt tolerant species with different economical potentials. Further experiments and studies on these successful plants in a larger scale, for economical production of forage, food, feed, and wood production is recommended.

Introduction

There are 380 million hectares of saline soil on the globe. Iran is located between 25 and 40 North latitude and is predominantely an arid and semi-arid country. About 15 per cent or 25 million hectares of the country is affected by salinity. According to data published by UNESCO, Iran ranks 3rd in Asia only after Russia and China and ranks 5th in the world in total salt affected land areas. Reclamation of salt affected land is being practiced using different combinations of leaching and drainage. Beside the high cost of this method, this solution has some other difficulties such as removal of drained water and environmental issues. Use of salt tolerant plant species while utilizing saline ground water is more effective alternative for this purpose.

Halophytes are plant species that grow naturally in saline habitats. They are represented by several thousand species of grasses, shrubs and trees, which are distributed throughout different ecosystems; from coastal areas to mountains and lowland deserts. Halophytes can be utilized for rehabilitation of degraded lands, combating desertification, fuel wood and timber production, shade and shelter, industrial crops, and animal fodder. Thus, halophytes production with proper management can make a significant contribution to mankind and to human food security. Millions of hectares of soils are too salty for normal crops and pasture lie idle where halophytes may be grown (Breckle, 1986; Le Houerou, 1993; Malcolm, 1994; Squires, 1994).

This experiment was conducted to compare salt tolerance of 21 salt tolerant species for their ability to reduce soil salinity, produce forage, fodder, fuel and wood.

Materials and Methods

The study site was located at Cha-Afzal Research Station 90 km from Yazd in central Iran. The study area's climate is harsh dry and desert-like. Annual rainfall and the evaporation rates are (60-80 mm) and 4000 mm respectively. Annual temperature ranges from –16 to 45°C. About 90 per cent of the land surface is heavily affected by salinity. In order to perform management practices to select the best salt-tolerant crops and lower salt accumulation in top soil, 10000 pots, of different salt-tolerant species both local and from Pakistan (Table 5.4) were sowed in the green house of Nuclear Research Centre for Agriculture and Medicine (Karaj). These seedlings were grown to 8 inches tall and were transplanted to the site. In Feb. 1999, a number of seedlings were also obtained from the nursery of natural resources organization, and the total of 21 salt–tolerant species were selected to test in the site of Yazd. Survival rate of different salt tolerant species are presented in Table 5.5. Ten hectares of the land of Chah- Afzal salinity research station were devoted to this activity. It is worth to mention that the land was divided to plots, with area about 5000 m² surrounded by *Tamarix* trees.

Table 5.1: Chemical Properties of Water Available for Irrigation at the Selected Site of Yazd During February 1999 till October 1999

	Date		
	2-1999	*6-1999*	*10-1999*
EC × 10^6	9217	8104	7039
pH	7.4	7.3	7.1
S.A.R	17.7	17.1	16.3
S.I.	1.46	1.42	1.05
S.A.R. adj	44	41	33.4
Classificaion	C4-S4	C4-S4	C4-S4

Table 5.2: Soil Characteristics of the Selected Site of Yazd at Initial Stage

Depth		*Sample 1*		*Sample 2*	
	Unit	*0-35 cm*	*35-70 cm*	*0-35 cm*	*35-70 cm*
EC	dSm-1	126.38	50.64	200.2	75.2
pH	paste	8.2	8.1	8.2	8.2
P.W.	per cent	17.75	22.53	21.37	26.8
S.A.R.	-	283	116	544	247.2
E.S.P.	per cent	95	90	98	94.1
E.S.R.	-	21	8.9	44	16.8
Texture		L	S.L.	Clay	C.L.

P.W. = Soil moisture

S.A.R. = Sodium Absorption Ratio

E.S.P. = Exchangeable Sodium Percentage (Na/CEC × 100)

S.P. = Saturation percentage

E.S.R. = Exchangeable Sodium Ratio (Na/CEC - Na)

S.I. = Saturation Index = 8.4–(pK2 – pKsp) + p (Ca + Mg) + P (CO_3 + HCO_3)

S.A.R.adj = Sarirr [1.0 + (8.4–pHc)]

The quality of water available for irrigation is presented in Table 5.1 and soil characteristics of the experimental site are given in Table 5.2. In order to measure soil water content, monitoring its changes with time and make suitable schedules for irrigation, ten access tube (in two replications) were installed up to about 180 cm soil depth, in 5 selected plots (Those could resist and tolerate against highly salt accumulation and severe condition of Yazd i.e *Haloxylon aphyllum, Elaeagnus angustifolia, Pistacia vera, Eucalyptus camaldulensis* and *Atriplex canescens*). The neutron probe CPN 503DR calibrated and have been used for irrigation scheduling. Irrigation was started when the soil water depletion was about 50 per cent of volumetric soil water content at field capacity. Amount of irrigation water calculated based on difference between soil moisture prior to irrigation and the soil moisture at field capacity.

Table 5.3: Some of the Soil Characteristics 3 Years After Cultivation of Salt Tolerance Species at 0–30 cm Depth at Chah Afzal Research Station

Plot	*pH*	*EC*103*	*SAR*
Punica grantum	8.1	9.1	3.47
Elaegnus angustifolia	8.03	8.88	12.89
Acacia ampliceps	8.12	9.65	12.75
Pistachia vera	8.2	12.7	21.78
Atriplex lentiformis	8.2	11.87	20.319
Atriplex halimus	8.2	15.81	24.28
Haloxylon aphyllum	8.04	30.4	53.92
Eucalyptus	7.8	8.68	13.3
Tamarix	8.08	10.8	16.8

Table 5.4: List of Plant Species Cultivated at Chah Afzal Research Station (1999-2002)

1.	*Punica grantum*	12.	*Atriplex lentiformis*
2.	*Elaegnus angustifolia* L.	13.	*Atriplex halimus* L.
3.	*Fraxinus oxyearpa*	14.	*Atriplex canescens*
4.	*Olea europaea* L.	15.	*Haloxylon aphyllum*
5.	*Ulmus minor*	16.	*Tamarix ramosissima*
6.	*Ailanthus glandulosa*	17.	*Tamarix aphylla*
7.	*Robinia pseudoacacia*	18.	Wheat (13 varieties)
8.	*Eucalyptus camaldulensis*	19.	Barley (Three varieties)
9.	*Pistacia vera* L.	20.	Triticalle (Three varieties)
10.	*Acacia ampliceps*	21.	Rye (Three varieties)
11.	*Atriplex nummularia*		

Table 5.5: Survival Rate of the Salt-tolerant Species Cultivated at Chah Afzal Research Station, Yazd, Iran

Plant Name	*No. of Seedlings Planted*	*% Survival 1 Year After Plantation*
Punica grantum	612	49
Elaegnus angustifolia L	357	70
Acacia ampliceps	463	66
Pistacia vera L.	208	92
Eucalyptus camaldulensis	259	39
Haloxylon aphyllum	408	86
Tamarix	340	88
Atriplex	1020	87

Soil salinity was measured twice (before planting and after 3 years of halophyte growing). For studying the dynamics of underground water some water samples were collected at regular intervals from the site and from several wells located 1-2 Km around the site. The samples were analyzed at PINSTEC, Islamabad- Pakistan via isotope analysis methodology.

Results and Discussion

The results of soil EC, pH and SAR changes given in Figures 5.1–5.3 shows that electrical conductivity and sodium absorption ratio at upper layer of soil profile have decreased due to leaching of sodium salts and chloride ions. Figure 5.2 demonstrates that soil pH has been improved. This attributes to cultivation of salt tolerant species which added organic matter to the soil. Organic acids, the by product of O.M. biological decomposition lowered the soil pH. The results showed that out of 21 selected salt tolerant plants, 12 species, and sub-species performed adequately. These species have potential uses for orchard products (Pistachio, Pomagranate, and wild olive), for forage (*Acacia, Atriplex, Sesbania, Kochia* and *Kallar grass*), and for wood (*Eucalyptus*, wild olive, *Haloxylon* and *Tamarix*). Plantation of salt tolerant species and irrigation had profound effect on reduction of soil pH, EC, and SAR. Results of neutron meter study revealed that Eucalyptus and Sesbania consumed 2.5 times more water than Pistachio, wild olive, *Atriplex, Haloxylon, Tamarix, Acacia* and *Kochia*, and 2 times more than pomegranate. Plantation of species with low water requirement for arid region of Chah Afzal is more desirable (Figure 5.4).

In order to screen the most promising species among salt-tolerant species for the local conditions, adaptation trials are necessary. Such trials were successfully implemented in this project. The site served very well to demonstrate the capabilities and usefulness of Biosaline Agriculture to local farmers and research scientists. The application of neutron probe as a mean of nuclear technique for irrigation management was also successfully demonstrated. However, due to its high cost, and the high level of expertise required for operation, neutron meter can be used mainly by research scientists and research institutes as a reliable tool for irrigation management studies. Based on the resutls of this adaptation trial study, *Punica grantum, Elaegnus angustifolia,*

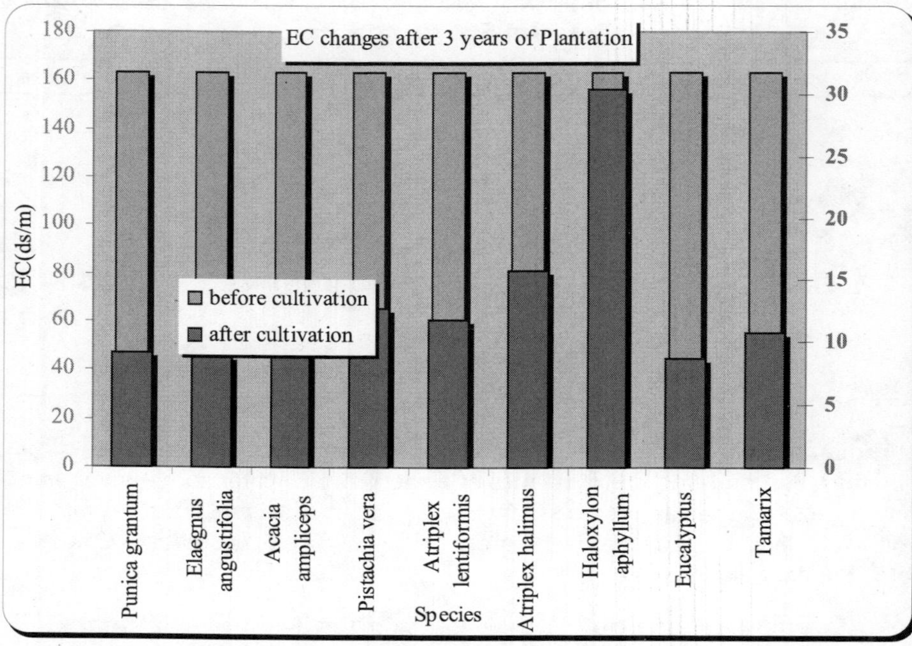

Figure 5.1: EC Changes

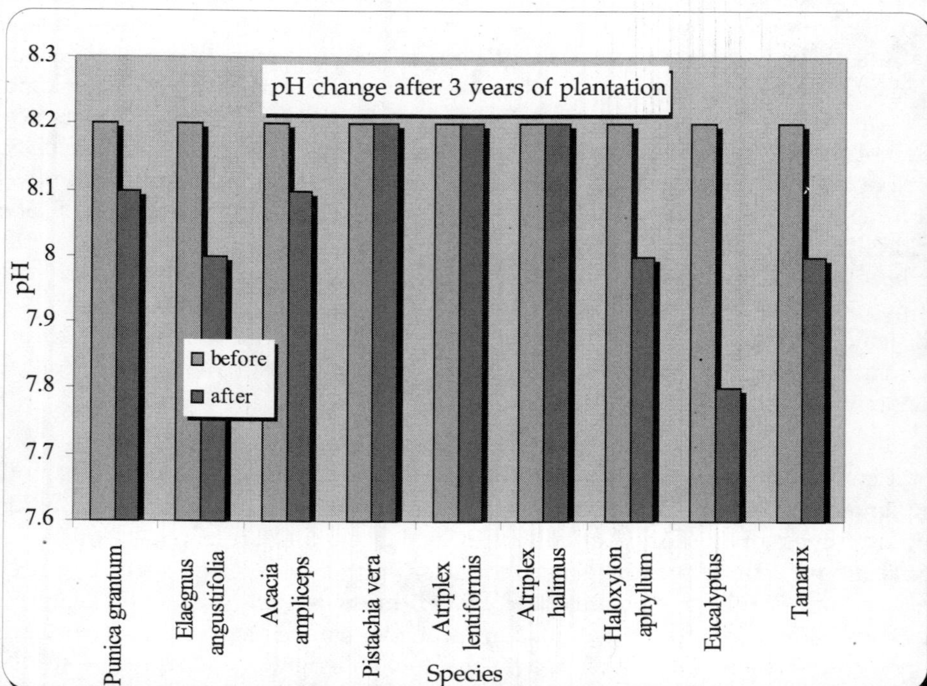

Figure 5.2: pH Changes

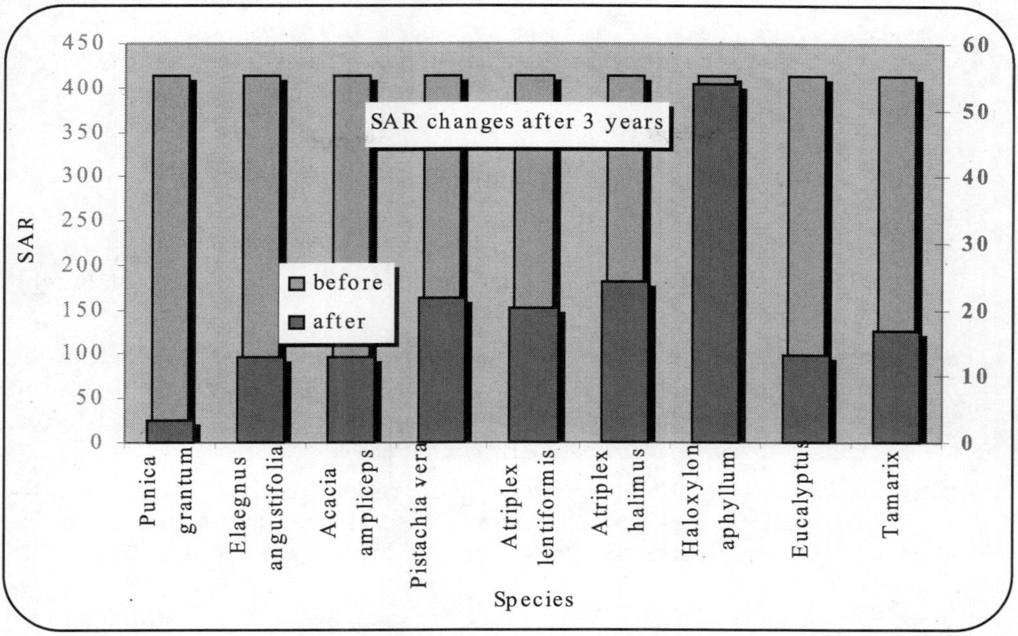

Figure 5.3: SAR Changes

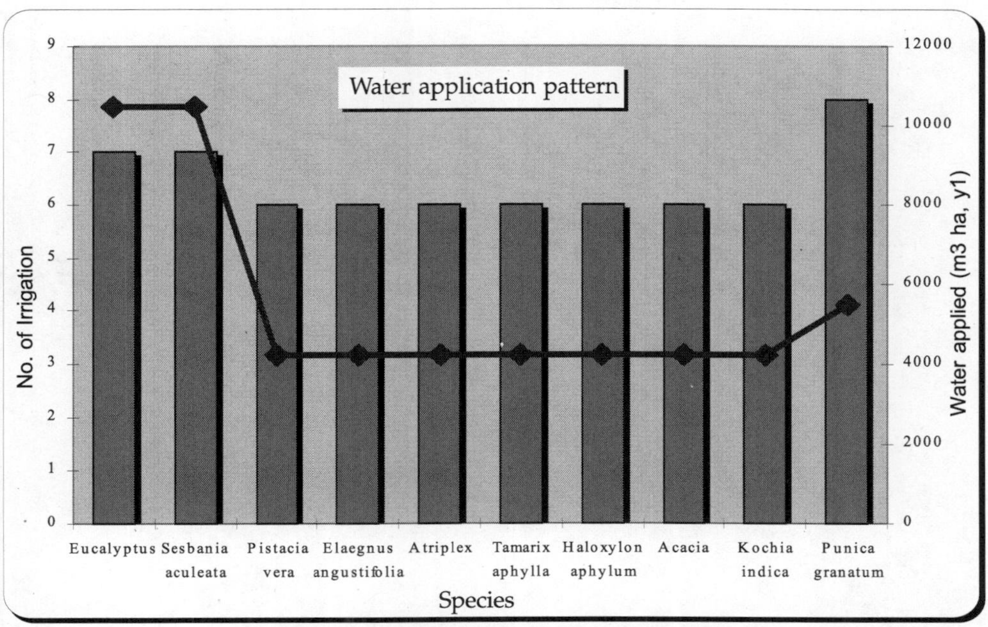

Figure 5.4: Water Application Pattern

Acacia ampliceps, Pistacia vera, Eucalyptus camaldulensis, Haloxylon aphyllum, Tamarix spp. and *Atriplex* were the most suitable salt-tolerant species among the ones tested for the local conditions. However, more salt-tolerant species should be investigated. The appropriate agronomic practices for maximum production of these species need to be studied further.

References

Breckle, A.W. 1986. Studies on halophytes from Iran and Afghanistan. II. Ecology of halophytes along salt gradients. Proceedings of the Royal Society of Edinburgh, 89B:203-215 pp.

Le Houerou, H.N. 1993. Salt-tolerant plants for the arid regions of the Mediterranean isoclimatic zone. In H. Leith and A. Al Masoom, eds., Towards the rational use of high salinity tolerant plants. Tasks in Vegetation Science, 27: 403-422, Kluwer Academic Publishers, Dordrecht, The Netherlands.

Malcolm, C.M. 1994. Use of halophyte forage for rehabilitation of degraded lands. Ch. 4, in V.R. Squires and A.T. Ayoub, eds., Halophytes as a resource for rehabilitation of degraded lands. Kluwer Academic Publishers, Dordrecht, The Netherland.

Squires, V.R. 1994. Overview of problems and prospects for utilizing halophytes as resource for livestock and for rehabilitation of degraded lands. Ch. 1, in V.R. Squires and A.T. Ayoub, eds., Halophytes as a resource for rehabilitation of degraded lands. Kluwer Academic Publishers, Dordrecht, The Netherlands.

6

Water Uptake by Roots and Crop Salt Tolerance Under Brackish Irrigation as Affected by Root Environment

U.W.E. Schleiff

Independent Expert for Irrigation and Salinity, Fertilizers and Crops, Soils and Environment, D-38289 Wolfenbuettel, P.O. Box 1934, Germany

E-mail: schleiff.uwe@t-online.de

ABSTRACT

In many countries of the dry areas water resources of good quality are required for human consumption and industrial development. Consequently irrigated agriculture and horticulture are exposed to increasing pressure to expand the use of brackish and even high-saline waters for crop production. Experience shows that successful and sustainable application of lower quality waters such as brackish or saline waters requires special management practices to realize the agronomic potential and avoid environmental stress. The adaptation of irrigation management to saline growth conditions is often unsatisfactory, as our understanding on the key factor for crop salt tolerance– the process of water uptake by roots from saline soils - still is far from being satisfactory. The paper analyzes the present concept for rating the salt tolerance of irrigated crops following the 4-layer-profile after FAO and USDA, where the vertical movement of solutes is considered. This concept does not reflect the effects of the lateral salt movement which dominates between water applications to cover the plant water supply. Field experiments with onions under irrigation with brackish waters of increasing salinity and under strongly arid climate (Saudi Arabia) have shown a continuous increase of soil salinization in the rhizospheric soil (soil adhering the roots) as compared to the bulk soil (soil far from roots). The presented model calculation demonstrates the effect of root hair length and rhizospheric soil volume on salt concentrations in the rhizospheric soil solution. The effects of root morphology in conjunction with soil properties are considered as an important link to complete our understanding on crop salt tolerance with respect to optimize irrigation technique and breeding of more tolerant crops. Under controlled greenhouse condition the fine roots of young rape plants (equipped with many long root hairs) were about 5-times more efficient than the thicker roots of onions (equipped with fewer short root hairs) in their water uptake rate from saline soils.

Introduction and Definitions

In the dry areas of the world, agriculture and horticulture are exposed to increasing pressure to develop brackish water resources for irrigation and to reduce the consumption of fresh water, which are required for human consumption and industrial development. In Oman for example the over-pumping of good quality groundwater along the Batinah-Coast has caused a continuous intrusion of seawater into the groundwater, resulting in severe soil salinity and expanding problems for crop growth, even for palm trees (Ahmed *et al.*, 2004). The Government of Oman is now developing a strategy to limit agricultural production in coastal areas and to transfer the production of selected crops on suitable land inside the country, where brackish groundwater resources can be developed for crop irrigation (MRMEWR/OMAN *et al.*, 2005). Similar strategies to expand the use of brackish and treated waste waters for irrigation are under discussion in many countries of arid regions (Dichtl, 2002).

As there is no internationally accepted clear definition of the term 'brackish water', I will give my understanding applied for this paper. The term 'brackish water' was originally used for waters found at the mouth of rivers, where sweet river water is mixed with saline seawater. Under such conditions the salinity of brackish waters may vary in wide ranges following the tides. As the term is not clearly defined with respect to salinity for irrigation, we proposed for a regional study in Oman the rating of brackish groundwater salinity as given in Table 6.1.

Table 6.1: Brackish Water Classification for Irrigated Agriculture (Proposed for brackish groundwater resources of Oman, 2005)

Unit	*Fresh*	*Brackish*			*Saline*
		Mildly	*Moderately*	*Strongly*	
dSm^{-1}	<2	2–5	5–10	10–15	>15
g/l	<1.5	1.5–3.5	3.5–7	7–10	>10

Figure 6.1 presents this definition graphically for crops differing in their salt tolerance, but is completed for halophytes. The growth of halophytes is included as

they are sometimes taken into consideration for excessive salinity levels, *e.g.* re-use of saline drainage waters. Their agronomic value is still under discussion (Lieth, 1999).

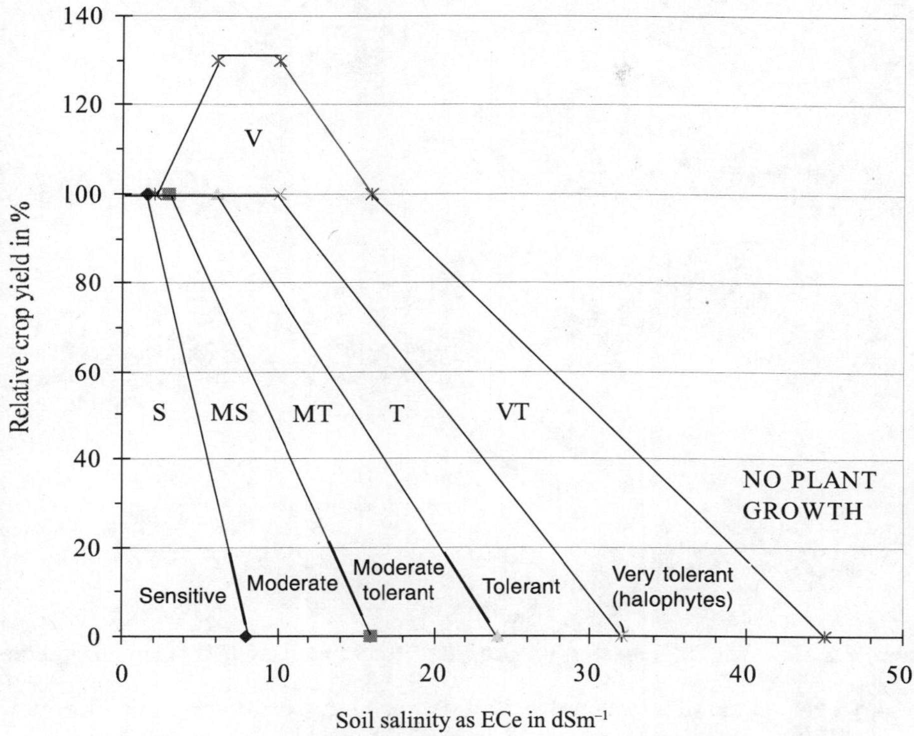

Figure 6.1: Rating of Relative Crop Salt Tolerance After FAO (Ayers and Westcot, 1985) completed for Halophytes (after Schleiff, 2003)

Analysis of FAO-Concept for Rating of Crop Salt Tolerance

The rating of crop salt tolerance after FAO considers the root zone as a 4-layered soil profile, the top layer of 0-25 per cent, the middle layers of 25-50 per cent and 50-75 per cent, and the bottom layer 75-100 per cent as presented in Figure 6.2. To cover the water demand of plants the water depletion follows the pattern 40/ 30/ 20/ 10 per cent (0.4/ 0.3/ 0.2/ 0.1 ET = Evapo-transpiration), which reflects the decreasing rooting density: 40 per cent of the ET demand is taken from the top 25 per cent of the rooted soil layer, 30 per cent, 20 per cent and 10 per cent from the 2nd, 3rd and bottom layer. Consequently the share of water for salt leaching (LF = leaching fraction) and controlling soil salinity decreases with soil depth. As a result the root zone salinity is not considered as homogeneous, but increases with soil depth. Rating of crop salt tolerance is based on this concept and all ECe-values given for relative yield losses of crops refer to the average root zone salinity calculated from the 4-layer soil salinity. For special cases (*e.g.* high frequency irrigation) this concept can be adapted by modification of the water depletion pattern.

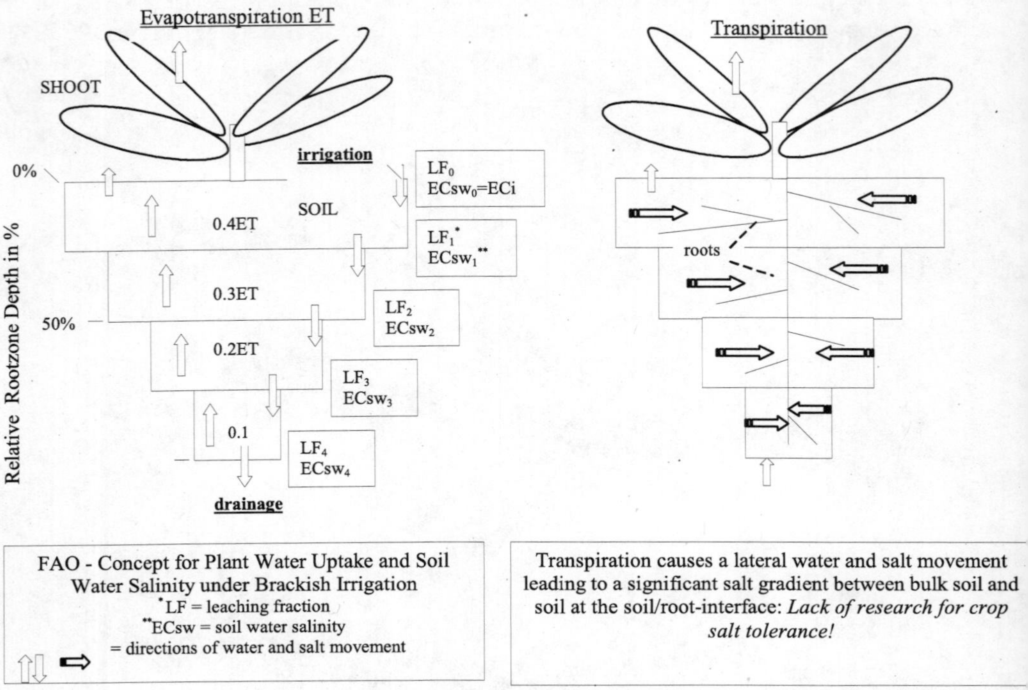

Figure 6.2: Concept for Research on Crop Salt Tolerance Under Brackish Irrigation

However, as pointed out on the right half of Figure 6.3 there is one important aspect of water and salt movement under brackish irrigation that is not taken into consideration. It is the lateral movement of saline soil solution to cover plant water requirement and which dominates between two water applications, when the soil water content drops below field capacity. Stimulated by transpiration, which is usually stronger under arid weather conditions, a flow of saline soil solution is directed to the water absorbing root surface. The soil water is taken up by roots, but in saline soils most salts remain in the soil solution surrounding the roots. As shown earlier the uptake of damaging salts by most crops can be considered as nearly negligible in this context (~<5 per cent of soil solution salinity by most crops; Schleiff, 1982a, b).

Salt Dynamic Around Irrigated Roots

The following results from a field investigation carried out under extremely arid conditions in Saudi Arabia with onions show that these considerations are not just theory, but can be even found also under the difficulties of field experiments. Onions were irrigated for around 4 months with brackish waters of different salinity levels ranging from about 2.3 dS m^{-1} to 8.0 dS m^{-1}, which corresponded to Cl-concentrations from 13 to 52 meq/l (Figure 6.3). The plots were irrigated once (winter time) and twice (summer time) a week with an application rate of 50 mm. Soil samples were taken from the rooted bulk soil (soil far from roots) and soil adhering the roots (rhizospheric soil) at two occasions:

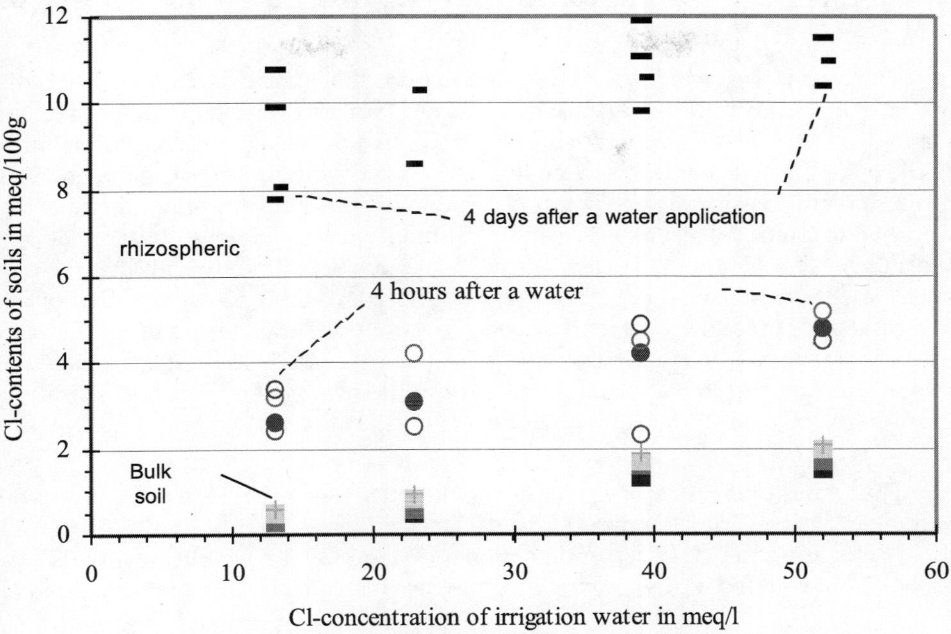

Figure 6.3: Chloride Contents of Bulk and Rhizospheric Soil Under Brackish Irrigation (Schleiff, 1981b)

1. 4 hours after a water application, when the soil water still exceeded field capacity
2. 4 days later, a few hours before the next water application

As shown in Figure 6.3, the Cl-contents of the bulk soil followed the Cl-concentration of the irrigation waters and increased from about 0.5 meq/100g to nearly 2 meq/100g soil. However the Cl-contents of the rhizospheric soil exceeded the Cl-contents of the bulk soil even immediately after the water application with values in the range from about 3 to about 5 meq/100g soil. Thus the water application of 50 mm did not level the Cl-contents of the bulk and rhizospheric soil. The following period of 4 days transpiration caused a Cl-accumulation in the rhizospheric soil, which was 2 to 3 times higher than immediately after the water application and ranged from 8-9 to 10-12 meq/100g soil (Schleiff, 1981b).

The results from the field experiment have clearly shown that soil salinity is not only heterogeneous in the vertical dimension as supposed in the FAO-concept, but there is also a significant gradient in the lateral dimension. This gradient is not considered in the present concept for rating the crop salt tolerance, but doubtless is of particular importance for plant water supply and consequently crop salt tolerance (Schleiff, 2005).

Modelling of Soil Solution Salinity as Affected by the Volume of the Root Hair Zone

The conclusion from the field measurements is that salinity stress for plant roots under irrigation is rapidly varying within short periods and can be much stronger than expected from calculations or measurements of the average soil solution salinity. The level of salt concentrations in the rhizospheric soil solutions that can be achieved through plant transpiration depends on several factors. One important factor is the water potential of the leaves as already reported for maize, barley, wheat and sugar beets (Schleiff, 1981a, 1982b, 1983, 1984). Numerous experiments with many crops have clearly shown that it is the salt concentration or osmotic water potential of the root surrounding solution, which the roots are actually exposed to and that is decisive for the water uptake by roots. However the level of the osmotic soil water potential the roots were exposed to was usually taken from the average soil solution salinity, a gradient of soil solution salinity between bulk and rhizospheric soil was not considered.

The results of the following model calculation presented in Figure 6.4, however, points out the considerable differences of the effective soil solution salinity around roots following a period of water depletion and depending from the morphology of roots. The simplified model calculation compares the effect of a water depletion of

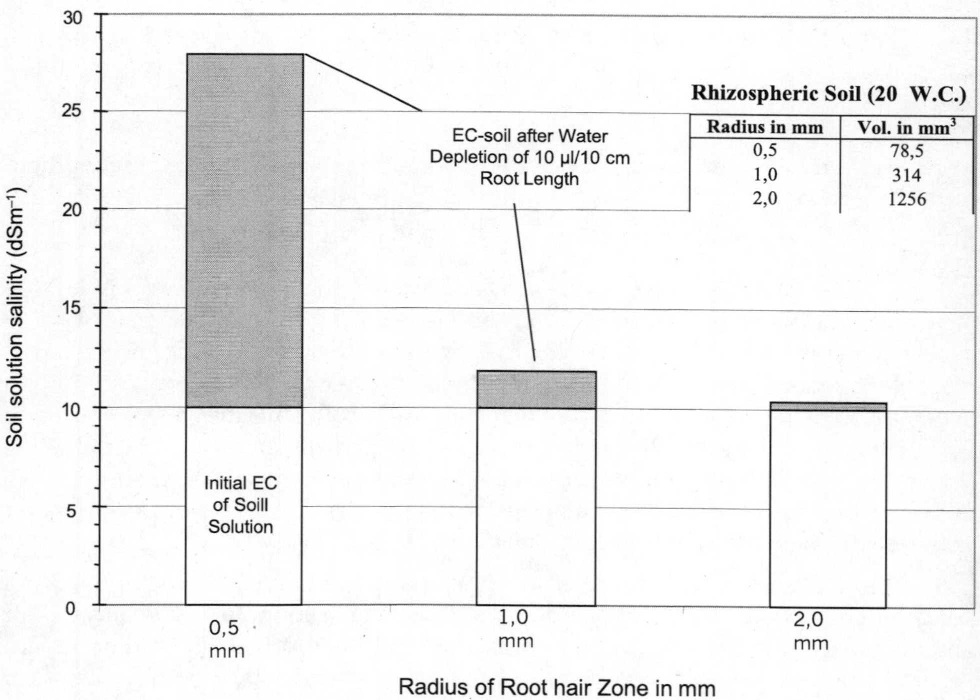

Figure 6.4: Effect of Water Depletion by Roots of Different Morphology on Rhizospheric Soil Water Salinity

10 µl/10 cm root length by roots, which are equipped with root hairs of 0.5, 1.0 and 2.0 mm length and form a rhizocylinder volume of 78.5, 314 and 1256 mm^3. The columns of Figure 6.4 show a large difference in the increase of the rhizospheric soil solution salinity. In case of short root hairs (0.5 mm length) the water depletion affects an increase of soil solution salinity from 10 to 28 dSm^{-1}, in case of medium root hair length (1.0 mm) from 10 to 12 dS m^{-1} only and in case of longer root hairs (2.0 mm) the increase can be neglected. These ECe-values close to the roots correspond to osmotic soil water potentials of about -0.9, -0.4 and -0.35 MPa and may differ greatly with respect to their effect on water uptake by roots.

Even when the presented model calculation may suffer from some simplifications of minor relevance (*e.g.* diffusion effects, water flow from bulk into rhizospheric soil, ion uptake), it clearly attracts our attention on the substantial contribution of root morphology for the salt concentrations in the soil solution near roots, which the roots are really exposed to and from where water uptake occurs.

Comparison of Water Uptake from Saline Soils by Roots of Different Morphology

A first experimental indication on the relevance of root morphology for the efficiency of water uptake from saline soils and consequently crop salt tolerance is presented in Figure 6.5. It is shown that under equal climatic conditions for the shoot, equal soil water contents and soil solution salinity water uptake rates by rape roots was much higher than by onion roots. Water uptake rates by onion roots were only about 5 to 6 ml/(g root-DM/h), but close to 30 ml/(g root-DM/h) for rape roots, when the osmotic potential of the soil solution was around -0.5 MPa (ECsw ~14 dSm^{-1}). This observation coincides with findings on the efficiency of nutrient acquisition by various crops. It is well known that roots of onions are much less efficient in their uptake of Nitrate, Potassium and Phosphate than roots of rape (Jungk, 2002). This

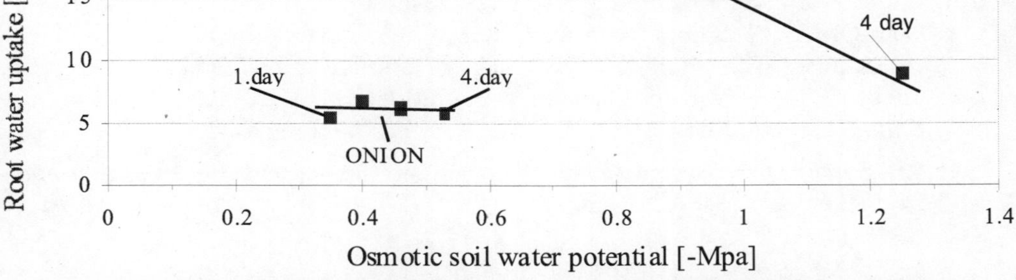

Figure 6.5: Efficiency of Water Uptake from Saline Soil Solutions by Roots of Onion (Short root hairs) and Rape (Long root hairs) (Schleiff, 1985)

fact is related to the differences in the soil volume of the rhizocylinder formed by the same root mass, which is in the case of rape about 20-times larger when compared with onion roots. It is most unlikely that this large difference in the water uptake rates is mainly due to differences in shoot properties, for example osmotic adaptation to saline soils or resistance of stomata, but these factors will be involved too.

That modifications of roots under saline soil conditions principally play an important role to adapt to salinity is not just a theoretical consideration, but can be even watched in the field and was reported earlier for field grown sweet potatoes. Roots growing in saline soils were significantly finer and more numerous than roots under non-saline conditions. Energy produced by leaves was invested in the extension of the root system instead of tuber formation (Schleiff, 2005).

Conclusions

In the last decades much research has been done to extend our understanding of the physiological and biochemical complexities of plant salt tolerance at the cellular as well as at the whole plant level, but the output measured at the elevation of crop salt tolerance in the field seems to be nearly negligible. We should ask ourselves for the reason of this little progress in spite of the large investment. Personally I guess that many of the experimental set-ups to study salinity stress under well-controlled conditions such as water cultures are part of the problem, as these conditions have nothing to do with the reality the plants are exposed to in the field. There is a high probability that under the well-controlled artificial growth conditions of experiments too many growth factors of dominant importance in the field are excluded and consequently the value of the results is very limited with respect to natural or field conditions. From my experience and results presented in this paper I want to underline that there is an urgent need to understand processes within soils, above all interactions between roots, soils, salts and water uptake, as they have been overlooked in the past. Their quantitative understanding is most promising for a significant progress in research for crop salt tolerance, even when the experimental work and data evaluation will be trickier as compared to water culture experiments. However some important steps to develop adequate experimental set-ups were already presented (Schleiff, 1987a, b, 2005).

References

Ahmed, M., H. Esechie and A. Al-Ajmi. 2004. Biosaline agriculture: perspectives from OMAN. Presented at a workshop on Biosaline Agriculture, June 2004, Wageningen, The Netherlands Bernstein N. and Kafkafi U., 2002: Root growth under salinity stress. In: Plant roots–the hidden half. Ed. Waisel Y., Eshel A., Kafkafi U.; 3rd edition, Marcel Dekker, N.Y.–Basel - Hongkong.

Ayers, R.S. and D.W. Westcot. 1985. Water quality for agriculture. Irrigation and Drainage Paper 29, FAO, Rome.

Dichtl, N. 2002. Wiederverwertung von Abwasser. ATV-DVWK-Bundestagung 2002; ISBN 3-936514-06-2; 239-253.

Hu, Y. and U. Schmidhalter. 2004. Limitation of salt stress to plant growth. In: Plant Toxicology, fourth edition, eds. Hock and Elstner; Marcel Dekker, N.Y., 191-224

Jungk, A.O., 2002. Dynamics of nutrient movement at the soil-root interface. In: Plant roots–the hidden half. Ed. Waisel Y., Eshel A., Kafkafi U.; 3rd edition, Marcel Dekker, N.Y.–Basel - Hongkong

Lieth, H. 1999. Development of crops and other useful plants from halophytes. Proceedings: saline irrigation: halophyte production and utilization, 1-20, Agadir, Morocco

MRMEWR-OMAN/Parsons Intern./Schleiff, 2005. Study on the development of brackish groundwater resources of Oman (unpublished report)

Pasternak, D., Y. De Malach and I. Borovic. 1986. Irrigation with brackish water under desert conditions. VII. Effect of time of application of brackish water on production of processing tomatoes. J. Agric.Water Manage.12: 149-158

Rhoades, J.D., A. Kandiah and A.M. Mashali. 1992. The use of saline waters for crop production. FAO Irrigation and Drainage Paper 48, Rome

Riley, D. and S.A. Barber. 1970. Salt accumulation at the soybean (*Glycin max*. L.*Merr*.) root-soil interface. Soil Sci. Soc. Amer. Proc. 34: 154-155

Schleiff, U. 1981a. Determination of maximum salt concentration in the rhizospheric soil solution due to transpiration of young maize plants. Mitt. d. Dtsch. Bodenkundl. Gesellsch. 32: 1981

Schleiff, U. 1981b. Osmotic potentials of roots of onions and their rhizospheric soil solutions when irrigated with saline drainage water. Agricultural Water Managem. 3: 317-323

Schleiff, U. 1982a. Dynamics of salts in the rooted soil and its significance for the water supply of crops - an overview. Z. f. Kulturtechnik u. Flurbereinigung 23: 38-49

Schleiff, U. 1982b. Maximum salt concentration in the rhizospheric soil solution of young sugar beets in dependence from the salt adaptation of their shoots. Plant and Soil 66: 397-404

Schleiff, U. 1983. Water uptake of barley roots from rhizospheric soil solution of different salt concentrations. Irrigation Science 4: 177-189

Schleiff, U. and G. Schaffer. 1984. The effect of decreasing soil osmotic and soil matric water potential in the rhizosphere of a loamy and a sandy soil on the water uptake rate of wheat roots. Z. Acker- und Pflanzenbau 153: 373-384

Schleiff, U. Salztoleranz von Kulturpflanzen unter besonderer Berücksichtigung der Salzgehalte des wurzelnahen Bodens. Unpublished Report 1985

Schleiff, U. 1986. Water uptake by barley roots as affected by the osmotic and matric potential in the rhizosphere. Plant and Soil 94: 143-147

Schleiff, U. 1987a. A vegetation technique to study the water uptake by roots from salinized rhizospheric soils. Z. Pflanzenernähr. Bo-denk. 150: 139-146

Schleiff, U. 1987b. A comparison of transpiration rates of young rape plants from salinized soils of different texture. Z. Pflanzen-ernähr. Bodenk. 150: 403-405

Schleiff, U. 2003. Handbook 'Salinity and Soil Fertility Kit'–a portable field laboratory for soil, water and plant analysis. See under http://www.salinity.de; self-published, Wolfenbuettel; ISBN 3-00-016639-4

Schleiff, U. 2005. Research aspects for crop salt tolerance under irrigation with special reference to root environment. In: Research Accents in Agricultural Chemistry, Special Issue/Sonderheft FAL Agricultural Research, Braunschweig, 83-94; and also under http://www.salinity.de

7

Investigation of Crop Production Potentiality of Saline Lands

M. Dehghani, G. Hadarbadi, A. Nasrabady, Mohamad and E. Kyhan

Agricultural Jihad Organization, Southern Khorasan Province, Iran

ABSTRACT

Annually about 2250 million cubic meter of water are drained in Zohre river and thousands hectares of deltaic lands of this river are left useless. In this paper investigation of crop production potentiality of saline lands is discussed. For this purpose soil and water parameters were analyzed in relation with native plants (Wheat, Barley and Palm), based on some experimental measurements. As a result, temperature and evaportranspiration, precipitation, water quality and quantity, native plants, the crops water requirement, leaching fraction and salt increment in the soil due to irrigation, underground water level and gypsum need for leaching were analyzed. The study shows that salty lands of Zohre River Delta could be amended by water of river.

Introduction

Salt-affected lands are mostly found in coastal areas. However, this is a problem but its magnitude and intensity have been increasing. The rapid growth of human population urged scientists to find new resources such as poor quality and marginal soil and water resource (75 per cent of world's population are residing in developing countries which possess only 55 per cent of the world's arable land). Soil salinity is considered as an environmental hazard and one of the factors causing desertification in arid, semi-arid and dry sub-humid areas both dryland farming and irrigated regions. It causes decreasing or losing agricultural productivity.

Nearly 10 per cent of the total world land surface is covered with different types of salt-affected soils. At present, there are nearly 954 million hectares of saline soils on the earth's surface. More than 80 million hectares of such soils are in Africa, 50 million hectares in Europe, 357 million hectares in Australasia, nearly 147 million hectares in Central, North and South America. Similarly, a large bulk of about 320 million hectares of land in South and South East Asia is under the grip of salinity. Estimates indicate that the world as a whole is losing at least 3 ha of arable land every minute due to soil salinization/sodification.

In Iran, salt-affected regions include the vast parts of eastern, central and southern areas. It is estimated that 15 per cent of the total land surface of Iran and about 30 per cent of its plains are affected by salinity. Salinity of soil and water resources is a serious threat in many parts of the country. Approximately 11 per cent of the total land area of the country is used for crop production. Of this area, annually, about 12.7 Mha is cultivated (irrigated plus rainfed) and nearly 5.8 Mha is left as fallow. An estimated area of 32 Mha of unused land of Iran is potentially suitable for crop production.

Amelioration of these soils needs a source of calcium that can replace the exchangeable sodium. These soils are generally classified to 3 categories, including saline, sodic and saline-sodic. There are several methods for soil reclamation, including utilizing of various chemicals (gypsum application followed by excessive irrigation), leaching and biological methods (growing salt tolerant crops). Recently researchers are recommending phytoremediation (biosaline agriculture) methods.

Annually about 2250 million cubic meter of water (fresh, marginal and brackish water) are drained in Zohre River. On the other hand thousands hectares of deltaic lands of this river are left useless. Main objective of this research were: evaluation and investigation of using soils and water recourses around this river in sustainable agriculture. The evapotranspiration, precipitation, water quality and quantity, native plants properties, the crops water requirement, leaching fraction and salt deposition in the soil due to irrigation, water level and gypsum requirement for leaching were estimated.

Materials and Methods

The study site is located in Hendijan Delta, which is close to Persian Gulf coastal lands (Figure 7.1).

In order to evaluate agriculture and plants growth capabilities in Hendijan delta soil and water parameters were analyzed in relation with native plants based on some experimental measurements. Random samples of soil were taken from 0-15, 15-30, and 30-60 cm depths. Soil properties were assessed in laboratory and their chemical and physical parameters determined. Meanwhile, the evapotranspiration, precipitation, water quality and quantity, native plants properties, the crops water

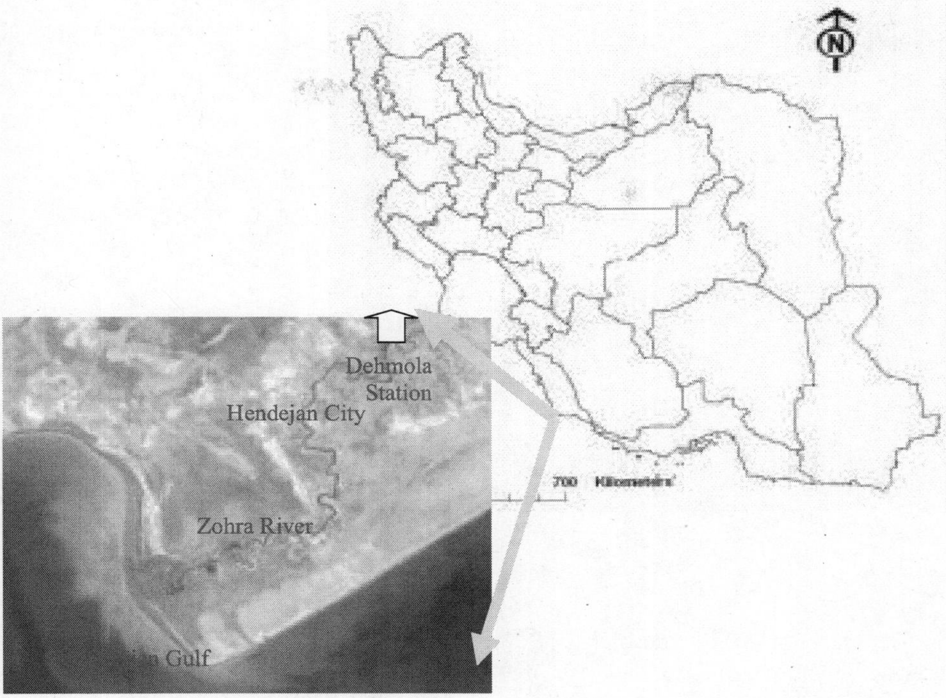

Figure 7.1: Study Area

requirement, leaching fraction and salt accumulation in the soil due to irrigation, water level and gypsum content which is necessary for leaching were also calculated.

All spatial data (soils sampling) were entered in to a geographic information system (GIS) and prepared ISO chemical maps for all parameters and 3 mentioned depths. The maps show the project area has saline-sodic lands and for improvement the quality of soils, we need to ameliorate substances such as gypsum (It is commonly used and cost effective).

Results and Discussion

Temperature and Evapotranspiration

Using Dehmola meteorological station data and analyzing temperature and evapotranspiration in a long-term period (36 years), properties like absolute maximum, minimum, average temperature and evapotranspiration were calculated (Figures 7.2 and 7.3). The results show that maximum and minimum monthly evapotranspiration were 329.75 and 77.8 mm respectively (Figure 7.4).

Precipitation

Using long term data of Dehmola meteorological station (36 years) and reliable rainfall data, monthly average of rainfall was computed (Figure 7.5). The result

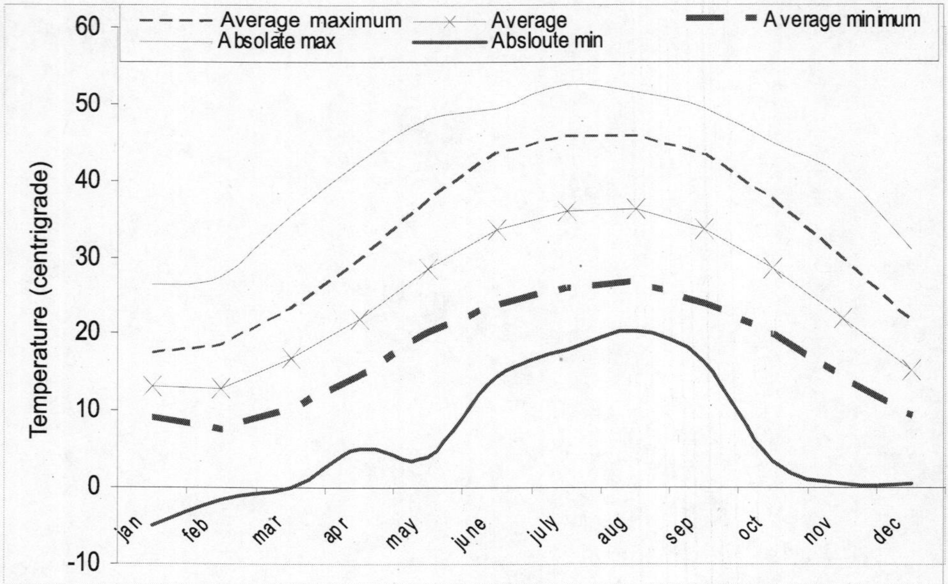

Figure 7.2: Temperature Parameters in Study Area

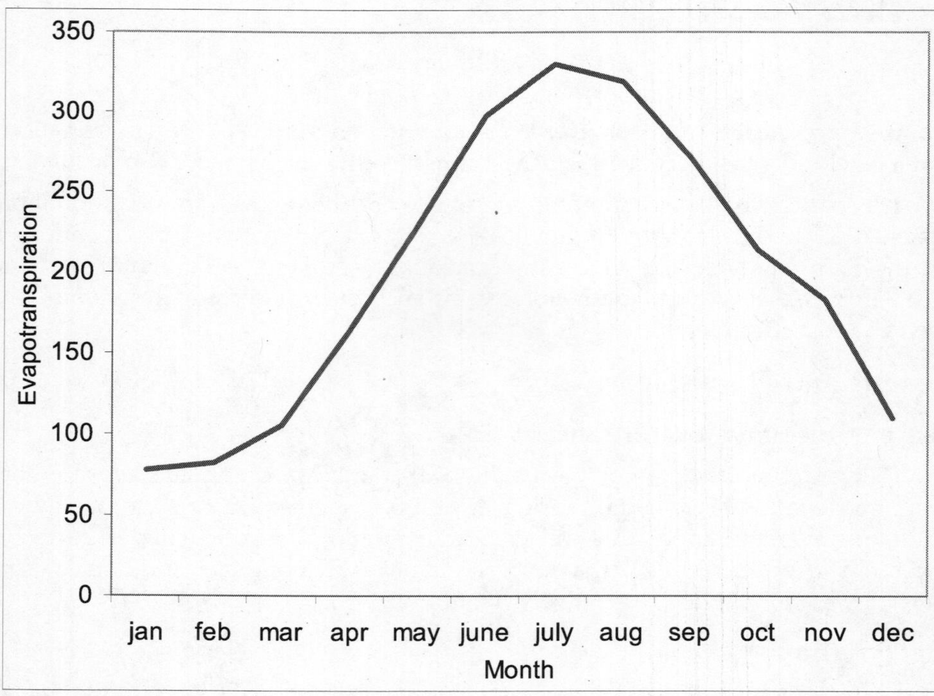

Figure 7.3: Monthly Average of Evapotranspiration

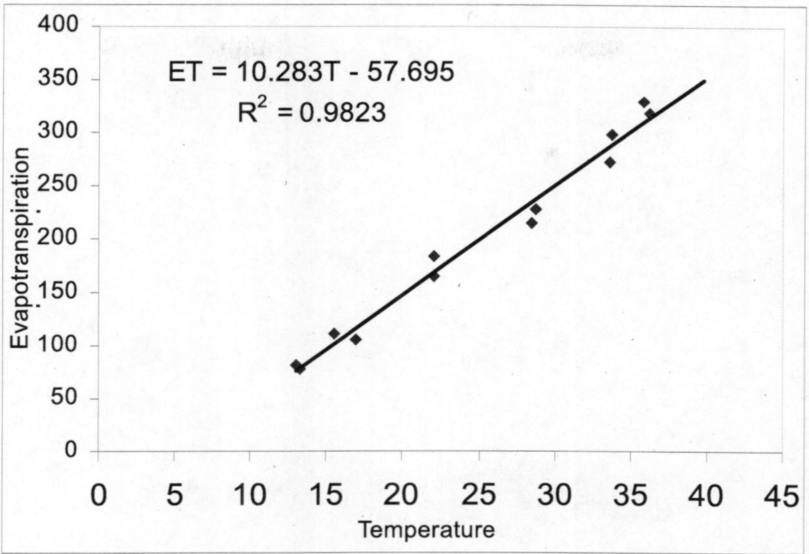

Figure 7.4: Temperature and Evapotranspiration Relation in Study Area

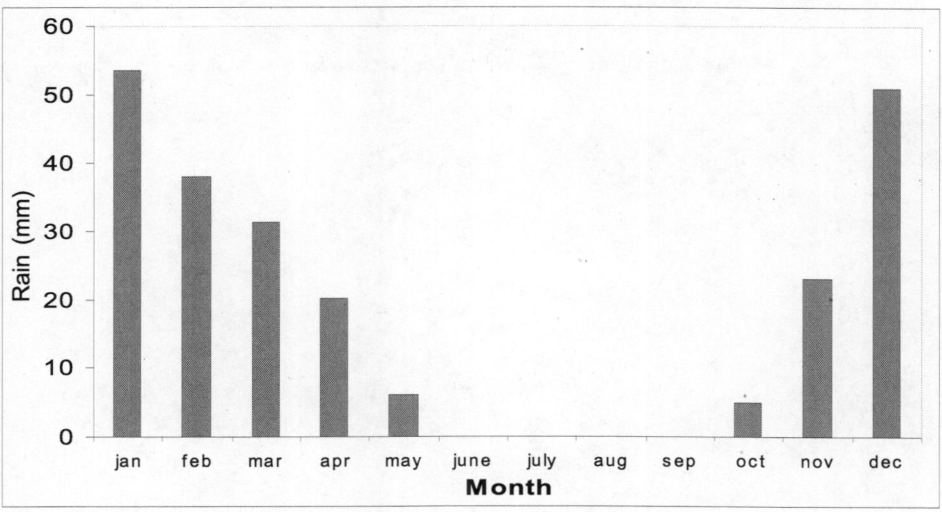

Figure 7.5: Monthly Average of Rainfall in Study Area

indicates that annual average rainfall and coefficient of variable were 228 mm and 37 per cent respectively.

Discharge

Monthly average volume and flows discharge and different return periods including 2, 5, 10, 25, 50 and 100 years and discharge-class curve were calculated (Figures 7.6–7.9 and Table 7.1).

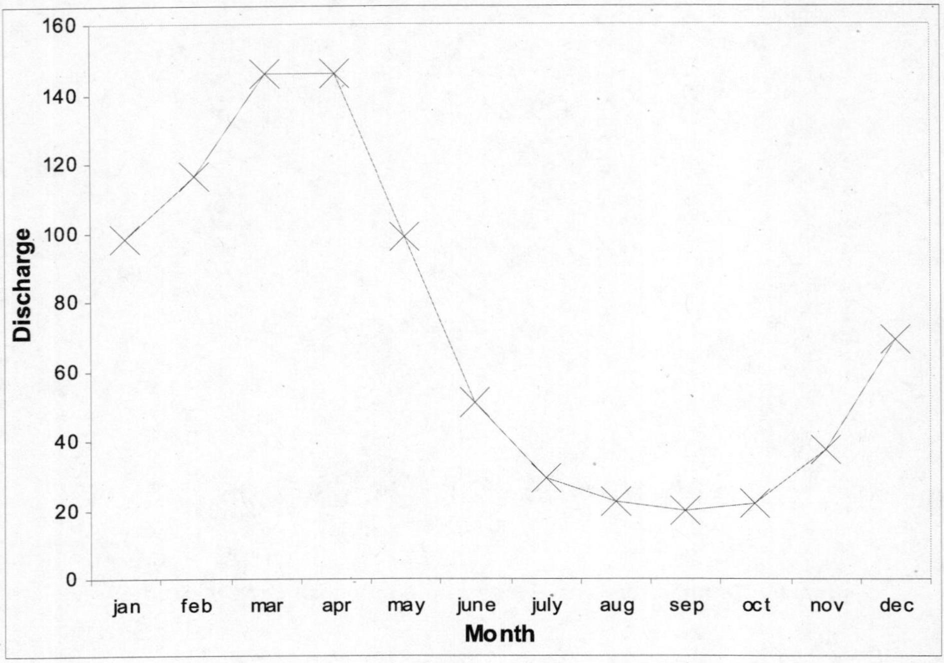

Figure 7.6: Monthly Average Discharge in Study Area

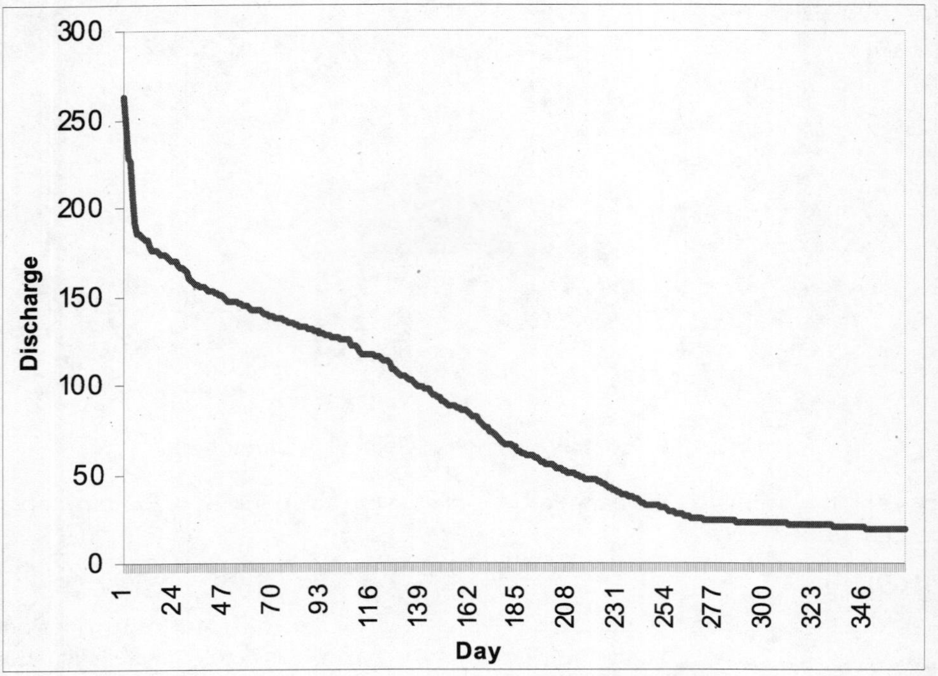

Figure 7.7: Discharge-Class Curve in Study Area

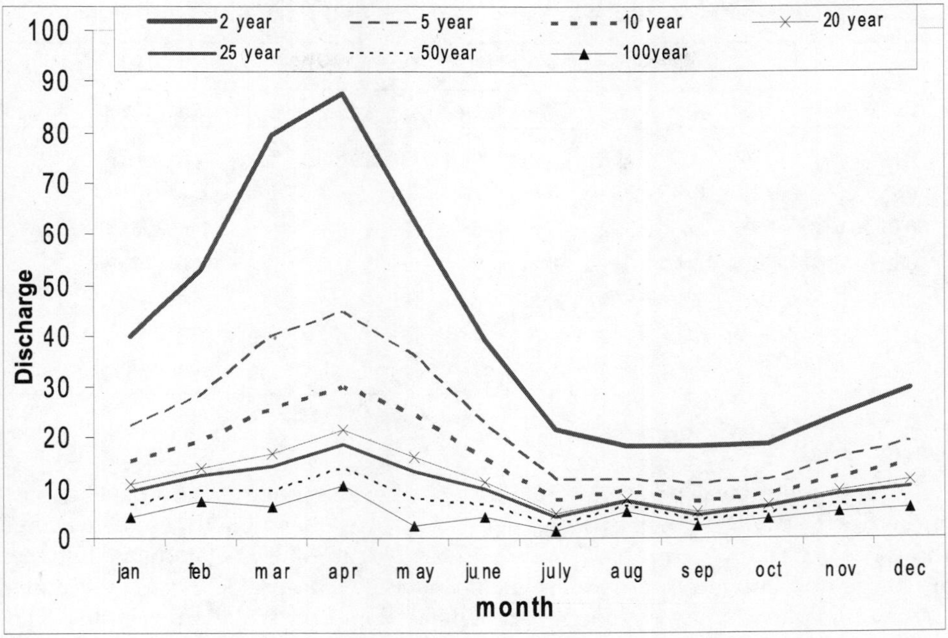

Figure 7.8: Low Flow Curve in Different Return Period

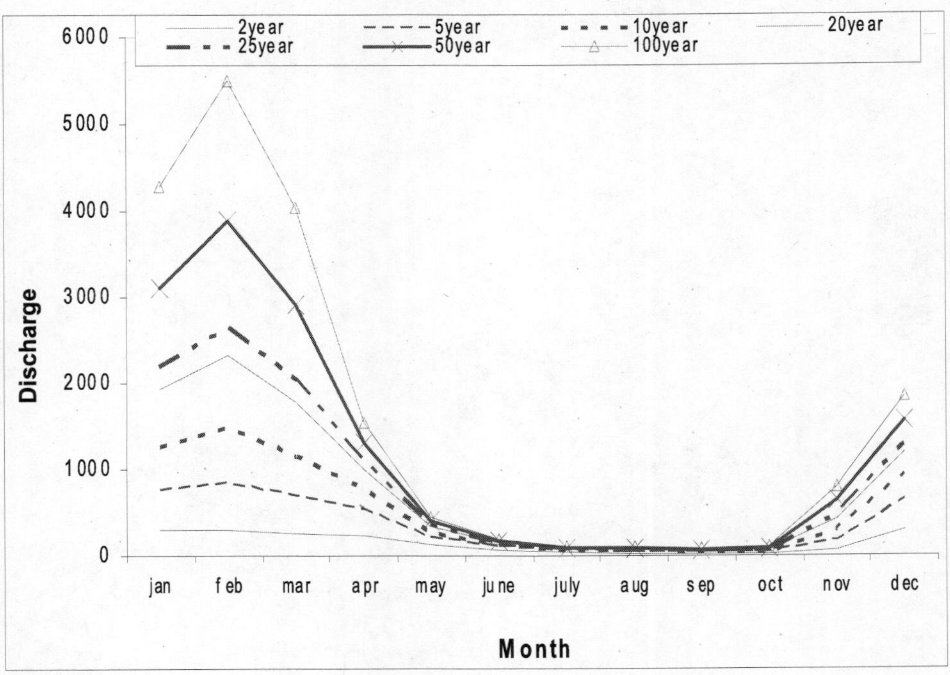

Figure 7.9: Peak Flow Curve in Different Return

Table 7.1: Monthly Average Volume Discharge and Different Return Period

Volume ($*10^6$ m^3)												Return Period
Jan	*Feb*	*Mar*	*Apr*	*May*	*June*	*July*	*Aug*	*Sep*	*Oct*	*Nov*	*Dec*	
254	301	378	390	263	136	78	60	53	57	97	178	2
486	644	729	674	392	216	122	98	81	83	160	357	5
675	972	978	863	460	268	151	120	98	99	211	526	10
882	1373	1222	1045	515	317	178	139	112	113	266	732	20
954	1519	1300	1103	531	333	186	145	116	117	284	807	25
1190	2033	1542	1282	578	379	211	160	127	128	346	1069	50
1450	2645	1783	1462	620	424	235	174	138	139	413	1379	100

Water Quality

Water quality was determined by using Dehmola sampling records. Chemical properties (EC, SAR, TDS, pH, Mg^{++}, Ca^{++}, Na^+, Cl^-, HCO_3^-, and SO_4^{--}) were measured and it has been shown in Table 7.2 and Figures 7.10 and 7.11. Relationship between discharge and chemical parameters including EC, SAR and TDS were estimated (Table 7.3). Water quality has been classified based on Wilcox diagram (Figure 7.12).

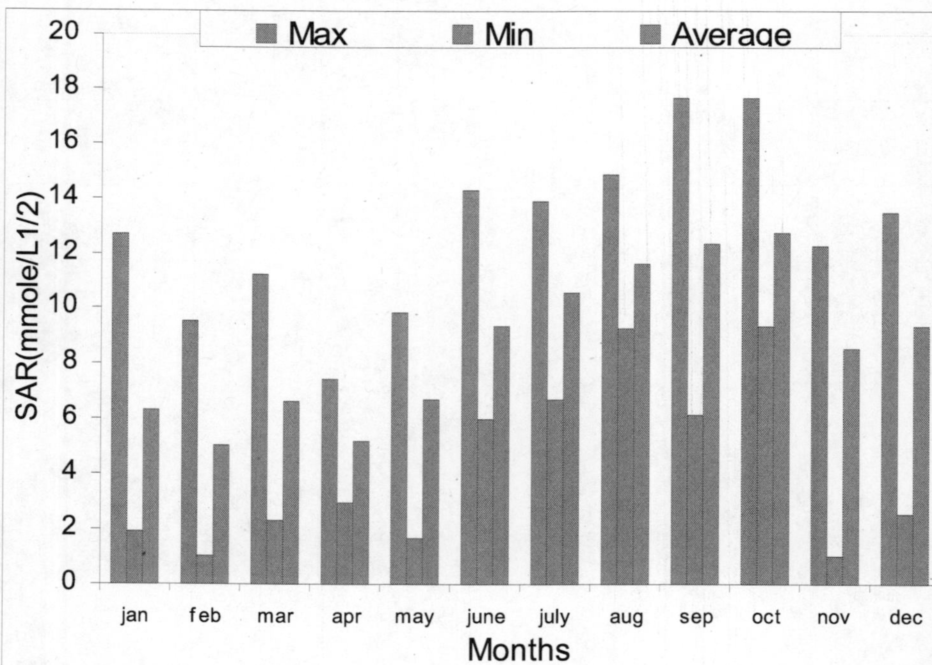

Figure 7.10: SAR Value per Month in Study Area

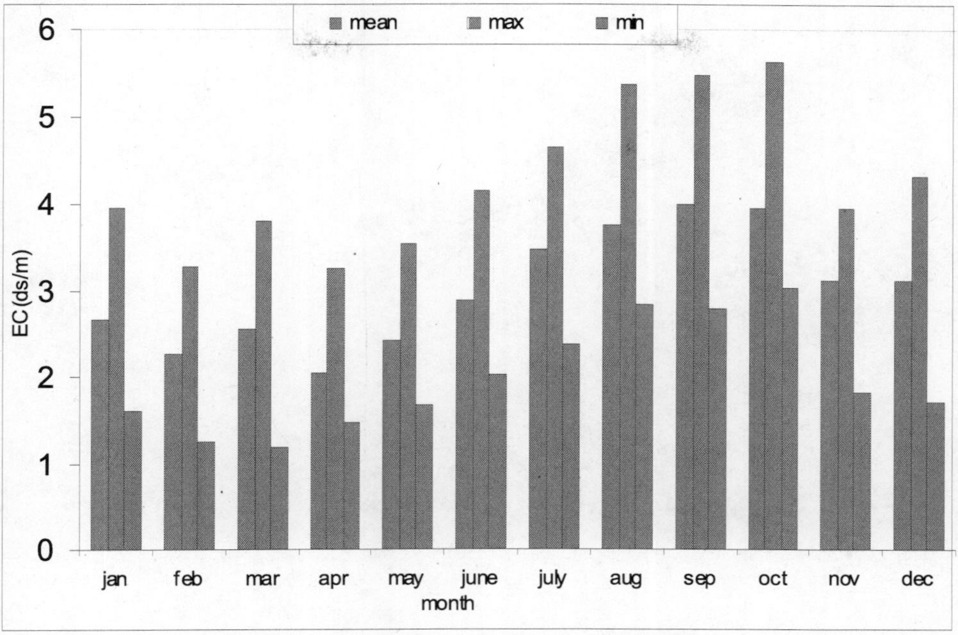

Figure 7.11: EC Value per Month in Study Area

Table 7.2: Average Water Quality Properties

TDS (mg/l)	*EC (dSm^{-1})*	*pH*	*Meq/l*						*SAR*	*Parameters*
			HCO_3^-	*Cl^-*	*SO_4^{--}*	*Ca^{++}*	*Mg^{++}*	*Na^+*		
1.9	3	7.9	2.3	20.3	7.8	7.3	3.2	19.7	8.7	Mean annually

Table 7.3: Relation Between Quality and Quantity of Water

Formula	*Correlation*
$SAR = 47.581Q^{-0.4413}$	$R^2 = 0.7325$
$EC = 8.4806Q^{-0.266}$	$R^2 = 0.7397$
$TDS = 0.6314EC - 15.207$	$R^2 = 0.8727$
$TDS = 4.9749Q^{-0.2412}$	$R^2 = 0.6578$

Water Requirement

Calculated water requirement for native plants by using monthly effective rainfall, crop coefficient (K_c), growth stage and evapotranspiration parameters (Figures 7.14 and 7.15) were assessed.

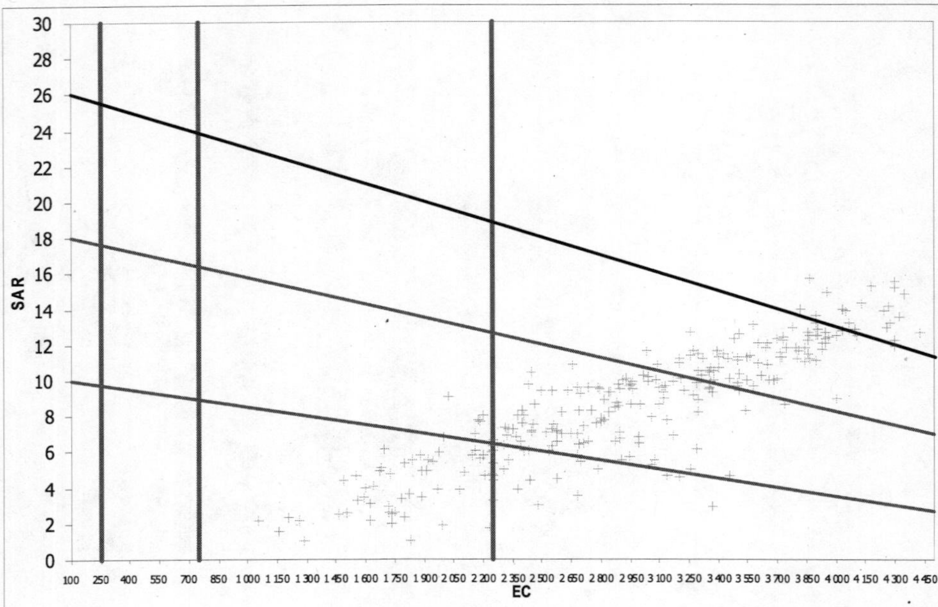

Figure 7.12: Wilcox Diagram

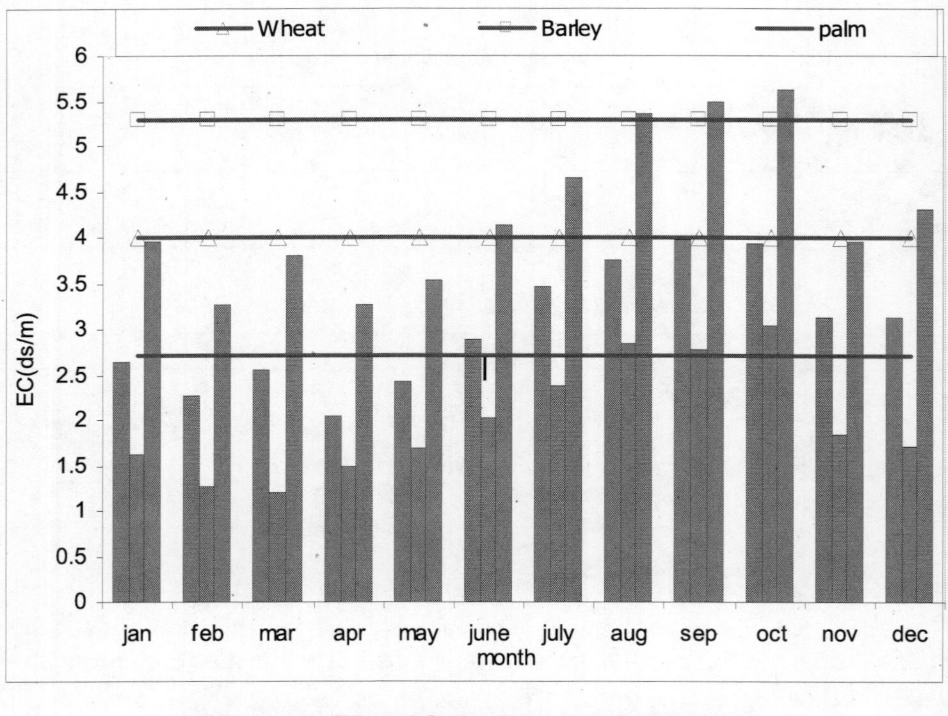

Figure 7.13: EC and Saline Tolerance Relationship

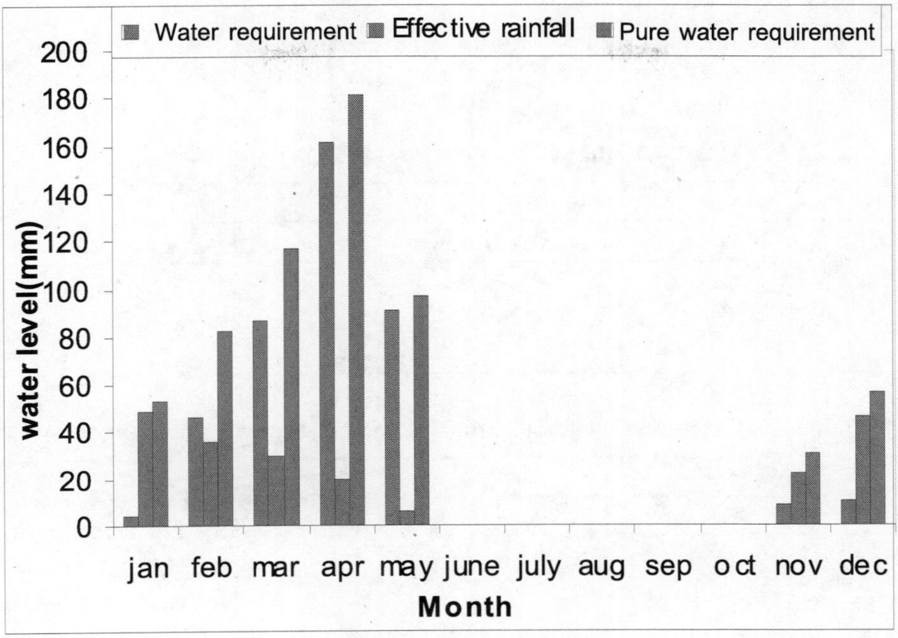

Figure 7.14: Monthly Water Requirement for Wheat and Barley

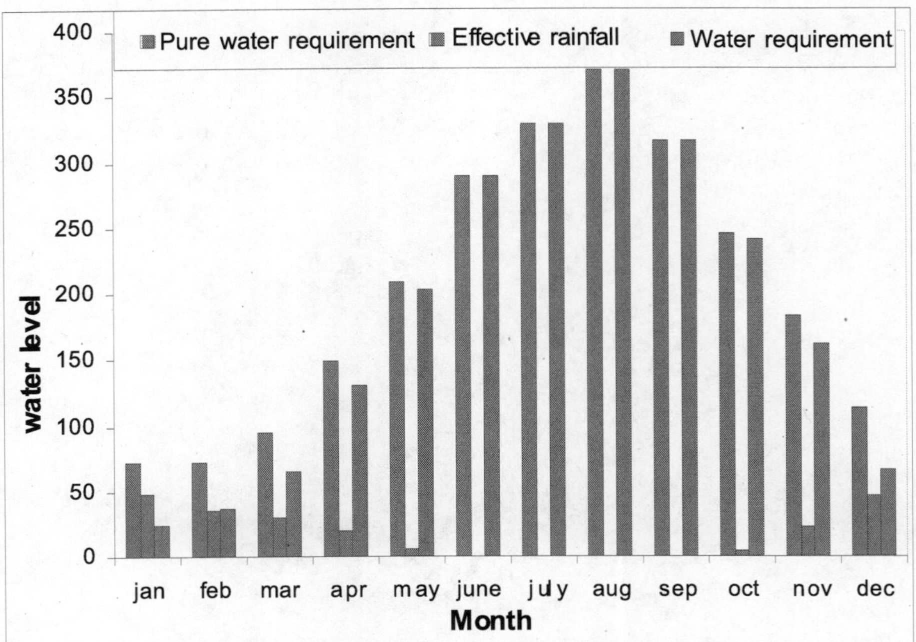

Figure 7.15: Monthly Water Requirement for Palm

Leaching Fraction

Added salt to soils by irrigation were calculated by using TDS and water requirement (Table 7.4), on the other hand leaching fraction for native plants in different months were computed (Table 7.5).

Table 7.4: Salt Accumulation in Growth Period (Ton/ha)

Month												Plant
Jan	*Feb*	*Mar*	*Apr*	*May*	*June*	*July*	*Aug*	*Sep*	*Oct*	*Nov*	*Dec*	
0.4	0.6	1.1	1.7	2.9	5.0	7.6	9.0	7.7	5.8	3.3	1.4	Palm
0.1	0.8	1.4	2.1	1.3	*	*	*	*	*	0.2	0.2	Wheat & barley

Table 7.5: Leaching Fraction in Growth Period of Native Plants (per cent)

Month												Plant
Jan	*Feb*	*Mar*	*Apr*	*May*	*June*	*July*	*Aug*	*Sep*	*Oct*	*Nov*	*Dec*	
25.6	22.4	23.2	18.2	22.4	26.1	35.3	40.4	42.5	41.2	31.6	30.2	Palm
16	14.1	14.6	11.6	14.1	*	*	*	*	*	19.3	18.6	Wheat
11.6	10.3	10.6	8.5	10.3	*	*	*	*	*	14	13.4	Barley

Soils Chemical and Physical Analysis

Physicochemical properties related to soil quality were determined from 70 random soil samples designated for analysis of soil chemical properties (Figure 7.16). Electrical conductivity (EC), pH, anions (SO_4^{--}, CO_3^{--} and Cl^-) and cations (Mg^{++}, Ca^{++}, Na^+ and K^+), cation exchange capacity (CEC), exchangable sodium per cent (ESP), sodium absorption ratio (SAR) and soil texture were analyzed for 3 soil depths (0-15, 15-30 and 30-60 cm).

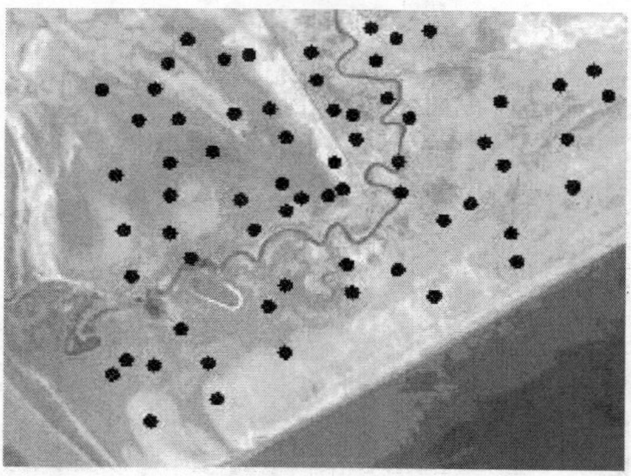

Figure 7.16: Soil Sampling Position in Study Area

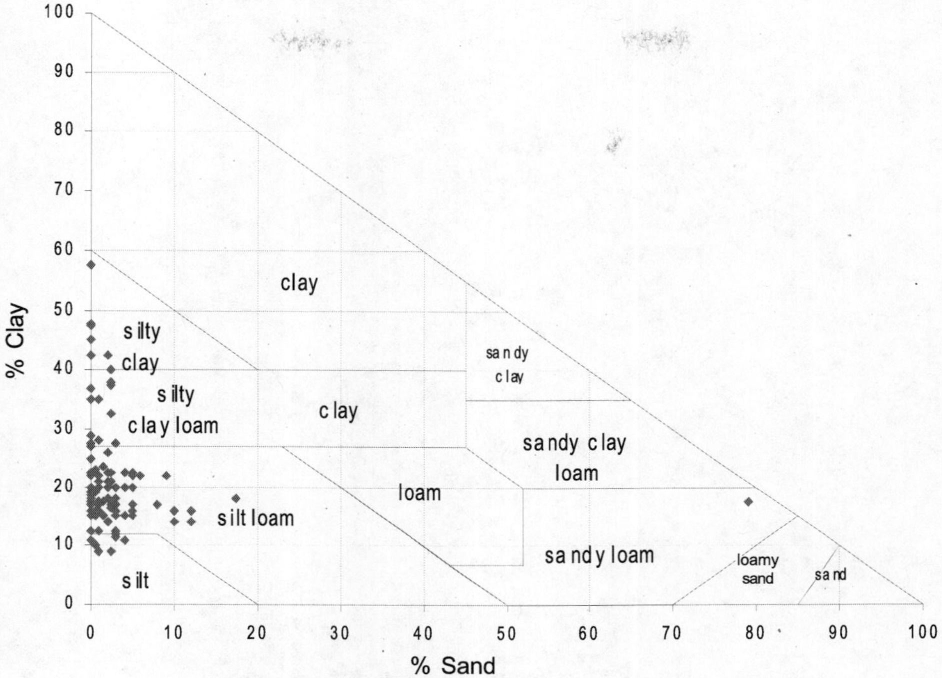

Figure 7.17: Soil Texture in Study Area

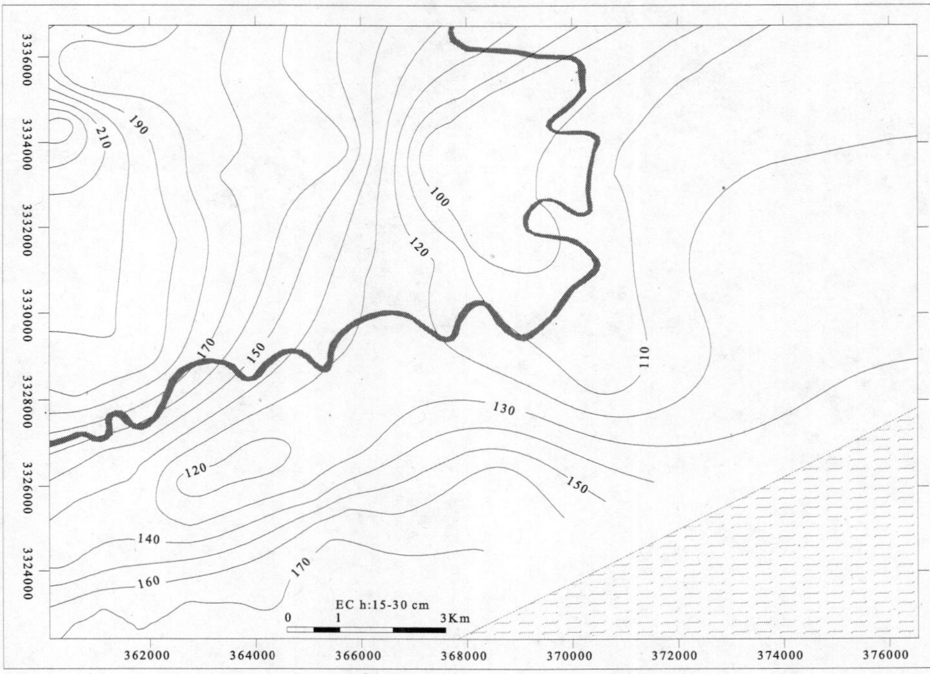

Figure 7.18: ISO EC Curve in 15–30 Depth

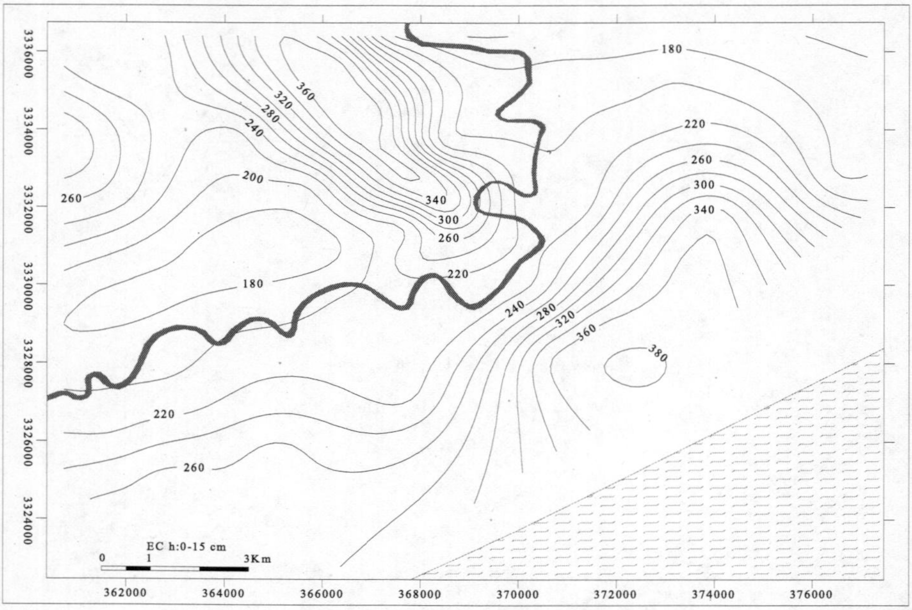

Figure 7.19: ISO EC Curve in 0–15 Depth

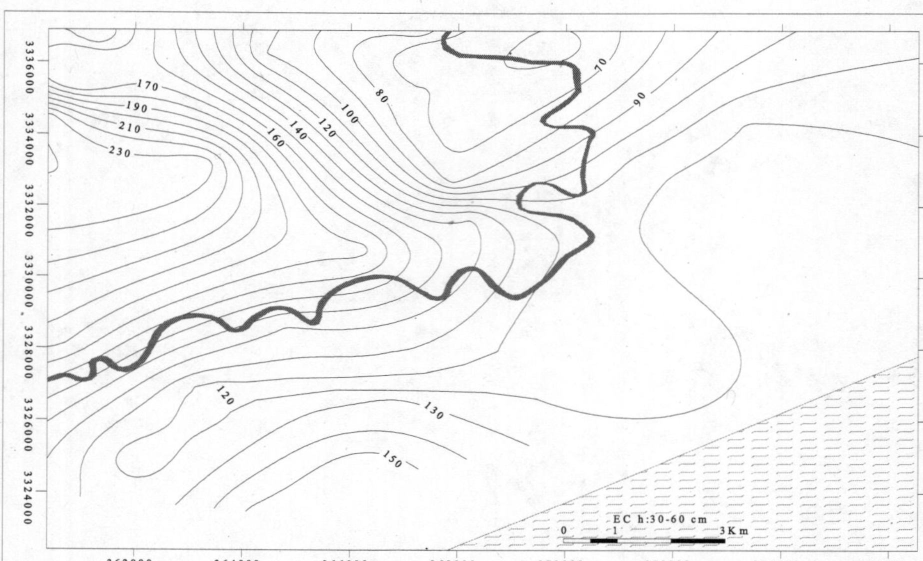

Figure 7.20: ISO EC Curve in 30–60 Depth

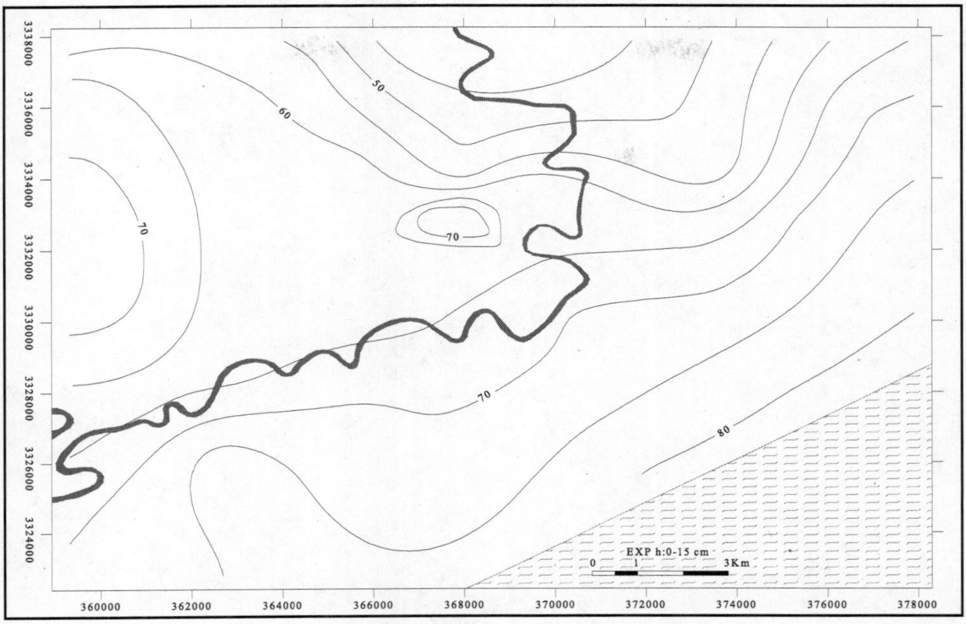

Figure 7.21: ISO ESP Curve in 0–15 Depth

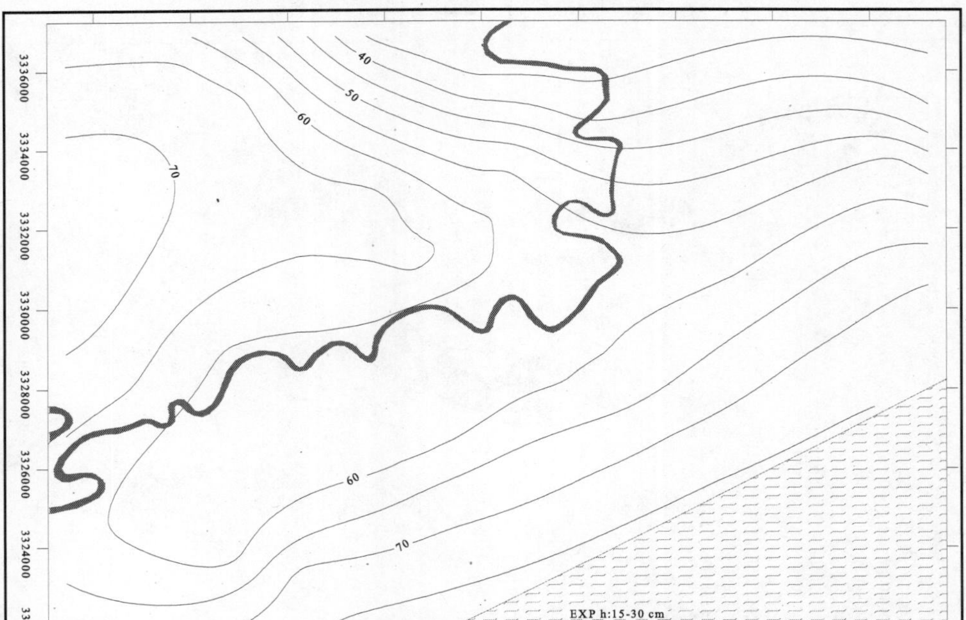

Figure 7.22: ISO ESP Curve in 15–30 Depth

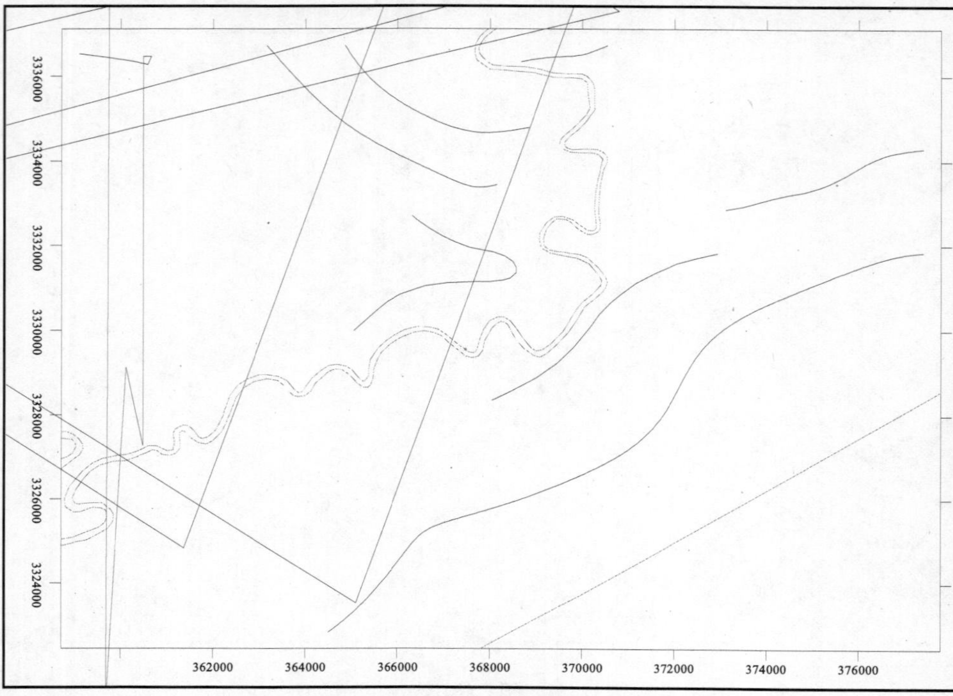

Figure 7.23: ISO ESP Curve in 30–60 Depth

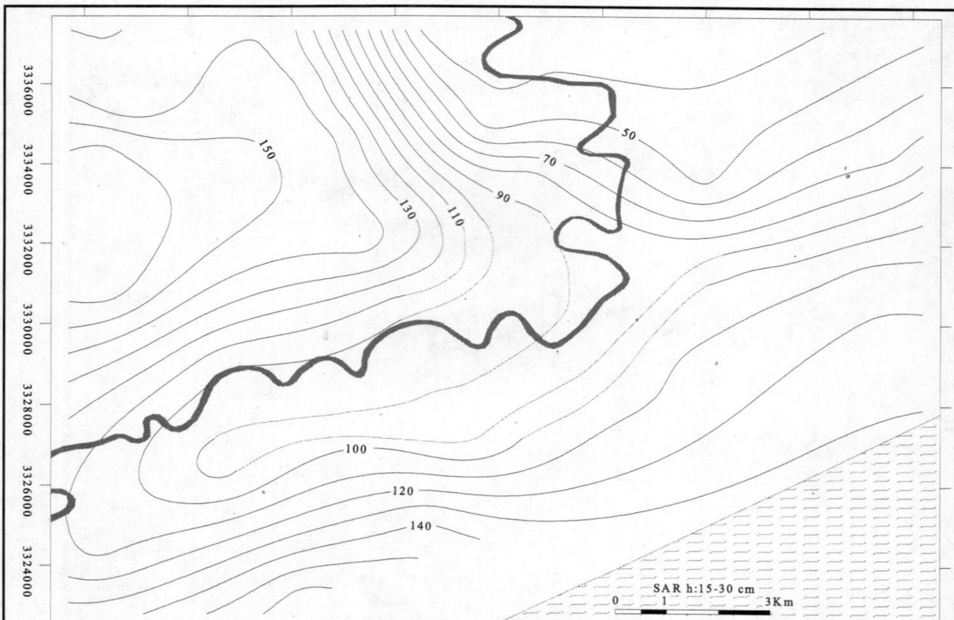

Figure 7.24: ISO SAR Curve in 0–15 Depth

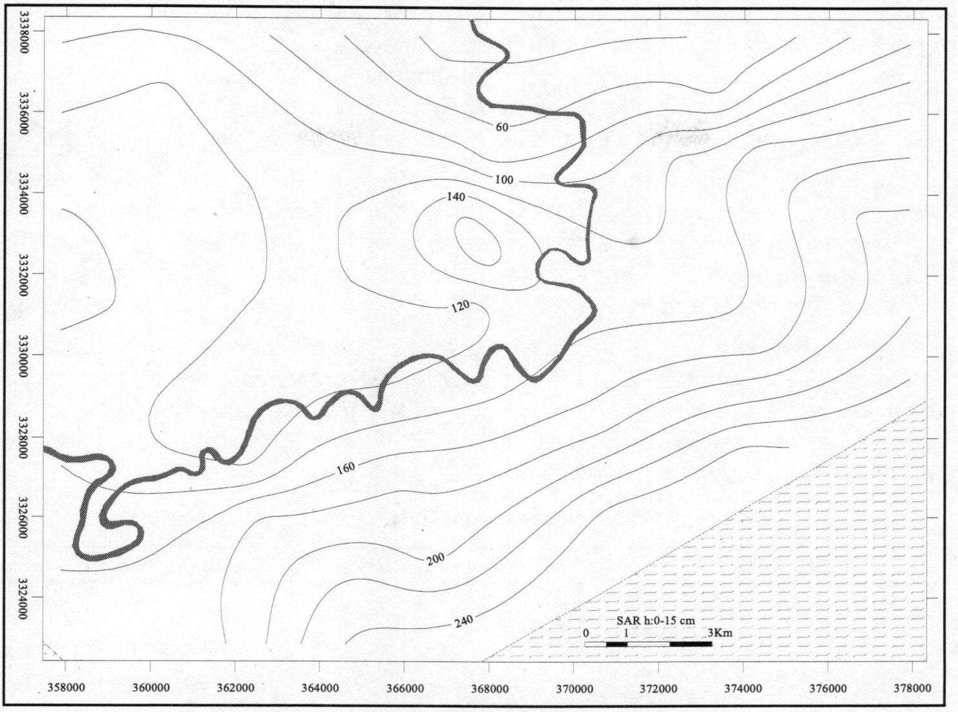

Figure 7.25: ISO SAR Curve in 15–30 Depth

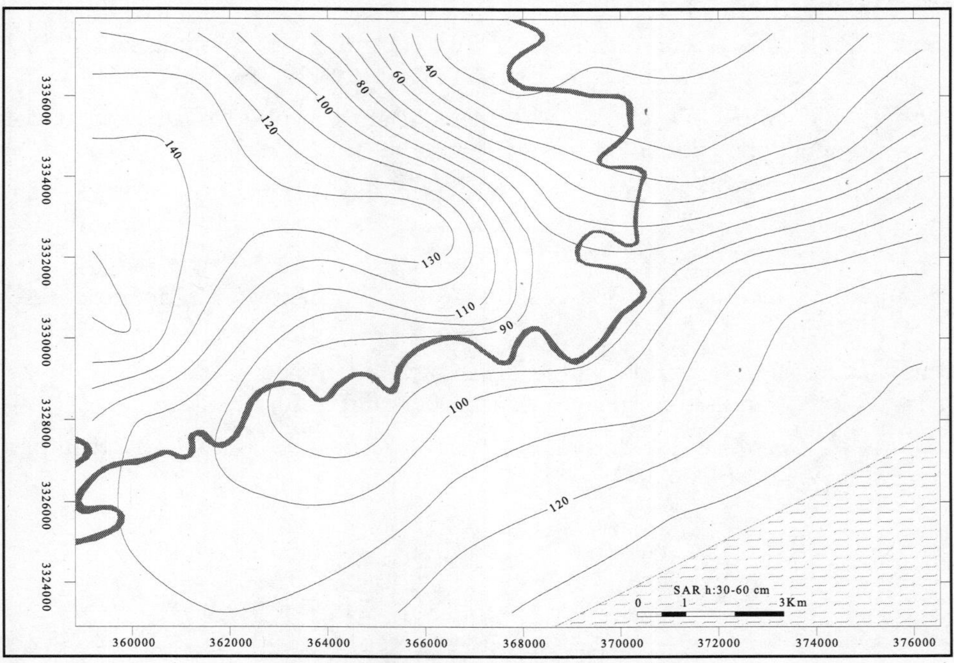

Figure 7.26: ISO SAR Curve in 30–60 Depth

Figure 7.17 shows that soil texture is silty-loam. All spatial data were entered in to a geographic information system (GIS) using ArcGis9.1 and sufer7. Maps of the soil chemical properties were prepared by interpolating the measurement at the 70 sample sites, using Kriging interpolation. The studies showed that kriging method is the best for mapping chemical properties of the soil.

The ESP, SAR and EC maps most closely reflect the spatial distribution of root zone salinity. These maps show that the soils are saline-sodic and the concentration of salinity in the soil surface is more than deeper layers (Figures 7.18–7.26). Regarding to CEC, bulk density, soil depth, initial ESP (35 and 60), final ESP (5) and water quality the quantity of required gypsum and water level has been defined (Table 7.6).

The results indicate that there are no problem in terms of water properties and climatology for agriculture and soil reclamation. We recommend that: 1-take out surface layer of soil, 2- using suitable drainage for soil leaching and crop plantation such as barley because this could reduce salt from soils after harvesting and its saline tolerance is more than other.

Table 7.6: Amount Water and Gypsum for Leaching and Reclamation

ESPi=60 ESPf=5		*ESPi=35 ESPf=5*		*Condition*
60	15	60	15	Leaching depth (Cm)
3.9	1	2.2	0.5	Water level (m)
33	8.2	18	4.4	Gypsum (ton per ha)

References

Alam, S.M., R. Ansari and M.A. Khan. 2001. Reclaiming saline/ sodic soil. http://www.pakistaneconomist.com/issue2000/issue19&20/i&e3.htm

Cooper, R.M. and G.D. Istok. 1998. Geostatistics applied to ground water contamination methodology, J. Environmental 114: 270-286.

Gallichand, J.G, D, Buck land, G.D. Marcotte and M.J. Hendry. 1992. Spatial interpolation of soil salinity and sodicity for a saline soil in southern Alberta, Canadian Journal of Soil Science 72: 503-516.

Irrigation and Drainage Paper No. 29. 1976. Water quality for agriculture, FAO publication.

Mughal, F.H. 2002. Irrigation- Induced Salinity (Environmental Impacts). http://www.pakistaneconomist. com/issue2000/issue19&20/i&e3.htm

Rhoades, J.D. 1999. Use of saline drainage water for irrigation. Journal of Agricultural Drainage 38: 615-657.

8

Ecosystems as Accelerators of the Energy Flow from the Earth Surface

Helmut Lieth

Environmental Systems Research, University of Osnabrueck, Germany

E-mail: helmum.lieth@t-online.de, helmut.leith@usf.uni-osnabrveck.de

ABSTRACT

Recent evaluations of species diversity in ecosystems are mainly concerned with number of species in relation to a variety of external and internal parameters. Species diversity is mainly considered as number of species per unit area, but the flow of energy is seldom considered among them. Depending on the size of the area concerned the data for number vs. surface area are plotted in log/log scale, log/linear scale, or even linear scale usually without concern for the size of the species.

The energy flow is considered in ecology on various scales and for various purposes. The common names for it are photosynthesis, respiration, net primary productivity, gross primary productivity, food chain, food consumption, predation, transpiration and so on. In all cases the functional parameters are related to biota. The possibility of considering these functions as part of a general principle of energy flow is only discussed in biometeorology, but even here the process of energy flow is mostly restricted to its impact upon weather and climate. The calculation of energy transfer from the earth to outer space is left to astrophysicists.

The development of mankind in biomass, number of people, utilization of fossil energy resources, and consumption of ecosystems resources require a careful analysis of the present conditions in the context of global energy flow. It is necessary for ecologists to participate in the discussion not only in the above-mentioned aspects, but also in a theoretical context.

In ecology, the recent discussions of ecosystems diversity are mainly related to species diversity. The maximum number of species per unit area or per unit ecosystem appears as the ultimate goal in such papers. Only a few discussions consider the energy flow of the biota in general, and practically none of them consider the total energy flow of an ecosystem including all aspects of energy reception, conduction and dissipation as part of an evolutionary principle.

In this paper we propose new research in this direction. We are concerned with the inclusion of the biological energy flow and especially its conduction speed as part of the general energy flow pattern between earth and outer space. We shall try to explain the biological energy flow as part of the "mass" in the Einstein equation $E = m \times c^2$ and its flow velocity as a modifying function of the c^2 depending on the temperature. Based on this equation we see the evolution of ecosystem's structure and function in part as the attempt of the biological evolution to maximise the total energy flow. This may be achieved by improving the total energy conduction and by developing different kinds of pathways for energy conduction. We propose, that from the various options available those will be preferably developed, which can conduct the energy under the existing environmental conditions most efficiently. The conduction options can be assessed through Kirchhoff's laws written for electrical currency. In analogy to these laws we can calculate the best ways of energy conduction if we include all avenues of energy conduction through any particular ecosystem on earth. This calculation will provide the goal for evolution of any ecosystem anywhere on the earth.

This includes besides the energy transfer from biomass production and decay under certain temperature levels the much larger energy transfer through water evapotranspiration and abiotic energy transportation properties of ecosystems compartments. It also includes the energy spent for the development of information units, which enhance the optimisation of energy allocation in the development of energy conducting structures across the entire ecosystem. This will help us to get a logical explanation for species diversity as well as for the structural diversity in ecosystems parameters, like compartment sizes, height of ecosystems, shoot/root ratio, herbaceous vs. woody tissue, perennial vs. annual species and a multitude of other properties. On a different level of ecosystem development, are the generation of information processing tissue and the information processing itself inside each ecosystem important?

The value and the need to look at the ecosystem in this way are stressed and new theoretical work is suggested. We need to describe the evolution of ecosystems not only as given facts but also as part of the development of the energy flow optimisation between the sun, our planet and outer space. In the context of information processing abiotic compartments are as important as biotic compartments. A series of model developments and concurrent field experiments are needed to analyse in which ways the energy conduction is optimised in the various ecosystems on earth. A special problem hereby is the question that which properties of an ecosystem are developed intrinsically and which is controlled by external forces?

Introduction

Several papers on biodiversity appeared recently under various aspects in ecological journals. Schreiner (2004), Turner (2004), Blackburn and Gaston (2004) take the increase of biodiversity for granted and discuss the pattern as it is usually done, following the paradigm expressed in island species diversity and endemism development, depending on area size and distance from source.

The fact that ecosystems are an assembly of producers and consumers in a food chain is well recognised in ecology, but is often discussed for taxonomically separated species groups as if they were independent from other parameters. So was it done in some of the papers mentioned above. This way of discussing species diversity may be interesting for understanding a particular section of the ecosystem, but from my point of view it misses one of the main goals of ecology, the explanation for the purpose of ecosystem structure and function for the system itself and for the place of the various ecosystems in the context of a universal law of evolution.

Only in the wake of the IBP studies of the 60 ies and 70 ies of the last century was emphasis placed upon the energy accumulated and digested by entire ecosystems. The papers by Odum (1977) became a classic with respect to the gross primary production as an input into the energy budget of ecosystems including man. My own contribution to this topic focussed on the net primary production and culminated in the global net primary productivity pattern as a surface property of the world vegetation (Figure 8.1).

Both, Odum and myself considered at that time only the biomass (carbon) related energy as the net amount of the energy flow through the entire ecosystem. Following that, Lieth and Box (1977) compared several global patterns of climatic parameters with the global NPP and GPP.

A full discussion of all different comparisons with climate and vegetation parameters is available in Lieth (1981). During these analyses it became evident, that major ecosystems functions were strongly correlated if not completely controlled by environmental parameters optimising the energy flow. The total energy flow from any ecosystems in the presence of sufficient water was mainly optimised by the vegetation. Its structure is built for the optimisation of energy flows in many forms. Additionally it provides the first set of information processing devices, which can be used by all other compartments of the ecosystems. When we compared the efficiency of solar energy fixation by the ecosystems in general, we found that it was surprisingly low in comparison to the entire energy conducting compartments. This led us to question the paradigm of that time that the ecosystems were evolving only for the benefits of the biota.

We try to explain in a few chapters our views and offer a number of problems for analytical and modelling work to compare the importance of energy flow efficiency management in different ecosystems. This will show us how information processing can improve the energy flow developed through gravity and water evaporation alone. This may explain the creation of new information systems by the biota with the help of nervous tissue.

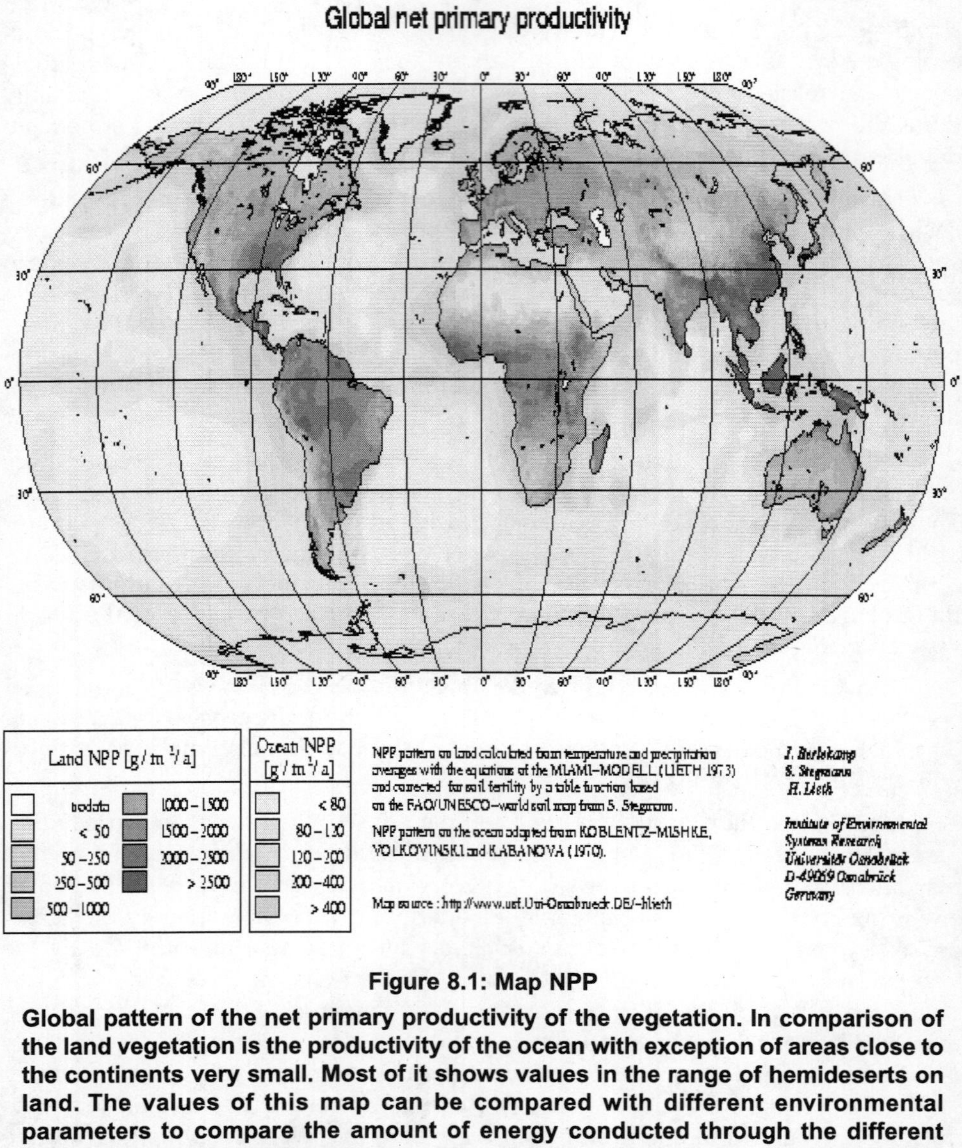

Figure 8.1: Map NPP

Global pattern of the net primary productivity of the vegetation. In comparison of the land vegetation is the productivity of the ocean with exception of areas close to the continents very small. Most of it shows values in the range of hemideserts on land. The values of this map can be compared with different environmental parameters to compare the amount of energy conducted through the different types of ecosystems. Wherever enough water is available the energy flows at this place along the use and evaporation of water.

Some Facts of the Global Energy Flow

We consider the maximization of energy flow and transfer to be a universal principle. It takes place on all levels of the organisation of matter according to the thermodynamic laws of open systems. Energy in the form of matter or attached to it will flow to places of lower energy content in various ways. The temperature difference between the earth surface and outer space is the force for it. The amount of energy transferable through a particular system compartment may be the stimulation to prefer its development with other possibilities.

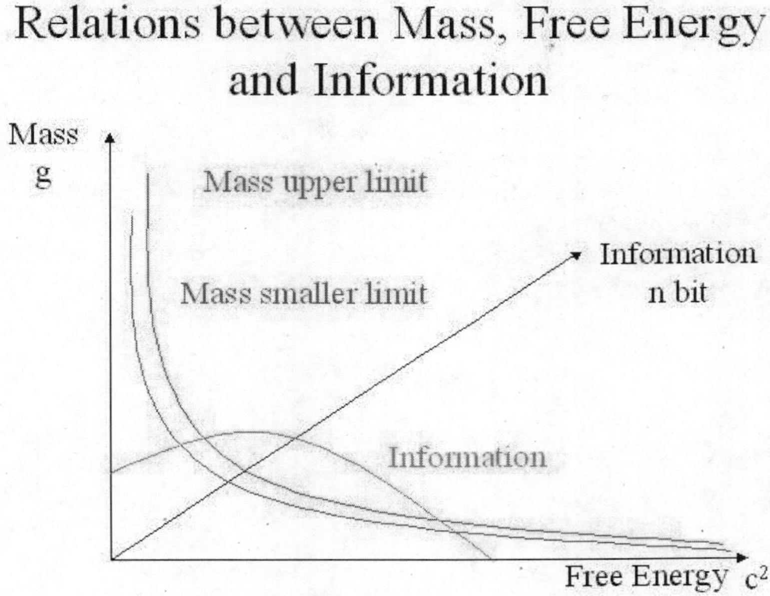

Figure 8.2: Relation Between Mass, Free Energy and Information comparing the Energy Flow in and to Outer Space in Relation to the Mass Size Carrying the Energy and the Information Content. Large masses in the universe travel slow and release energy with carriers of a mass size tending practically towards zero. This energy travels with the speed of light, while the larger masses travel much slower. All matter contains information. Big masses appear to have only information related to gravity, while the smaller masses carry many more types of information.

In order to understand the energy flow at the surface of the earth, we must forget initially the standard distinction between biological and abiological energy flow. Energy is absorbed everywhere on the earth surface, stored, conducted and dissipated according to the same laws of thermodynamics and energy conductance, valid in the same way for biomass as well as for rock and water. These laws are known since almost two centuries but rarely used in the analysis of ecosystems. Only a small part of it is converted into mass, the other part is used to increase the temperature of the total ecosystem, which then provides the force for energy conduction to compartments with lower temperature. The importance of temperature for all systems on earth is recently discussed in a sociological context by Markert *et al*. (2006). If we follow the universal energy flow pattern, as it is understood in principle by the astrophysicists, we can use for the biological energy flow the following diagram.

Mass can be understood as gravitational energy, as we know it from lake water in a higher position on the earth surface. Gravity can pull the water down and enables the utilization of the energy of the flowing mass for purposes other than the water transport itself.

In the case of biomass the mass itself is comparable to the lake water and radiation of any kind the flowing energy. Instead of gravity, provides the temperature difference between earth and outer space, the power needed to pull with lower gravitational energy gases and radiation away from the earth. It is like the release of small particles, which continuously flow from the large masses of the earth in several steps into outer space. On their way out these particles may be accelerated by solar energy absorbed by soil surface and air, by the centrifugal power of the earth movements and from the attraction of other big masses. This flowing energy will be combined with the heat conducted from the interior of the earth and converted in the low atmosphere into other forms of energy or air mass properties.

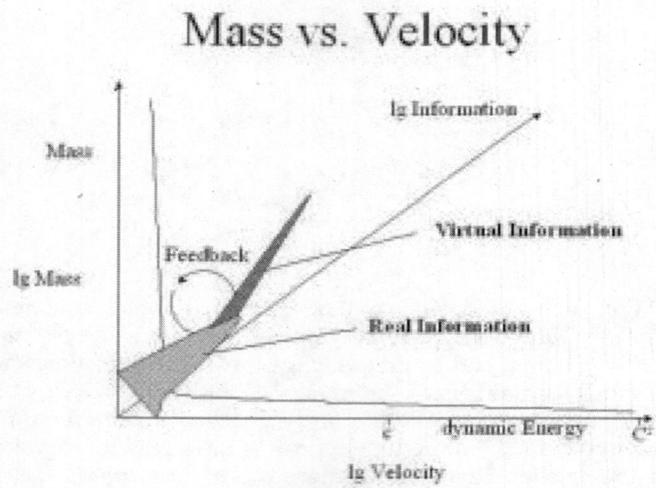

Figure 8.3: Mass vs. Velocity

In Figure 8.2 we postulate that lower masses contain more information than big masses. In this range of mass sizes is obviously information processing devices evolved which can accelerate the energy flow. The information bits in some of these mass sizes can generate information types, which have no relation to the mass. In the cases where this occurs we must distinguish between real and virtual information. The importance of this distinction is discussed by Lieth (2005).

We know the conversion of solar energy usually under the names photosynthesis, transpiration, growth, temperature increase, re-radiation and so on. This can be compared to the energy utilization in industry where energy from fossil fuel or nuclear fuel is converted into structures and functions useful for man. Man tries to optimize these utilizations in various ways, improving the efficiency of the conversion as well as improving the structure of the system itself. This activity follows the general

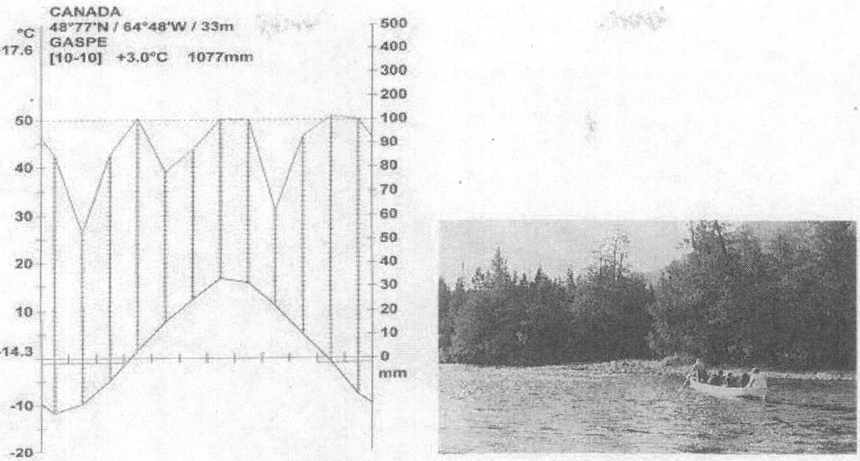

Figure 8.4: Climate diagram Canada (Gaspe)

The climate diagram of Gaspe represents the boreal coniferous forest, here shown from a forest at the Gaspe Peninsula in Quebec/ Canada. This forest type conducts energy in summer in large quantities through the evaporating foliage. In winter, when water is frozen, the system is covered with snow, which can reflect the incoming radiation much faster than the small amount, which can evaporate.

principle of universal energy flow maximisation. The natural development of ecosystems follows this principle as well. In order to enable the ecosystems compartment to convert energy into flow improving structure, they use a substantial part of it for the building of information systems, which allow them to process the energy, flow partly for their own benefit and partly for the purpose of the global energy dissipation *e.g.* relative humidity, wind, dispersal mechanisms etc. The present condition of mass size, internal and surface temperatures provide together the power to accelerate "below quantum size particles" as energy away from the big mass of the "planet earth" into outer space against the gravitational power of the planet. An important aspect is, however, the transfer of radiant energy into masses of any kind and vice versa. This aspect is not yet recognized in ecology, but might be possible if we can calculate the energies absorbed or released according to Einstein´s formula. It might give us a better understanding why the energy conversion efficiency of photosynthesis is so low in comparison with mechanical energy conversions. It requires much analytical work if we want to understand some of the astonishing facts of information processing within and along the structural and functional properties of the ecosystem and how much energy it takes to develop these systems in comparison with the energy needed for structural elements and/or evaporating water.

In biometereology we explain the energy budget of any place on the earth surface in the following way: The incoming solar energy is reflected from the bare soil or is used in and above the soil for temperature increase, on open waters mostly into water vapour. This is visible on the daily and annual march of temperatures on climate

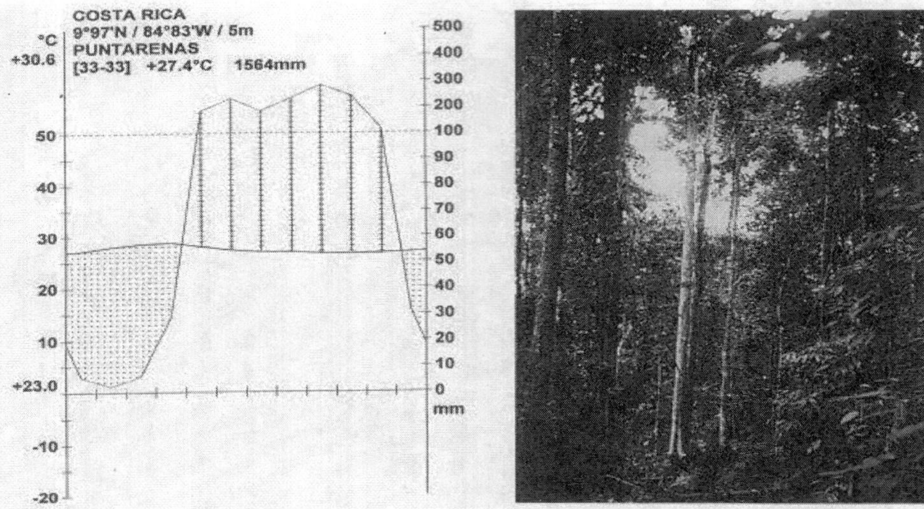

The evergreen tropical rainforest grows into 50- to 70-m tall stands of trees of many species from many genera. On fertile soil like here in the Osa peninsula in Costa Rica year-round temperatures of about 27°C and rainfall of up to 3000 mm annually provide conditions for fast growth.

Figure 8.5: Climate Diagram Costa Rica

The climate diagram of a place on the Osa Peninsula in Costa Rica shows a system, which receives a lot of rain during certain parts of the year. At this time we notice a slight drop in temperature, which is related to the evaporation of water from the vegetation. The majority of the energy received at the surface of the earth is carried outward through the conversion of liquid water to water vapour. This process is largely improved by a vegetation cover. The amount, which cannot be evaporated under the existing temperature range, flows as liquid water by gravity eventually towards the ocean.

measuring stations all over the world. This is quite different in contrasting environments with and without vegetation. See climate diagrams in Figure 8.4, 8.5 and 8.6, taken from Lieth *et al.* (1998).

The conversion of radiation to heat at the surface of the earth is comparable to the energy flow in industry and for the water flow. The difference in the principle is only that the example of the lake water stays within the mass principle, the industry example converts mass to energy but is implemented by external information inputs, whereas the energy flow through the ecosystem is first used to accumulate and even convert energy into mass and then converts this mass back to energy, which is conducted by the atmosphere to outer space.

Our postulate is that the ecosystem compartments are evolving always new means of energy flow as long as there is a force present which can improve the energy flow. Temperature increase as well as biomass accumulation provide such energy potential.

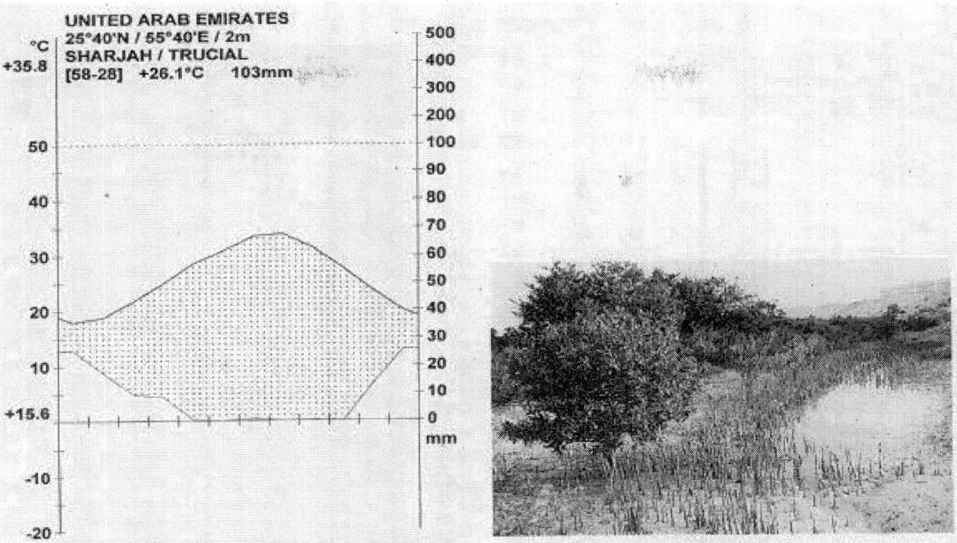

Flat and muddy coastlines above and below the mean tidal sea level are the signals to which in the tropical zone mangrove ecosystems respond. The picture was taken at the shore of Abu el Abbyat in Abu Dhabi Emirate. The mangrove ecosystems are among the most productive biological systems on earth, which demonstrates clearly that salinity itself is not a general stress factor. Most of the plants in this system require sodium chloride for proper development. In the temperate zone such locations are occupied by salt marshes, void of trees.

Figure 8.6: Climate Diagram United Arab Emirates

The climate of the United Arab Emirates represents desert conditions. The low water regime prevents the development of a vegetation cover around most of the year. Only near the ocean coast is vegetation. Here the evaporation of water from vegetation can improve the flow of energy onto the air. The vegetation at this belt consists mainly of mangrove species, which can tolerate the elevated salt content of the water.

The major problem in understanding the connection of the universal energy flow from radiation energy to biomass production to respiration and water evaporation, to atmospheric processes of air mass movement and cloud formation on the base of the Einstein equation is the fact that we have on earth the sharp distinction between energy transfer between lager masses and energy conversion from and to very small masses which can flow with the speed of light. It is necessary in all future research to detect the positions in the ecosystem energy flow, where this occurs. Figure 8.7 shows a simplified model of such a flow with the indication of points where the conversion between mass flow and radiation flow occurs. We need in ecology the calculation of conversion rates between these two types of energy transfer.

The Evolution of Ecosystems for the Energy Flow from the Earth Surface to Outer Space

When we look at the ecosystem from this point of view, we will soon see that biodiversity is a natural force to maximise the energy flow between the combined

Simplified model of the present avenues of energy flow across the ecosystem on earth.

Energy Sources	Internal heat	Solar radiation		
Soil surface compartments	Conduction to soil, to air	Conversion		Information
	conduction turbulence air lifting	water evaporation	soil water temperature evaporation	biotic compartments conversion to biomass CO_2 reduction oxidation food chain
	temperature radiation	accumu-lation of biomass		
		degradation storage	conduction improvement	
sink	transfer to outer space	fossil fuel sediments	more avenues higher speed	

Figure 8.7: Simplified Models of Water Flow Avenues in Ecosystems. For more explanation see the text.

energy volume (heat) created from the indigenous energy of the interior of the earth and the additional energy absorbed by incoming radiation. The internal heat energy of the planet is mainly responsible for the average temperature on the earth surface, being around 14℃. The incoming energy from solar radiation is added onto the indigenous heat provided by the hot earth interior. Its daily and annual fluctuations create the undulating temperature patterns on the earth surface. We postulate that both, the outcoming as well as the incoming energy flow is responsible for the evolution of the ecosystems on earth. Both are important parts of the energy flow to outer space. This is reflected in the calculation of the biometeorologists. The equation they negotiate is usually neglecting the irradiation of the globe itself. It only tries to balance the various parts of the energy release from the surface of the earth against the incoming solar radiation including the radiation from the blue sky.

The values for the parameters in this equation are very different in contrasting ecosystems. In Table 8.1b some estimates are shown as percentage of the energy flow conducted by each type of energy conduction. The table shows that E carries the largest amount over the ocean and tropical forests, O and R the largest amount over the desert and P appears in all systems very small.

To find more details of the energy flow developments will increase our understanding of the ecosystems structure and function. Especially the reason for the development of information processing to improve the energy flow is not sufficiently understood.

Table 8.1a: Simplified Equation of the Energy Processing in an Ecosystem

The equation reads (meaning of the symbols is shown in table 1a):

I = O + R + A + E + P + S

Energy balance for Meteorology

I–Solar input

O–Direct radiation into outer space

R–Reflection

A–Advection (air mass exchange)

E–Evapotranspiration of Water

P–Photosynthesis–Biomassproduction

S–Storatge or loss of sensible heat

Table 8.1b: Estimates of Percentage of the Energy Flowing through each Avenue Distinguished in Equation a. These shows, that deep water have no vegetation at the surface. One of the reasons for this may be the fact that a vegetation cover will not improve the speed or the amount of energy transported which an open water surface can achieve

Vegetation Type	*O*	*R*	*A*	*E*	*P*	*Total (%)*	*Temperature Difference Day/ Night (°C)*
			(%)				
Dry desert	10	60	30			100	80-90
Tropical rainforest	5	8	10	75	2,00	100	0-10
Grassland	5	10	25	60	0,5	100,5	10-40
Ocean	5	10	15	70	0,01	100,01	0-10

Conclusions and Suggestions

The first lesson we learn from such analysis is the fact that the ecosystems did not evolve the food chain for the sole purpose of feeding animals and men, but to accelerate the conduction of the heat from the planet earth, plus the absorbed incoming solar energy from the earth to outer space. We assume, however, that the biota has the possibility to create information processing devices, which optimise this flow partially for their own benefit. One of the most prominent features in this respect is the evolution of the best possible homeothermy of the entire ecosystem and the lowest energy input into its structural development. The response to environmental constraints includes the biotic as well as the abiotic compartments of the system. This appears to be controlled for the biotic compartments on the genomic level and for the abiotic compartments by following the physical forces, which provide the fastest energy flow.

For these tasks all levels of the individual taxa as well as the combination of taxa in the ecosystems are involved. The results of such studies may have a great influence on our understanding of nature conservation and future land use options under the present population pressure.

The different amounts of energy fixation on the globe and the different amounts of water converted from liquid to gas or ice must have locally large consequences for the pressure upon the solid earth crust as well as for weight and volume of the global pool of air and water. These are intrinsic parameters for the function of ecosystems. The external controls for the result of these functions are the environmental parameters existing at the place of any ecosystem type. The difference in weight and energy release or fixation is rarely being calculated. We propose that such calculations should be attempted in the near future because it might give us more insight in the global climate variabilities which are presently becoming a major problem for the human society.

Acknowledgements

We thank the Arnold-Sommerfeld-Gesellschaft for the possibility to present this paper at one of their seminars in Leipzig. We also thank Dr. Meiss for his assistance to get this paper published. Several members of the Sommerfeld-Gesellschaft provided critical discussions. Frau Brigitte Herzog deserves special thank for her assistance to type the manuscript.

References

Bais, S. 2005. Die Gleichungen der Physik, Meilensteine des Wissens. 96 p. Birkhaeuser, Basel.

Blackburn, T.M. and K.J. Gaston. 2004. Macroecology. Basic and Applied Ecology. Vol.5, pp. 385-387.

Del Monte-Luna, P., B.W. Brook, M.J. Zetina-Rejon and V.H. Cruz-Escalona. 2004. The carrying capacity of ecosystems. Global ecology and biogeography vol.13, Nr.6, pp. 485 - 495.

Lieth, H. and E. Box. 1977. The gross primary productivity pattern of the land vegetation: A first attempt. Tropical ecology 18, pp. 109–115, Varanassi, BHU press.

Lieth, H., J. Berlekamp, S. Fuest and S. Riediger. 1999. Climate Diagram World Atlas CD 1, Leiden, Backhuys Publishers.

Lieth, H. 1981. Der Energievorrat der Pflanzendecke und seine Nutzung. pp.141–177, In: Hosemann, G. and Finckh, E., Herausgeber: Auf der Suche nach Energiequellen. Erlanger Forschung Reihe B, Bd.12, Erlangen.

Markert, B., H. Lieth, P. Mencke-Gluekert, M. Hofsang and S. Fraenzle. 2006. Zur Existenz eines ganz starken anthropischen Prinzips. 60 p. Norderstedt, Books on Demand GmbH.

Odum, H.T. 1971. Environment power and society. 331 pp., New York, Wiley and Sons, Inc.

Scheiner, S.M. 2004. A melange of curves–further dialogue about species-area relationships. Global ecology and biogeography vol.13, Nr. 6, 479–484.

Turner, J.R.G. 2004. Explaining the global biodiversity gradient: Energy, area, history and natural selection. Basic and applied ecology, vol.5, 435–448.

9

Chemical Composition and Digestible Parameters of Various Halophytes

A. Riasi[1]* and M. Danesh Mesgaran[2]

[1]*Department of Animal Science, Faculty of Agriculture, University of Birjand, Iran*

[2]*Department of Animal Science, Faculty of Agriculture, Ferdowsi University of Mashhad, Iran*

**E-mail: riasi@wali.um.ac.ir, riasi004@yahoo.com*

ABSTRACT

A wide range of plants are used as forage in ruminant nutrition. Each plant has a unique morphology and physiology that gives it specific adaptation, growth, and forage quality features. However, halophytes have an important role as maintenance feeds during periods of fodder shortage in summer and winter and can provide supplemental or emergency feed for ruminants (especially sheep, goats and camels) in dry and saline environments. Chemical composition and digestible parameters of halophytes have been reported by several researchers and there is wide variation between the nutritive values of halophytes. This review summarizes the chemical composition and digestible parameter of *Kochia scoparia*, *Atriplex sp.* and *Suaeda sp.* as forage in ruminant nutrition.

Introduction

Halophytes constitute a significant part of the local flora in central Iran and in many cases represent supplementary or emergency feed during the prolonged seasons of drought. These forages are grazed by sheep, goats and camels and usually contain sufficient crude protein and (or) carbohydrate components to have significant nutritional potential. However, many factors can affect their chemical composition and nutritive value. There has been increased interest in cultivating some of the halophytic plants as forage in the salty agricultural regions for improving animal production in Iran.

Kochia *(Kochia scoparia)* is a hardy, drought-resistant plant widely used as emergency forage for livestock. Madrid *et al.* (1996) reported that Kochia could be substituted in diets fed to goats as proper forage during the warm season in Mediterranean regions. Protein content and digestibility of Kochia has been reported to be comparable with alfalfa (Sherrod, 1973). Many species of Atriplex are excellent livestock fodder because of their favourable crude protein content (El Shaer, 1997; El-Shatnawi *et al.*, 2003). However, there is wide variation in the mineral content of Atriplex due to the difference in their salt accumulation, geographical regions and age. Hopkins and Nicholson (1999) showed that finishing lambs fed *Atriplex nummalaria*, supplemented with either hay or grain produced acceptable meat based on both objective and subjective criteria compared with lambs fed alfalfa. Biomass of Suaeda is able to replace conventional grass hay in fattening diets for sheep (Swingle *et al.*, 1996).

The aim of this paper is reviewing information about the chemical composition and digestible parameter of some halophytic forage for ruminants.

Chemical Composition

Chemical analyses can provide valuable information about the actual chemical constituents influencing digestion, unlike in vitro methods. A routine system of chemical characterization of forages for ruminants must include determination of fiber, lignin and protein and critical to these is an accurate dry matter (DM) determination. Some characterization such as ash, macro and micro minerals, and anti-quality factors is dependent on the researcher's objectives.

Chemical composition of *Kochia scoparia*, *Atriplex arcuata* and *Suaeda dimorphestegia* were determined by different researchers (Table 9.1). The results shown that there is wide variation in chemical composition of halophytes and among the plant species, which could be related to the stage of harvesting, leaf/stem ratio or genetic variation, and availability of salts in the soil solution (Benjamin *et al.*, 1995; El-Shatnawi *et al.*, 2003). According to the Table 9.1, the mean content of crude protein (CP), organic matter (OM), neutral detergent fibre (NDF) and active detergent fiber (ADF) of Kochia (13.5, 86.2, 53.5 and 33.5 per cent, respectively) was higher than Atriplex (12.1, 79.6, 51.0 and 31.5 per cent, respectively) and *Suaeda species* (8.7, 67.3, 37.5 and 23.6 per cent, respectively). These data showed that the halophytes were considered as poor quality roughage since they contained lower levels of CP and higher levels of cell wall constituents. In general all halophytic species contained sufficient levels of minerals in comparison with those recommended for ruminants by the Nutritional Research Council (NRC) (2001), and Kearl (1982).

Table 9.1: Chemical Composition of *Kochia* sp., *Atriplex* sp. and *Suaeda* sp. (%)

Halophyte Species	*CP*	*OM*	*EE*	*NDF*	*ADF*	*ash*	*Ca*	*P*	*Na*	*K*	*Mg*	*Cu*	*Se*	*Reference*
Kochia sp.														
K. scoparia	9.1	–	–	–	–	11.1	0.7	0.2	–	–	0.5	–	–	(8)
K. scoparia	11.7	–	1.3	57.1	–	11.6	–	–	1.8	0.9	0.1	0.004	–	(9)
K. scoparia	19.4	–	–	46.9	32.1	–	–	–	–	–	–	–	–	(21)
K. scoparia	20.6	83.5	2.9	51.8	26.7	16.5	–	–	–	–	–	–	–	(22)
K. scoparia	13	85.2	–	68.2	44.9	–	–	–	–	–	–	–	–	(28)
K. scoparia	7.4	89.8	1.4	43.4	29.7	10.2	0.9	0.2	2.1	1	0.2	0.003	0.0005	(34)
Mean	13.5	86.2	1.9	53.5	33.5	12.4	0.8	0.2	2	1	0.3	0.004	0.0005	
Atriplex sp.														(1)
A. nummularia	10.3	–	–	45.5	28.9	23	0.6	0.2	7.2	2	0.5	–	–	(4)
A. nummularia	13.8	79.8	–	46.9	26	–	1.8	0.5	4.9	–	–	–	–	(5)
A. nummularia	16.9	75.4	–	44.2	–	–	3.1	0.1	4.2	–	–	–	–	(6)
A. nummularia	17.8	74.5	–	44.5	22.5	–	1.4	0.2	4.7	2.9	0.9	0.004	–	(9)
A. imorphostegia	6.2	83.1	0.9	61.1	–	16.9	–	–	–	–	–	0.0001	–	(11)
A. halimus	13.2	–	2.3	65.3	40.8	23.1	1.7	0.3	3.9	0.6	0.3	–	–	(38)
A. barclayana	10.2	–	4	46.9	27	23.8	–	–	4.1	1.8	–	0.003	–	(34)
A. imorphostegia	8.5	85.4	1.1	53.7	43.5	14.6	0.7	0.2	5.6	1.6	0.2	0.003	0.0007	
Mean	12.1	79.6	2.1	51	31.5	20.3	1.6	0.3	4.9	1.8	0.5		0.0007	
Suaeda Sp.														
S. arcuata	7.5	67.5	2.4	47	–	32.5	–	–	6.8	1.4	0.2	0.008	–	(9)
S. fruticosa	12.1	–	5	38.2	29.1	14.2	4	0.2	5.7	1.5	0.6	0.0001	–	(11)
S. esterora	7.6	–	–	36	19.4	34.1	–	–	10.2	1.4	–	–	–	(38)
S. arcuata	7.5	67	2.5	28.7	22.3	33	1.4	0.1	7.2	1.5	0.2	0.008	0.002	(34)
Mean	8.7	67.3	3.3	37.5	23.6	28.5	2.7	0.2	7.5	1.5	0.3	0.005	0.002	

Atriplex arcuata and *Suaeda dimorphostegia* accumulated higher Na^+ (4.9 and 7.5 per cent, respectively) and K^+ (1.8 and 1.5 per cent, respectively) compared with *Kochia scoparia*. Animals can tolerate large amount of Na^+ and K^+ intake (Bayoumi *et al.*, 1990) and the kidneys normally regulate excretion of these minerals in accordance with variations in dietary intake (Pond *et al.*, 1995). However, the excessive amounts of Na^+ and K^+ could change physiological status of animals (McDowell, 1992). Magnesium (Mg) content in halophytes (0.1 to 0.9 per cent) was recorded to be enough for dry and pregnant animals according to the nutritional recommendation (0.13 per cent) by Kearl (1982). Riasi (2006) showed that Se content of *Suaeda arcuata* (22.0 mg/kg) was higher than *Kochia scoparia* (5.4 mg/kg) and *Atriplex dimorphostegia* (6.7 mg/kg). It seems likely that, using of *Suaeda arcuata* as organic Se supplement for animals could be important in deficient areas.

Digestible Parameters

The *in vitro*, *in situ*, and *in vivo* procedures have been used for evaluating different plant materials in nutrition of ruminants. However, *In vitro* and *in situ* data on ruminal and post-ruminal disappearance of low-quality forages, especially halophytes, are limited.

In vitro and *in situ* data

Numerous laboratory methods (gas production, enzymatic, three-step, daisyII, and dual flow continuous culture procedures) are currently used for estimating ruminal and intestinal disappearance of forage protein. These methods are simple to conduct and reasonable in cost, compared with the *in vivo* methods (Gargallo *et al.*, 2004; McNiven *et al.*, 2002; Stern *et al.*, 1997). *In situ* digestion coefficients have been also used in an attempt to develop a system that can predict not only feed nutritive value but consumption as well (Ørskov, 2000).

Haddi *et al.* (2003) determined digestibility of five halophytes by gas production technique. They reported that among halophytes, the chenopodiaceae family had better fermentation parameters. There was an interaction between NDF, ADL, CP, and ash of halophytes and the gas production kinetic. In general cell wall, lignin, and acid detergent insoluble ash content of forages are important factors which influence kinetic of microflora into the rumen during the *in vitro* procedures (Van Soest, 1993). Riasi and Danesh Mesgaran (2004) compared ruminal degradability of DM and CP of some halophytes using *in situ* procedures. They reported that the DM rapidly degradable fraction of *Suaeda arcuata* (0.53) was higher than *Kochia scoparia* and *Atriplex dimorphostegia* (0.31 and 0.39, respectively). In addition, *Kochia scoparia* had higher DM slowly degradable fraction (0.37). The dry matter digestion coefficients of Atriplex in this study were supported by Ben Salem *et al.* (2002). *Kochia scoparia* had a lower rapidly degradable CP fraction (0.35) compared with A*triplex dimorphostegia* and *Suaeda arcuata* (0.50 and 0.55, respectively); therefore the amount of its true protein and undegradable protein should be high. According to Kaitho *et al.* (1998), Atriplex had about 50 per cent rapidly degradable CP fraction, but Ben Salem at al. (2002) reported that about 70 per cent of Atriplex N was rapidly degradable. Danesh, Mesgaran and Stern (2005) determined the ruminal, post-ruminal and total tract CP disappearance of various feeds (grain, by-products and halophytes) using *in situ* and *in vitro* methods. They observed that Atriplex had the lower ruminal disappearance of CP (0.27) compared with Suaeda and Kochia (0.34 and 0.51, respectively). Variation in the ruminal disappearance of DM and CP may be attributed to differences in NDF, ADF, ADL (Van Soest, 1994; Yan and Agnew, 2004), ash (Benjamin *et al.*, 1995) and tannins (Reed *et al.*, 1990) of feeds. It also observed that halophytes had the lowest ruminal protein disappearance (0.41) and no differences (P>0.05) were found between the procedures regarding the ruminal, post-ruminal and total tract of protein disappearance. The total tract protein disappearance of *Ariplex sp.*, *Suaeda sp.* and *Kochia sp.* (0.62, 0.64, and 0.71, respectively) was lower than alfalfa hay (0.83). It is concluded that *Kochia sp.* had higher nutritive value than that the other species of halophytes (Danesh Mesgaran and Stern, 2005).

An experiment was conducted on *Kochia scoparia* and *Atriplex dimorphostegia* using daisyII and three-step procedures (Riasi *et al.*, 2005) (Table 9.2). Results showed that ruminal CP disappearance of Kochia was lower (P<0.05) than Atriplex after 12 h incubation in the rumen during daisyII method, but there was no difference (P>0.05) between forages after 16 h incubation during three-step method. Post-ruminal CP digestion was not different between Kochia and Atriplex when using the three-step procedure (33.0 and 35.0 per cent, respectively) or the daisyII procedure (58.0 and 60.0, respectively). No differences (P>0.05) was observed between total tract CP digestion of Kochia (86.6 per cent) and Atriplex (88.6 per cent) using two procedures. The inconsistency between these results and previous study (Danesh Mesgaran and Stern, 2005) may be attributed to the genetic variation of halophytes.

Table 9.2: Ruminal, Post-ruminal and Total Tract Disappearance of CP (% of DM) of *Kochia scoparia, Atriplex dimorphostegia* Using Three-step and Daisy II Technique (Riasi *et al.*, 2005)

Forages	*Three-step*			*DaisyII*		
	Rumen[1]	*Post Rumen*	*Total Tract*	*Rumen*[2]	*Post Rumen*	*Total Tract*
Kochia scoparia	80.0	33.0	86.6	72.4b[3]	58.0	88.4b
Atriplex dimorphostegia	82.5	35.0	88.6	78.5a	59.6	91.3a

[1] Crude protein disappearance from Dacron bags after 16 h incubation in the rumen.

[2] Crude protein disappearance from Dacron bags after 12 h incubation in the rumen.

[3] Means in the same column with different letters are significantly different (P<0.05).

In vitro continuous culture fermentation offers several advantages compared to *in vivo* studies including no confounding effect of endogenous contribution of N, decreased protozoan purine contribution, and constant feed intake, out flow rates, and pH (Stern *et al.*, 1997). Riasi *et al.* (2005) tested nutritive value of two halophytes (*Kochia scoparia* and *Atriplex dimorphostegia*) using dual flow culture system. These researchers reported that there was no difference (P>0.05) between ruminal disappearance of DM and CP of Kochia and Atriplex. But, Atriplex had higher (P<0.05) apparent and true digestibility of OM (405.0 and 450.0 g kg^{-1}, respectively) compared with Kochia (357.0 and 398.0 g kg^{-1}, respectively). In addition, NDF digestibility of Atriplex (411.0 g kg^{-1}) was higher ($P<0.05$) than that of Kochia (348.0 g kg^{-1}) into the fermenters. The result showed that total VFA production into the fermenters was lower for Kochia compared with Atriplex. There was no differences ($P>0.05$) for molar proportions of acetate and propionate after feeding Kochia or Atriplex.

In vivo data

Direct *in vivo* measurement would appear to be the most desirable way to determine rates and extents of ruminal degradation of forage proteins. However, *in vivo* procedures require considerable investment of resources; each experiment involves use of several ruminants with cannulae in various sections of the gastrointestinal tract (Fahey Jr, 1994).

There are some reports about *in vivo* apparent nutrient digestibility and voluntary feed intake of halophytes. Madrid *et al.* (1996) fed goats with various levels of Kochia hay and wheat straw (0:100, 25:75, 50:50, 75:25 and 100:0). The DM and OM digestibility of diets and dry matter intake had no significant differences ($P>0.05$) between treatments. However, apparent digestibility of N, cellulose, NDF and ADF was reduced as Kochia level increased. In an experiment the nutrient digestibility of *Atriplex sp.* was higher than other halophytes and it assumed to be due to high level of salts in Atriplex (Benjamin *et al.*, 1995). Benjamin *et al.* (1995) reported the *in vivo* DM and OM digestibility of *Atriplex nummalaria* (73.5 and 58.7 per cent, respectively) and *Atriplex canescens* (62.0 and 46.7 per cent, respectively). Swingle *et al.* (1996) tested digestibility of various halophytes by *in vivo* methods. They concluded that there was no difference between DM and OM digestibility of halophytes. However, fiber digestibility of Suaeda was lower than the other forages.

Riasi *et al.* (2005) determined the apparent nutrient digestibility of Kochia, Atriplex and alfalfa using difference method in Baloochi ewes (Table 9.3). In agreement with Hinojosa *et al.* (1985), nutrient digestibility of the halophytes (except ash) were significantly ($P<0.05$) lower than that of alfalfa. The digestibility of DM and CP of Kochia and Atriplex in this trial was lower than previous *in situ* and *in vitro* evidences (Danesh Mesgaran and stern, 2005). There was no significant difference ($P>0.05$) in apparent digestibility of DM, OM, CP, EE and ash between the halophytes. However, the NDF digestibility of Kochia (325.0 g kg^{-1}) was significantly ($P<0.05$) lower than Atriplex (445.0 gkg^{-1}). Smith *et al.* (1972) reported that the proportion of soluble DM, lignin and hemi-cellulose may account for differences in forage cell wall digestibility. Ash digestibility of the halophytes was significantly ($P<0.001$) higher than alfalfa. As has been mentioned before the halophytes have high content of ash and soluble minerals. Hence, these forages are more susceptible to ash digestibility in ruminants.

According to El Shaer (2003), sheep became adapted to salt bush and increased their intake of forage over 3-5 month period. Mixed diet containing halophytes were also consumed by sheep and goats. Ben Salem *et al.* (1985) compared the voluntary feed intake of *Atriplex nummularia* by sheep (273±48 g d^{-1} or 10.8±3.4 g $kg^{-0.75}$) and goats (229±65 g d^{-1} or 12.7±9.3 g $kg^{-0.75}$). They concluded that Atriplex had the best palatability between six halophytes species. Riasi *et al.* (2006) replaced 50 per cent of alfalfa with *Kochia scoparia* or *Atriplex dimorphostegia* in sheep feed. The result showed that the voluntary dry matter intake of T1 (Kochia + alfalfa) and T2 (Atriplex + alfalfa) (468 and 486 g d^{-1}, respectively) was lower ($P<0.05$) than T3 (alfalfa) (628 g d^{-1}). Mean ruminal pH (T 1= 7.53, T 2= 7.52, T 3= 7.46) and ammonia-N concentrations (T 1= 10.25, T 2=9.89, T 3= 9.71 mg/100ml) were similar for all treatments at different times after feeding (0, 0.5, 1, 2, 3, 4, 6, 8 hours) ($P>0.05$). The N retention was higher ($P<0.001$) in ewes fed T 3 (2.03 g d^{-1}) compared with T 1 (1.07 g d^{-1}) and T 2 (0.93 g d^{-1}).

Table 9.3: Apparent Nutrient Digestibility of Kochia, Atriplex and Alfalfa (g kg^{-1}) (Riasi *et al.*, 2005)

Nutrients [1]	*Apparent Digestibility*			*Contrasts* [2]	
	Kochia	*Atriplex*	*Alfalfa*	*Ko vs At*	*Ko + At vs Alf*
DM	468.2	486	626.1	NS [3]	*
OM	405.1	425.2	625	NS	***
CP	531.4	476	679.3	NS	*
EE	435	412	593	NS	**
NDF	325	445	501	*	**
Ash	927	931.1	669.4	NS	***

1. DM: dry matter; OM: organic matter, CP: crude protein; EE: ether extract; NDF: neutral detergent fibre.
2. *Ko:* Kochia scoparia*; At:* Atriplex dimorphostegia*; Alf:* Alfalfa.
3. NS: none significant; *: $P<0.05$; **: $P<0.01$; ***: $P<0.001$

Conclusion

There was wide variation in chemical composition and nutritive values of different species of halophytes and among the each species. However it seems these plants have good potential as forage for ruminants especially as maintenance feed during periods of fodder shortage in summer and winter and can provide supplemental or emergency feed for ruminants in arid and semi-arid regions. Greater attention is needed to use halophytic forages (especially Kochia and Atriplex) as a resource for ruminant feeding, particularly when higher quality feeds were too expensive or not available in sufficient amounts. More research is needed for the best stage of harvesting of forages, the effect of halophytes on animal production and animal physiology.

References

Aganga, A. A., J. K. Mthetho and S. Tshwenyane. 2003. *Atriplex nummularia* (old man saltbush): A potential forage crop for arid regions of Botswana. Pak. J. of Nut. 2: 72-75.

Bayoumi, M. T., H. M. El Shaer and A. fawzia. 1990. Survival of sheep and goats fed salt march plants. J. Arid Environ. 18: 75-78.

Benjamin, R. W, Y. Lavie, M. Forti, D. Barkai, R. Yonatan and Y. Hefetz. 1995. Annual regrowth and edible biomass of two species of Atriplex and Cassia sturtii after browsing. J. Arid Environ., 29: 63-84.

Ben Salem, H., A. Nefzaoui and L. Ben Salem. 1985. Sheep and goat preferences for Mediterranean fodder shrubs. Relationship with the nutritive characteristics. INRA–Tunisia Rue Hedi Karray, 2049.

Ben Salem, H., A. Nefzaoui and L. Ben Salem. 2002. Supplementation of Acacia cyanophylla Lindl. Foliage-based diet with barley or shrubs from arid areas (*Opuntia ficus-indica f. inermis* and *Atriplex nummularia* L.) on growth and digestibility in lambs. J. Anim. Feed Sci. and Technol. 96: 15-30.

Ben Salem, H., A. Nefzaoui and L. Ben Salem. 2004. Spineless cactus (*Opuntia ficus-indica f. inermis*) and oldman saltbush (*Atriplex nummularia* L.) as alternative supplements for growing Barbarine lambs given straw-based diets. Small Rum. Res. 51: 65-73.

Calsamiglia, S. and M. D. Stern. 1995. A three step *in vitro* procedure for estimating intestinal digestion of protein in ruminants. J. Anim. Sci. 73: 1459-1465.

Cohen, R. D. H., A. D. Iwaasa, M. E. Mann, E. Coxworth and J. A. Kernan. 1989. Studies on the feeding value of *Kochia scoparia* (L.) Schrad. Hay for beef cattle. Can. J. Anim. Sci. 69: 735-743.

Danesh Mesgaran, M., A. Riasi and M. D. Stern. 2004. Chemical composition, and in vitro and in situ protein digestibility of some halophytes located in central Iran. Proc. Brit. Soc. Anim. Sci. 242.

Danesh Mesgaran, M. and M. D. Stern. 2005. Ruminal and post-ruminal protein disappearance of various feeds originating from Iranian plats varieties determined by the in situ mobile bag technique and alternative methods. Anim. Feed Sci. Technol. 118: 31-46.

El Shaer, H. M. 1997. Sustainable utilization of halophytic plant species as livestock fodders in Egypt. International Conference on "water management, salinity and pollution control toward sustainable irrigation in the Mediterranean region". Valenzano-Bari, Italy.

El Shaer, H. M. 2003. Potential of halophytes as animal fodder in Egypt. H. Lieth (ed.), Cash crop halophytes: Recent studies, 111-119.

El-Shatnawi, M. K. J. and A. Y. Abdullah. 2003. Composition changes of *Atriplex nummularia* L. under Mediterranean arid environment. African J. of Range and Forage Sci. 20: 253-257.

Fahey Jr, G. C. 1994. Forage Quality, Evaluation and Utilization. American Society of Agronomy. Madison, Wisconsin.

Gargallo, S., S. Calsamiglia and A. Ferret. 2004. Development of an *in vitro* technique to determine intestinal digestion of protein supplements by daisyII incubator. J. Dairy Sci. 87: 512.

Haddi, M. L., S. Filacorda, K. Meniai, F. Rollin and P. Susmel. 2003. *In vitro* fermentation kinetics of some halophyte shrubs sampled at three stages of maturity. Anim. Feed Sci. Technol. 104: 215-225.

Hinojosa, R. D., H. E. Kiesling and G. P. Lofgreen. 1985. Digestibility of diets and performance of growing feedlot steers fed alfalfa and kochia roughages. J. Anim. Sci. (Suppl.), 63: 427 (abstract).

Hopkins, D. L. and A. Nicholson. 1999. Meat quality of wether lambs grazed on saltbush (*A. nummalaria*) plus supplements or Lucerne (Medicago sativa). Meat Sci. 51: 91-95.

Kaitho, R. J., I. V. Nsahlai, B. A. Williams, N. N. Umunna, S. Tamminga and J. Van Bruchem. 1998. Nitrogen in browse species: ruminal degradability and post-ruminal digestibility measured by mobile nylon bag and *in vitro* techniques. J. Sci. Food Agric. 76: 488-498.

Kearl, L. C. 1982. Nutrient requirements of ruminants in developing countries. Utah Agricultural Experimental Station, Logan, Utah, USA.

Kirkpatrick, J. G., R. G. Helman, G. E. Burrows, D. V. Tungeln, T. Lehenbauer and R. J. Tyrl. 1999. Evaluation of hepatic changes and weight gains in sheep grazing *Kochia scoparia*. Vet. Hum. Toxicol. 41: 67-70.

Madrid, J., F. Hernandez, M. A. Oulgar and J. M. Cid. 1996. Nutritive value of *Kochia scoparia* L. and ammoniated barley straw for goats. Small Ruminant Res. 19: 213-218.

McDowell, L. R. 1992. Minerals in Animal and Human Nutrition. Academic Press, San Diego, 524 pp.

McNiven, M. A., E. Prestlokken, L. T. Mydland and A. W. Mitchell. 2002. Laboratory procedure to determine protein digestibility of heat-treated feedstuffs for dairy cattle. Anim. Feed Sci. Thechnol. 96: 1-13.

Nutritional Research Council (NRC).2001. Nutrient Requirements of Dairy cattle, 7th revised Ed. National Academy of Science, Washington D.C.

Ørskov, E. R. 2000. The *in situ* technique for the estimation of forage degradability in ruminants. In: Given, D. I., E. Owen, R. F. E. Axford, and H. M. Omed (Eds) Forage Evaluation in Ruminant Nutrition. CABI publishing. pp. 175-188.

Pond, W. G., D. C. Church and K. R. Pond. 1995. Basic Animal Nutrition and Feeding. John Wiley and Sons, New York, USA.

Rankins Jr, D. L., G. S. Smith and D. M. Hallford. 1991. Altered metabolic hormones, impaired nitrogen retention, and hepatotoxicosis in lambs fed *Kochia scoparia* hay. J. Anim. Sci. 69: 2932-2940.

Reed, J. D., H. Soller and A. Woodward. 1990. Folder tree and straw diets for sheep: intake, growth, digestibility and effects of phenolic on nitrogen utilization. Anim. Feed Sci. Tech. 30: 39-50.

Riasi, A. and M. Danesh Mesgaran. 2004. *In situ* dry matter and crude protein degradability of halophytes located in central Iran. Proc. Brit. Soc. Anim. Sci. 243.

Riasi, A., M. Stern, M. Danesh Mesgaran and M. Ruiz Moreno. 2005. Ruminal and post ruminal crude protein digestion of halophytes forages (*Kochia scoparia, Atriplex dimorphostegia*) determined by various procedures. Proc. U. S. A. Joint Annual Meeting. 663.

Riasi, A., M. Danesh Mesgaran, H. Nasiri Moghadam and A. R. Heravi Moosavi. 2005. The nutrient digestibility (in vivo) of halophytic forages (*Kochia scoparia* and *Atriplex domorphostegia*). Proc. Brit. Soc. Anim. Sci. in Thailand. T53.

Riasi, A., M. Danesh Mesgaran and M. D. Stern. 2005. Ruminal digestibility and fermentation parameters of *Kochia scoparia* and *Atriplex domorphostegia* using dual flow continuous culture system. 2 the Research Seminar of Sheep and Goat. Institute of Animal Research, Tehran, Iran, 51-52.

Riasi, A. 2006. Chemical composition, nutritive value and some physiological effects of halophytes in Baloochi sheep. PhD thesis, Faculty of Agriculture, Ferdowsi University of Mashad.

Sherrod, L. B. 1973. Nutritive value of *Kochia scoparia*. III. Digestibility of kochia hay compared with alfalfa hay. J. Dairy Sci. 56: 923-926.

Smith, L. W., H. K. Goering and C. H. Gordon. 1972. Relationships of forage compositions with rates of cell wall digestion and indigestibility of cell walls. J. Dairy Sci. 55:1140-1147.

Stern, M. D., A. Bach, and S. Calsamiglia. 1997. Alternative techniques for measuring digestion in ruminants. J. Anim. Sci. 75: 2256-2276.

Swingle, R. S., E. P. Glenn and V. R. Squires. 1996. Growth performance of lambs fed mixed diets containing halophyte ingredients. Anim. Feed Sci. Technol. 63: 137–148.

Van Soest, P. J. 1993. Cell wall matrix interactions and degradation-session synopsis.. In: Forage Cell Structure and Digestibility. ASA-CSSA- SSSA, 677 S. Segoe Road, Madison, WI 53711, USA, P. 377.

Van Soest, P. J., 1994. Nutritional Ecology of the Ruminant, 2nd ed. Comstock Publishing Associates, Cornell University Press, Ithaca, NY.

Yan, T. and R. E. Agnew. 2004. Prediction of nutritive values in grass silage: II. Degradability of and dry matter using digestibility, chemical composition, and fermentation data. J. Anim. Sci. 82: 1380-1391.

Kirkpatrick, J. G., R. G. Helman, G. E. Burrows, D. V. Tungeln, T. Lehenbauer and R. J. Tyrl. 1999. Evaluation of hepatic changes and weight gains in sheep grazing *Kochia scoparia*. Vet. Hum. Toxicol. 41: 67-70.

10

Genetics of Tolerance to Water Stress and Salinity in Crop Plants

B. Sharma

Division of Genetics, Indian Agricultural Research Institute, Delhi – 110 012, India

E-mail: drbrsh@yahoo.co.in

ABSTRACT

Millions of hectares of cropped area around the world are inflicted with the scourge of salinity and water deficiency to various degrees. It is not an exaggeration that there will never be a shortfall in agricultural production even if the below average yields of today can be ensured in these permanently sick stretches. Crop failures are common and even low productivity levels extremely uncertain depending on the weather conditions in different years. Salt affected soils can be retrieved to some extent by chemical treatment, even though it involves large investments over long periods. But this can be achieved only with assured availability of fresh water. The problem is further compounded where the water is also saline, and severe restrictions are placed even in soils with normal reaction. This restricts the applicability of soil reclamation measures in the perpetually water-deficient areas. Developing crop varieties with enhanced tolerance to salinity and drought through genetic manipulation could be a reliable and cost effective approach to facilitate crop cultivation in the problem areas. This certainly does not preclude the employment of other measures to improve soil health and crop management. The information generated so far on the mechanisms and genetic control of salt and drought resistance gives hope in this regard. It is known for a long time that salt and drought tolerance are complex traits, have a large number of components, each under a different but partially overlapping system of genes, and therefore, difficult to handle in a breeding programme. Nevertheless, recent advances in genetics and biotechnology give a hope in this elusive area. It is common knowledge that arid regions invariably have the associated problem of salinity. Yet several plant species have been growing there in the wild. This is the most glaring proof to suggest that these plants have evolved, over millions of years, their own system of handling two serious handicaps simultaneously. Since all biological processes are under genetic control, there has to be a set of genes causing resistance to the two major abiotic stresses, drought and salinity, either individually or both simultaneously. So gene scouting was undertaken on a massive scale, and has yielded encouraging results. Advances in molecular genetics and biotechnology have confirmed that it would be possible to find solution to these insurmountable problems through accumulation of genes for resistance. Sources for such genes have been found in virtually all organisms from microbes to man. Many genes and gene products have been identified, mapped, tagged with molecular markers, and cloned. Successful attempt have been made to transfer these genes across species. The transgenics thus created, can go as direct varieties and also serve as new donors for further breeding.

Introduction

Mechanisms of Drought Tolerance in Plants

A plant manages to survive under conditions of water scarcity with the help of several inherent mechanisms. The most important among them are escape, avoidance, and tolerance or resistance of various degrees (Levitt, 1972).

Escape

An early maturing genotype can escape damage due to moisture stress if it completes the major part of its reproductive life cycle before the drought strikes. It is established that duration of plant life, *i.e.* earliness or late maturity, is under genetic control (Turner, 1979). Mutations for early or late flowering are induced frequently, and genes for early or late maturity have been described and mapped in several cultivated plants. Rapid growth and development of plants leading to early onset of the reproduction phase involves mobilization of assimilates into the grain at a fast pace. As a result, an early plant is able to complete its life before its survival and reproduction is threatened by water shortage. Search for early genotypes and induction of early mutations can be an effective approach in the management of crops in drought affected areas, especially if water shortage toward crop maturity is a recurrent phenomenon.

Avoidance

Several structural mechanisms operate in the plant kingdom that can help in meeting the threat of damage to plant life. They exist in different plant species or are present even in the varieties of same plant species with varying degree of expression. Like all other properties, these traits are also conditioned by genetic factors which can be exploited to enhance drought tolerance in the common cultivated plants. A few examples are listed below:

Water Storage in the Tissues

Plants with fleshy stems and leaves are excellent example of how a plant can store water in its body for future use. Collectively, such species are called succulent plants. Many desert plants like opuntia and cacti are endowed with this property, although all succulent plants are not confined to arid regions alone.

Development of Root System

The roots play a crucial role in maintaining the desirable water level in all other organs where metabolism needs to be supported by regular water supply. A deep root system can extract water from lower levels when the surface soil has dried. Efficiency of root system is also decided by its overall spread and extent of branching. Invariably, the length of roots is associated with shoot length in the same plant species. The late maturing genotypes of a species also tend to have deeper and vigorous root system. One reason for this association is that the late maturing strains get more time to grow before they mature. Therefore lateness works against the escape mechanism.

Hydraulic Conductance

Plants vary in their ability to mobilize water within their body. A plant that can push water upward after absorption by the roots from the soil will have an advantage in the struggle against drought.

Reduced Water Loss

Plants have several mechanisms to conserve water under stress. These mechanisms have been shown to vary in different plant species (O'Toole and Moya, 1978; and Ekanayake *et al.*, 1985). Number of stomata per unit area, efficiency of stomatal closing, lenticular control, rolling and folding of leaves are such examples. Genotypes with smaller leaves are likely to loose less water through transpiration. Varieties with smaller leaves can be created by mutation.

Several grasses which have neither succulent leaves nor deep root systems cover large areas in desert and semi-desert regions. Obviously, other metabolic mechanisms help these plants to survive under conditions of severe drought and water scarcity.

Tolerance

Drought tolerance (or resistance) is a genetically controlled phenomenon which consists of a large number of components. It appears that the mechanisms imparting drought tolerance work in a coordinated manner. The multiplicity of metabolic processes appears to be under the control of a regulator gene. Therefore, in spite of the multigenic nature of drought tolerance it may appear to be monogenic if the master gene (regulator) exists in dominant and recessive state in different genotypes. This includes biochemical properties as well as structural differences. As will be discussed below, specific substances and enzymic processes help in water economy and enhanced water use efficiency. Among the well identified mechanisms are maintaining leaf turgor, osmotic potential, solute accumulation, and metabolic adjustment.

Thus, plant survival under moisture stress is a balancing act between turgor maintenance, osmotic adjustment, and solute accumulation. All these mechanisms work in tandem under genetic control. This raises hopes for genetic manipulation and gene mining followed by trans-species transfer through gene transformation (transgenics).

Genetics of Drought Tolerance

Little information is available about the genetic control of drought tolerance. This is probably due to the fact that drought tolerance is a multicomponent phenomenon in which individual factor may be playing a small role but they function in a coordinated manner under the control of regulatory systems of "master genes". This complexity of the whole system has led to the conclusion that drought tolerance is a polygenic trait in which "minor genes" determine the ultimate phenotype in additive manner. However, the picture that is emerging now suggests that, like any other trait, identifiable genes are responsible for the expression of individual components of drought tolerance. Through the coordinated function of these gene systems, the plant is able to manage its life functions under water scarcity. As mentioned above, morphological and physiological/biochemical traits associated with drought tolerance have been identified, and the process is continuing.

Genes Identified

Genes have been identified for some of the traits associated with drought tolerance. For example, a dominant gene increases root length. Another gene affects root number and causes profuse rooting in dominant state. Thick root tip is a recessive trait. The ability for leaf rolling is also controlled by a single gene. A single gene (only one so far) has been shown to control osmotic adjustment.

A gene which reduces the water requirement in rice has been discovered. It has been called Drl1. This gene has been mapped genetically and has been found to be linked with the genes for plant height, leaf pigmentation, hull colour, and awn formation. These morphological traits make it easy to decide the presence of Drl1. Its presence increases the suitability of a rice strain for cultivation in upland conditions. This gene has a pleiotropic effect on root development, which is also a property that helps water mining from deeper horizons. Presence of such genes, thus, helps the plant under water shortage through greater water absorption as well as water economy.

Two genes have been reported in cowpea which delay the death of the plant in total absence of water for long durations. The plants exposed to drought for more than a month are revived if water becomes available before the death point. The two genes act through different mechanisms of water conservation and keep the growing point green during the period of stress. Here again, the root system has better growth when the genes are present in dominant state. The pattern of root growth slightly differs under the influence of the two genes.

Genes have been identified for the synthesis of certain chemical substances that hold water like a sponge. These substances are not highly reactive and their accumulation in large quantities does not create significant load on the plant metabolism. They are collectively called "nontoxic compatible solutes". Examples of such products are fructan, trehalose, polyols, dehydrin, osmotin, proline, glycine betaine, polyamines, abscissic acid etc. Genes for biosynthetic pathways of these solutes have been identified. They seem to be present universally through the biological world, and have been cloned from bacteria, yeast, plants and even human source.

Biotechnology in Drought Tolerance

Identification of specific genes for drought tolerance on the one hand, and the availability of techniques for gene transfer across species borders through genetic transformation (Grover *et al.*, 1998) on the other hand, have created an opportunity that never existed before. The entire procedure of new plant varieties with enhanced drought tolerance can be summarized in a few steps:

1. Identify genes for drought tolerance in plants (mainly xerophytes, succulents), bacteria, yeast, human etc. Genes for survival under moisture stress in the crops already known for their drought tolerance, such as, safflower, horsegram, clusterbean, mothbean etc., wild annuals of desert like sewan grass, and tree species, *e.g.* date palm, *Prosopis cinararia*, casurina, and so on.
2. Mark these genes using molecular tags.
3. Gene cloning on a suitable vector and strengthening of their function by hooking on to a strong constitutive promoter.
4. Further strengthening of the genes for drought tolerance through directed mutagenesis.
5. Transfer of the tagged genes into the target plant through genetic transformation.
6. Verify the suitability of transformants for direct cultivation in the field under moisture stress.
7. Selection of plants carrying drought tolerance gene(s) in the segregating populations through marker assisted selection using the molecular marker that was used for tagging.
8. Hybridization of the drought tolerant transgenics with promising varieties under cultivation.
9. Drought tolerance can be further fortified by pyramiding more genes with similar function but acting through different metabolic steps.

Some of the better known genes for drought tolerance in different organisms are listed in Table 10.1.

Breeding for Drought Tolerance

Performance of Genotype

Appropriate genotypes as donors of the target trait have to be identified before embarking on any plant improvement programme. This is the first step toward success. For a drought tolerance breeding programme, it is essential that genotypes with good productivity potential are selected. Their yield performance must be assessed both under optimum moisture condition as well as moisture stress. Invariably, a genotype with high yield potential will give superior performance compared to poor yielding strains unless some specific factors come in play and distort the picture. Thus, generally speaking, yield performance is expresses itself at its own relative level under all conditions. The best strategy will be to expose the segregating populations to conditions of optimum moisture condition and stress. This will ensure selection of promising segregants under extreme environments. Simultaneous screening in diverse conditions will reduce the possibility of wrong selections which may not be confirmed for yield performance and stress tolerance in future generations. The progenies of all the plants selected will be divided in two parts each and again tested for confirmation of their yield potential and stress tolerance under moisture abundance and moisture deficiency. Progenies showing good performance in both situations will be confirmed performers for yield under moisture stress. The grain harvested from such progenies will be advanced to regular trials in the subsequent generations.

Table 10.1: Genes Identified for Transformation or Mutational Strengthening

Gene	*Source*	*Function*
Tps 1	Yeast	Trehalose synthesis
P5cs	Yeast	Pyrroline-5-carboxylate synthetase (involved in proline synthesis) [overexpressing P5cs in tobacco produced 10-18-fold more proline]
SacB	*Bacillus subtilis*	Levan sucrase (fructan synthesis) [tobacco]
betA	*E. coli*	Choline dehydrogenase
betB	*E. coli*	Betaine aldehyde dehydrogenase (glycine betaine synthesis) [tobacco]
ODC	Yeast and mouse	Ornithine decarboxylase
ADC	Oat	Arginine decarboxylase
SAMDC	Human	S-adenosyl-methionine decarboxylase (proline content)
SOD	Pea	Superoxide dismutase [tobacco]
hva1	Barley	Three LEA (late embryogenesis abundant) proteins [rice]

This simple procedure will definitely lead to development of high yielding stress tolerant varieties in a most direct manner. The key to success, however, is identification of parental genotypes with high yield potential and strong resistance to moisture stress. The other equally important element is to have reliable system of screening for drought tolerance–a procedure which allows only the tolerant plants to survive and virtually kills the drought sensitive plant at an early age (preferably seedling stage).

Breeding Strategy

A programme aiming at developing drought resistant varieties will attack the problem from two ends: (1) transferring drought tolerance from known sources of resistance to high yielding varieties which require ensured water supply in good measure, and (2) improving the yield potential of already existing drought tolerant varieties. Using such genotypes in a hybridization programme combined with appropriate screening technique will ensure success.

Although field screening for yield performance and drought tolerance are essential components of a breeding programme and cannot be compromised, it is highly desirable to confirm the drought tolerance of new selections through morphological, physiological, and biochemical tests. Some of the tests that can be applied at field scale are:

1. Infrared thermometry–measures efficient water uptake
2. Banding metriburin herbicide in soil
3. Hydroponic culture under stress for root growth
4. Psychrometric evaluation of osmotic adjustment
5. Diffusion porometry for water conductance
6. Mini-rhizotron technique for root penetration
7. Infrared aerial photography for detection of dehydration (*good for mass screening of varieties and populations)*
8. Carbon isotope discrimination for water use efficiency
9. Yield evaluation under drought conditions (real life testing)
10. Drought index–yield reduction under drought compared to under normal moisture availability
11. Visual observation for morpho-physiological traits
    (*a*) Leaf length
    (*b*) Leaf angle
    (*c*) Leaf rolling
    (*d*) Wilting symptoms
    (*e*) Stay green growth habit
12. Longevity under water stress:
    (*a*) seedling screening in cowpea
    (*b*) plant survival under stress

Salt Tolerance in Plants

Salt tolerance is branded as a genetically complex trait (Epstein *et al.*, 1980). The reason is that exposures to severe stresses like water shortage (and also excess), salinity or infections affect all life functions which are controlled by a galaxy of their own genes. Consequently, the genetic systems that are capable of protecting the various metabolic functions have to be numerous, and mostly regulatory in nature. Conventionally, such properties have been termed as physiological traits, although all physiological/biochemical functions are under control of specific genes. This is the basic premise which gives the hope that, like tolerance to moisture stress, salt tolerance can also be effectively tackled genetically.

Since long, salt tolerance was declared as a polygenic trait. This hypothesis did not help in developing a procedure to handle crop management in saline conditions organize breeding programmes for salinity resistance. Recent research has identified genes which impart tolerance to salinity to various degrees. This is also confirmed by the fact that several plant and animal species (*e.g.* marine organism) are able to survive and flourish under high salt concentrations. Collectively, such plants are called halophytes (salt loving) (Flowers, Hajibagheri and Clipson, 1986). Obviously, such organisms must have a system of avoiding or metabolizing salt, or a combination of both.

There is a wide range of tolerance to salinity, halophytes being at the top. Apparently, such plants carry genes and gene systems that cause different levels of tolerance. Experience has enabled the cultivators and plant scientists to categorize crop plants in accordance with their suitability for cultivation in soils with different salinity levels. Scouting for salt resistance genes can begin with such plants. Ultimately, halophytes will be the most valuable source of genes for salt resistance.

Genetically, salt tolerance can be enhanced by (1) alteration of a single limiting factor in a complex chain of reactions, and (2) change(s) in genes for signal molecules. It is known that irrigation increases salinization of soil (Ghassemi, Jakeman and Nix, 1995). Therefore only a combination of increased irrigation potential and higher salt tolerance of new varieties will ensure productivity in areas that are at the lowest ladder of crop productivity.

Genetics of Salt Tolerance

Since salt tolerance was designated as a quantitative trait, the genes identifies for this property have been called as "quantitative trait loci" or QTL. This expression has stuck even though all such genes have been mapped specific locations on chromosome and are inherited in typical Mendelian fashion. Since a large number of genes are involved creating the total effect of salt tolerance, this property is better described as a multigenic trait even while the genes identified are called QTLs. The so-called QTLs for salt tolerance have been identified in tomato and many other cultivated plants. Different QTLs for salt tolerance have also been demonstrated to be operating at different stages of plant development. At least two sets of QTLs for salt tolerance are described: those acting at seedling stage, and the others active during vegetative phase.

Approaches to Enhance Salt Tolerance

The procedure to strengthen salt tolerance will be same as described for drought tolerance. The two-pronged approach will involve enhancement of salt tolerance in the existing commercial varieties, improving yield potential of known salt tolerant genotypes. Strains from both these groups can be used for experimentation simultaneously. The steps involved can be summarized as follows:

1. Select salt tolerant genotypes existing in populations
2. Induce more variation for salt tolerance by mutagenesis

3. Interspecific hybridization with tolerant wild relatives
4. Develop halophytes as alternative crops
5. Identify genes for salt tolerance in halophytes
6. Gene tagging and cloning
7. Genetic transformation
8. Breeding for yield under saline conditions

For assured success it is important to verify salt tolerance of new recombinants, mutants or transgenics at field scale. Laboratory testing in Petri dishes or verification in pods frequently leads to erroneous conclusions. Confirmation in field trials should be treated as mandatory.

Biotechnology has made the job of identification and transfer of genes from alien sources easier. The source of gene(s) could be any organism (in addition to plants), including algae, invertebrates, fishes etc. thriving in sea water. Thirteen species have already been transformed with 40 genes. Interestingly, genes for compatible solutes are common for both, drought as well as salt tolerance. In fact, about two dozen genes are known to impart tolerance for cold, drought and salinity simultaneously. This makes the task of creating varieties with multiple resistances feasible. It is not necessary to transfer all these genes to achieve success. Even two good genes under the control of a strong constitutive promoter will create a stable and high level of resistance against three important abiotic stresses in a single attempt.

Species Successfully Transformed for Salt Tolerance

Instances of successful transfer of genes for salt tolerance from alien sources to plant (transgenics) are being reported at regular intervals. Transformation for salt tolerance has been reported in the following plants (Monforte, Asins and Carbonell, 1997):

Arabidopsis thaliana: Tomato

Brassica juncea: Alfalafa

Brassica napus: Tobacco

Citrus *(Carrizo citrange)*: Rice

Cucumis melo (melon): Brinjal (eggplant)

Diaspyros kaki: Potato

(Japanese persimmon): Wheat

Success in Transformation for Genes and Other Elements

Several metabolic factors and stable products have been identified which play a major role in salt tolerance. Organisms rich in such substances can serve as reliable donors of genes for salt (as well as drought) tolerance. A few examples are listed in Tables 10.1 and 10.2).

Table 10.2: Elements Identified for Salt Tolerance

Elements	*Function*
CDPK	Ca^{++} -dependent protein kinase
CAX	Ca/H antiporter
EhCaBP	Calcium binding protein
HKT1	High-affinity potassium transporter
HNX1	Proton sodium exchanger
RCI3	Rare cold inducible gene
RHL	Rice Hal2 - like
SPD	Sorbitol-6-phosphate dehydrogenase
Hal1	Yeast halotolerance gene
Hal2	Yeast halotolerance gene
Mn-SOD	Yeast mitochondrial superoxide dismutase

References

Akbar, M., G. S. Khush and D. Hille Ris Lambers. 1986. Proc. Intern. Rice Genet. Symp., May 1985, International Rice Research Institute, Los Banos, Philippines 399-409.

Armento-Soto, J. L., T. T. Chang, G. C. Loresto and J. C. O'Toole. 1983. J. Soc. Adv. Breed. Asia Oceania 15: 103-116.

Azhar, F. M. and T. McNeilly. 1988. Plant Breeding 101: 114-121.

Blum, A., J. Mayer and G. Golan. 1982. Field Crops Research 57: 137-146.

Blum, A., K. F. Schertz, R. W. Toler, R. I. Welch, D. T. Rosenow, J. W. Johnson and L. E. Clark. 1978. Agronomy Journal 70: 472-477.

Bohm, W. 1974. Z. Acker. Pflanzenbau 140: 282-287.

Bruckner, P. L. and R.C. Frohburg. 1987. Crop Science 27: 31-36.

Chen, T. H. H. and N. Murata. 2002. Current Opinions in Plant Biology 5: 250-257.

Clarke, J. M. and T. F. Townley-Smith, T. N. McCaig and D. G. Green. 1984. Crop Science 24: 537-541.

Descenzo, R. A. and S. C. Minocha. 1993. Plant Molecular Biology 22: 113-127.

Dure, l. 1992. Control of Plane Gene Expression (ed. D. P. S. Verma). CRC Press, Boca Raton 325-335.

E. Epstein, J.D. Norlyn, D. W. Rush, R. Kingsbury, D. B. Kelley and A. F. Wrana. 1980. Science 210: 399-404.

Ekanayake, I. J., J. C, O''Toole, D. P. Garritty and T. M. Masajo. 1985. Crop Science 25: 927-933.

Faquhar, G. D. and R. A. Richards. 1984. Australian Journal of Plant Physiology 11: 539-554.

Flowers T. J. 2004. Journal of Experimental Botany 55: 307-319.

Flowers, T. J. and Y. R. Yeo. 1995. Australian Journal of Plant Physiology 22: 875-884.

Flowers, T.J., M. A. Hajibagheri and M. J. W. Clipson. 1986. Quarterly Review Biology 61: 313-337.

Foolad, M. R. 1997. Plant Breeding 116: 53-58.

Foolad, M. R. 1999. Genome 42: 727-734.

Foolad, M. R., F. Q. Chen and G. Y. Lin. 1998. Theoretical and Applied Genetics 97: 1133-1144.

Foolad, M. R., L. P. Zhang and G. Y. Lin. 2001. Genome 44: 444-454.

Foolad, M. R., T. Stolz, C. Dervinis, R. L. Rodriguez and R. A. Jones. 1997. Molecular Breeding 3: 269-277.

Gay, A. P. 1986. Annals of Botany 57: 361-369.

Ghassemi, F., A. J. Jakeman and H. A. Nix. 1995. Salinization of Land and Water Resources, Human Causes, Extent, Management, and Case Studies. University of New South Wales, Sydney, Australia.

Gong, J. M., P. He, Q. A. Quian, L. S. Shen, L. H. Zhu and S. Y. Chen. 1999. Chinese Science Bulletin 44: 68-71.

Grover, A. *et al.*, 1998. Current Science 79: 689-696.

Hamill, J. D., R. J. Robins, A. J. Parr, D. M. Evans, J. M. Furze and M. J. C. Rhodes. 1990. Plant Molecular Biology 15: 27-38.

Holmstrom, K. O. *et al.*, 1994. Plant Journal 6: 749-758.

Holmstrom, K. O., E. Mantyia, B. Welin, A. Mandal, E. T. Palva, O. E. Tunnela and J. Londesborough. 1996. Natture 379: 683-684.

Kavi Kishore, P. B., Z. Hong, G. H. Miao, C. A. A. Hu and D. P. S. Verma 1995. Plant Physiology 108: 1387-1394.

Kodomi, Y. M., B. B. Singh, T. Terao, O. Myesss, Jr., J. H. Yopp and P. J. Gibson.. 1999. Indian Journal of Genetics 59: 317-323.

Levitt, J. 1972. Responses of Plants to Environmental Stresses. Academic Press, New York, USA.

Lyon, C. 1941. Bot. Gazette 103: 107-122.,

Mano, Y. and K. Takeda. 1997. Euphytica 94: 263-272.

Masgran, C., T. Altabella, R. Fascas, D, Flores, A.J. Thompson, R. T. Besford and A. F. Tiburcio. 1997. Plant Journal 11: 463-473.

Mitra, Jiban. 2001. Current Science 80: 758-763.

Moeljopawiro, S. and H. Ikechashi. 1981. Euphytica 30: 291-300.

Monforte, A. J., M. J. Asins and E. A. Carbonell. 1997. Theoretical and Applied Genetics 95: 284-293.

Morgan, J. M. 1980. Journal of Experimental Botany 31: 655-665.

Morgan, J. M. 1983. Australian Journal of Agricultural Research 34: 607-614.

Morgan, J. M. 1991. Australian Journal of Plant Physiology 18: 249-257.

Ndunguru, B. J., B. R. Ntare, J. H. Williams and D. C. Greenberg. 1995. Journal of Agricultural Science 125: 79-85.

Noh, E. and S. C. Minocha. 1994. Transgenic Research 3: 26-35.

O'Toole, J. C. and T. B. Moya. 1978. Crop Science 18: 873-876.

P. Breto, M. J. Asins and E. A. Carbonell. 1994. Theoretical and Applied Genetics 88: 395-401.

Pilon-Smith, E. A. H., M. J. M. Ebskamp, M. J. Paul, M. J. W. Jeuken, P. J. Weisbeek and S. C. M. Smeekens. 1995. Plant Physiology 107: 125-130.

Prasad, S. R., P. G. Bagali, S. Hittalmani and H. E. Shashidhar. 2000. Current Science 78: 162-164.

Quesada, V., S. Garcia-Martinez, P. Piqueras, M. R. Ponce and J. L. Micol. 2002. Plant Physiology 130: 951-963.

Robertson, B. M., A. E. Hall and K. W. Foster. 1985. Crop Science 25: 1084-1090.

Romero, C., J. M. Belles, J. L. Vaya, R. Serrano and F. A. Culianez-Maica. 1997. Planta 201: 293-297.

Sengupta, A., J. L. Heinen, A. S. Holaday, J. J. Barke and R. D. Allen. 1993. Proc. Natl. Acad. Sci. USA, 90: 1629-1633.

Singh, B. N. and D. J. MacKill. 1991. Proc. Second Internatl. Rice Genetics Symposium. IRRI, Los Bano, Philippines.

Tomar, J. B. and S. C. Prasad. 1996. Indian Journal of Agricultural Science 66: 459-465.

Tozlu, I., C. L. Guy and G. A. Moore. 1999. Genome 42: 1020-1029.

Turner, N. C. 1979. Stress Physiology in Crop Plants. Wiley, New York, USA: 343-372.

Ugherughe, P. O. 1986. Z. Acker. Pflanzenbau 157: 13-23.

Xu, D., X. Duan, B. Wang, B, Hong, T. H. D. Ho and R. Wu. 1996. Plant Physiology 110: 249-259.

Yancey, P. H., M. E. Clark, S. C. Hand, P. D. Bowlus and G. N. Somero. 1982. Science 217: 1214-1217.

11

Kochia (*Kochia scoparia*): To Be or Not To Be!

M.J. Al-Ahmadi[1] and M. Kafi[2]

[1]*Faculty of Agriculture, Birjand University, Birjand, Iran*
E-mail: mja230@yahoo.com
[2]*Faculty of Agriculture, Ferdowsi University of Mashhad, Iran*
E-mail: m.kafi@ferdowsi.um.ac.ir

ABSTRACT

Quick germination and growth of Kochia and its desirable drought and salinity tolerance indicate that it can be regarded as a valuable forage plant during shortage of conventional forage plants, particularly in arid and semiarid regions. Kochia can also provide a good source of forage through deficit irrigation or using saline waters and due to its high nutritional value can constitute an important component of ruminant diets in these areas. Cultivation of Kochia using saline waters for rehabilitation of saline areas that have been left barren can be regarded as an approach in sustainable and low-input agriculture, which needs to enhance genetical diversity of agricultural products, can help in food production for people settled in these regions.

Introduction

Salinization is a multidimensional problem that is caused by water, soil and biological factors, and resolving this problem would not to be simple. Traditional approaches of agronomic management and designing drainage systems and land rehabilitation, such as using conventional cultivars and/or soil and water manipulation to meet plant needs or genetical selection from present cultivars in order to improve plant products under adverse environmental condition, though could be useful but in many cases will face with nonchalance and disinclination by farmers, due to high costs and emerging secondary problems (Pessarakli, 1994; Siadat *et al.*, 1997). Sustainable utilization of saline water and soil resources needs reappraisal of natural resources and obsolete and/or under-exploited plants.

Agriculture in Iran has a history of thousands of years and since then most of the soil and water reservoirs have been exploited to produce different crops. Land surface in Iran comprises of arid and semiarid regions (about 90 per cent of total country area), and extensive surfaces of its lands (15 per cent of total country area) is affected by salinity and alkalinity, so the maximal expansion of saline soils, after Russia, India, and Pakistan, could be found in Iran (Pessarakli, 1994; Siadat *et al.*, 1997; Iranian Ministry of Agriculture, 1998). In most of the arid and semiarid regions of Iran more than 80 per cent of harvestable water is used for irrigation (Albrecht *et al.*, 1987), and during recent years salinity of this water has increased due to excessive harvesting and flow of saline water from surface water sources (Siadat *et al.*, 1997).

Forage and food production from halophytes is one of the sustainable utilization methods of soil and water resources in arid and semiarid agro-ecosystems and nourish settled populations in these regions. Domestication of salt tolerant plants (halophytes), which grow naturally in saline and arid vegetation sites, could be introduced as new crops cultivated under environmental stresses induced by salinity and aridity. These plants could be sensible alternative to conventional crops in many developing countries, including Iran. Thus, biosaline agriculture is an inevitable necessity particularly in arid and semiarid regions, which has received more attention recently. In fact, the constraint that has been already present against better utilization of these plants is absence of proper insight about possibility of using these plants (Glenn *et al.*, 1992; Aflakpui *et al.*, 1998).

Kochia species recently attracted the attention of researchers worldwide. These plants are drought and salt tolerant and are grazed in natural state by livestock (Zahran, 1993), and may provide a good source of forage by using saline water for irrigation in dry regions (Fuehring *et al.*, 1985). Farmers already cultivate *K. scoparia* as a drought tolerant forage crop in lands where planting other crops is difficult; thus, it has been called as "poor man alfalfa" (Undersander *et al.*, 1990). Its rapid growth during a relative shorter period and drought tolerance indicate that *K. scoparia* has the capacity to turn into important forage especially in semiarid regions (Sherrod, 1971). Characteristics like relative inexpensive production, adaptation to arid, saline and low fertile lands, low water requirement, and resistance to insects and diseases, has caught the attention of researchers during recent years to use this plant as a forage crop (Undersander *et al.*, 1990).

The economical cultivation expansion of halophytes like Kochia as a cash crop involves an understanding of biology, ecology, and superior management pattern of these plants in order to obtain optimal and sustainable production. The first prerequisite is the understanding of the plant's capabilities. The main objectives to study *K. scoparia* were to benefit from its halophytic nature and investigation of its cultivation possibility in regions that have a low productivity potential due to salinity or other specific environmental constraints.

Kochia Genus

The genus of Kochia contains 10-15 species that are spread over temperate regions of Northern Africa, Asia, Europe and south-western parts of North America (Gelin *et al.*, 2003). It seems that the species of Kochia genus are found in all countries (Young *et al.*, 1981). Out of them three species of *K. prostrata*, *K. americana*, and *K. indica* are present in desert areas. All these three species are perennial shrubs and are tolerant to saline-sodic conditions (Mullenix, 1998; Harrison *et al.*, 2000). However, according to Zahran (1993) *K. indica* is an annual bushy herbaceous plant, similar to *K. scoparia*. *K. indica* is a native of India and origin of *K. prostrata* and *K. scoparia* is Eurasia (Young *et al.*, 1981; Zahran, 1993). Polyploidy occurs in all *Kochia* species and male sterility has been found in *K. prostrata* (Mullenix, 1998). All species possess high diversity in structure and growth (Harrison *et al.*, 2000). It seems two Kochia species present in Iran, which demand considerable attention, are *K. scoparia* and *K. prostrata*.

Botany (*Kochia scoparia*)

Kochia, (*Kochia scoparia* L. Schrad) (Undersander *et al.*, 1990; Brevedan *et al.*, 1992; FEIS-Fire, 1996; Madrid *et al.*, 1996; Knight *et al.*, 2002) is a member of Chenopodiaceae (goosefoot) family (Zahran, 1993; Eberlin, 1996; Madrid *et al.*, 1996; Hager and Sprague, 2001; British Columbia Government, 2004). Glenn *et al.* (2003) in classification of Kochia species that exist in China, describe *K. scoparia* characteristics as "Herbs annual; Leaves narrowly complanate; Plants subglabrous or only inflorescences ferruginous villous; branches ascending, slender, margin of winglike appendages of perianth segments repand or incised". The word scoparia means "broom like". Other synonyms of this species are: *K. alata* Bates; *K. sieversiana* (Pallas) Mey; *K. trichophylla* (Stapf); *Bassia scoparia*; *Chenopodium scoparia* (L.) (Undersander *et al.*, 1990; FEIS-Fire Effects Information System, 1996). This plant also has numerous common names (Everitt *et al.*, 1983; Undersander *et al.*, 1990; Brevedan *et al.*, 1992; Choukr-Allah *et al.*, 1992; FEIS-Fire Effects Information System, 1996; Hager and Sprague, 2001; British Columbia Government, 2004), such as: Kochia, common Kochia, belvedere cypress, summer cypress, mock cypress, burring bush, fire weed, Mexican fire weed, iron weed, railroad weed, Hahakigi, Hai tham. Sha'R Banet. In addition, several different varieties of this species have been detected (FEIS-Fire Effects Information System, 1996): *K. scoparia* var. *scoparia* F., var. *cultiva* Farw, var. *pubescens* Moq, and var. *trichophylla* (Stapf.) Bailey.

It should be noted that *K. scoparia* is a species with high genetic diversity (Smith *et al.*, 1975; Mullenix, 1998; Mengistu and Messersmith, 2002), high variable appearance form leading to identify several forms, varieties and subspecies from it. In another reports two other of its known subspecies has been declared as *K. scoparia* var. *scoparia* and *K. scoparia* var. *subrillosa* Moquin- Tandon (Smith *et al.*, 1975; Eberlin and Fore, 1996; Mullenix, 1998; Gelin *et al.*, 2003). Although there is a wide genetic diversity in wild Kochia, any improved varieties still has not been developed. Generally, its intensely branched nature and hairs occur along leaves edges are the characteristics that help in differentiating *K. scoparia* from other species. The young

Kochia seedlings may confuse with *Chenopodium album* seedlings, but cotyledons of this plant are hairless (Wilson and Ng, 1975).

Kochia scoparia is a dicotyledon herbaceous erect annual forb (Everitt *et al.*, 1983; Undersander *et al.*, 1990; Brevedan *et al.*, 1992; Eberlin and Fore, 1996; FEIS-Fire Effects Information System, 1996; Madrid *et al.*, 1996; Mullenix, 1998). Plants of this species generally have a bushy growth tendency, which is very adaptable in their growth form (Zahran, 1993; Weiner and Fishman, 1994). Kochia has taproot and fusiform root system with high capacity to penetrate deep into the soil and spread horizontally (Undersander *et al.*, 1990; Zahran, 1993; Eberlin and Fore, 1996; FEIS-Fire Effects Information System, 1996; Mullenix, 1998; Gelin *et al.*, 2003; Unknown, 2001; British Columbia Government, 2004), which helps Kochia to tolerate drought conditions. This plant may range in height from 15 cm to more than 2 m depending upon the growth and competition conditions (Smith *et al.*, 1975; Shroyer and Erickson, 1987; Undersander *et al.*, 1990; Biswas, 1991; Zahran, 1993; Eberlin and Fore, 1996; FEIS-Fire Effects Information System, 1996; Madrid *et al.*, 1996; Mullenix, 1998; Gelin *et al.*, 2003; British Columbia Government, 2004); however, it has been observed that plant height did not differ in low density (600 $pl.m^{-2}$) (Galitzer and Oehme, 1979).

Kochia usually has strong, erect and highly branched main stem (Smith *et al.*, 1975; Undersander *et al.*, 1990; Zahran, 1993; FEIS-Fire Effects Information System, 1996; Hager and Sprague, 2001; Knight and Walter, 2002; British Columbia Government, 2004). Young stems are sub-glabrous below, but usually hairy towards the top (Knight and Walter, 2002; British Columbia Government, 2004) and often grooved on older plants (Undersander *et al.*, 1990; Zahran, 1993; Hager and Sprague, 2001), and its tip is pyramidal or rounded (British Columbia Government, 2004; FEIS-Fire Effects Information System, 1996). The color of stem varies from greenish to red and often both colours are present in the same individual plant (Hager and Sprague, 2001).

Stem of Kochia bears numerous lateral oblique upward branches (Smith *et al.*, 1975; Knight and Walter, 2002; Gelin *et al.*, 2003; British Columbia Government, 2004), which gives to plant a dense and compact feature. Stems and branches in *K. scoparia* are softer than *K. indica* (Zahran, 1993).

Succulent Kochia leaves (Zahran, 1993) are arranged alternatively on stems (Eberlin and Fore, 1996; Mullenix, 1998; Undersander *et al.*, 1990; Hager and Sprague, 2001; Knight and Walter, 2002; British Columbia Government, 2004). In general shapes of leaves are linear to narrowly lanceolate or linear (Undersander *et al.*, 1990; Zahran, 1993; Eberlin and Fore, 1996; Mullenix, 1998; Unknown, 2001; Knight and Walter, 2002; Gelin *et al.*, 2003; British Columbia Government, 2004). Leaves are 2–6 cm in length and 3-8 mm in width (Undersander *et al.*, 1990; FEIS-Fire Effects Information System, 1996; Gelin *et al.*, 2003; British Columbia Government, 2004). Kochia produces a constant amount of leaves per branch and leaf length is a simple function of number and length of produced branches (British Columbia Government, 2004).

Kochia has small, green and relatively inconspicuous and indistinguishable sessile flowers (Undersander *et al.*, 1990; Eberlin and Fore, 1996; Mullenix, 1998; Knight and Walter, 2002; British Columbia Government, 2004). Flowers are bisexual or female (Gelin *et al.*, 2003), and usually 1-3 flower found in a glomerule in axil of upper leaves and in terminal spikes (Undersander *et al.*, 1990; Eberlin and Fore, 1996; Mullenix, 1998; Unknown, 2001; Knight and Walter, 2002; Gelin *et al.*, 2003; British Columbia Government, 2004). Spikes are short and dense (Knight and Walter, 2002; British Columbia Government, 2004) and have distinct leaf-like, but reduced, glaucous bracts which give a thorny appearance to the flowering stem (Everitt *et al.*, 1983; Knight and Walter, 2002; British Columbia Government, 2004). Kochia flowers are incomplete, which allows cross-pollination occurrence that plays more important role to develop its resistance and adaptation characteristics (Hager and Sprague, 2001).

Kochia fruit is a small and brittle stellar shape sac, which is 2 mm in width and its pericarp is membranous and separated from seed known as utricle (Zahran, 1993; Eberlin and Fore, 1996; Gelin *et al.*, 2003). Each fruit contains one seed (Zahran, 1993; FEIS-Fire Effects Information System, 1996). Seeds of this species are small, with 1-3 mm in length (Everitt *et al.*, 1983; Mullenix, 1998; Zahran, 1993; Eberlin and Fore, 1996; FEIS-Fire Effects Information System, 1996; Gelin *et al.*, 2003). Seeds of perennial Kochia species are orthodox and their longevity is more than *K. scoparia* (Harrison *et al.*, 2000).

Ecology (*Kochia scoparia*)

Kochia is a salt excreting halophyte (Phelps and Bauder, 2002), which in some classifications has been assigned as facultative halophyte (Brevedan *et al.*, 1992); however, concerning increased tolerance of older plants to soil salinity than younger ones, Zahran (1993) has introduced other Kochia species (*K. indica*) as an obligated halophyte.

The origin of this plant is from Eurasia (Everitt *et al.*, 1983; Zahran, 1993; FEIS-Fire Effects Information System, 1996; Madrid *et al.*, 1996; Knight and Walter, 2002), particularly south and east of Russia (Knight and Walter, 2002; Unknown, 2001). Kochia also grow abundantly in Mediterranean regions (Madrid *et al.*, 1996). Bioclimatic dispersal of Kochia is in dry areas with precipitation between 100 to 400 mm (Hager and Sprague, 2001) and has also been found at elevations up to 2000 m from sea surface (FEIS-Fire Effects Information System, 1996).

Kochia is a versatile species (Knight and Walter, 2002) and is capable to spread along long distances rapidly. This plant is highly adaptable to different environments and its ability to drought tolerance enables it to spread quickly. It is adapted throughout Europe and some regions of temperate Asia, Africa, and North and South America (Gelin *et al.*, 2003) and is found in gardens, farm margins, cultivated fields, grasslands and pastures, disturbed soils, roadside, edges of waterways and barren lands (Gelin *et al.*, 2003; Knight and Walter, 2002). Some adaptations of Kochia such as tumbleweed characteristic and tolerance to salinity and drought have possibly helped to its extending to regions that are more arid.

Kochia is a relatively cold tolerant species. Generally it is not influenced by cold or early spring frost (Undersander *et al.*, 1990; Eberlin and Fore, 1996; Mullenix, 1998). It germinates very early in spring (Eberlin and Fore, 1996) and its seedlings are partly (Mullenix, 1998) or completely (Eberlin and Fore, 1996) freezing tolerant. However, it has been reported that if average temperature goes lower than 15.5℃, Kochia does not produce flowers and seeds. In general, due to C_4 metabolism, cold temperatures alleviate net assimilation rate (NAR) and dry matter production of Kochia and reduce its competitive ability severely. In evaluation of interferences of Kochia with wheat and barley, Fischer *et al.* (2000) found that in warmer temperatures Kochia had a higher NAR than wheat and barley, but its NAR reduced as the temperature decreased.

Kochia has an extended tolerance to a variety of soils and establishes very easily in saline, alkaline, neutral, acidic soils and establish where other plants are unable to sustain (Miyamoto *et al.*, 1992, Unknown, 2001). Although wild Kochia is mostly found in saline and alkaline soils (Eberlin and Fore, 1996; FEIS-Fire Effects Information System, 1996; British Columbia Government, 2004) but it has not been extended to regions with very acidic soils (Undersander *et al.*, 1990). This plant also has adapted to saline soils containing high nitrate levels (Edwing and Dobrowolski, 1992; cited in Madrid *et al.*, 1996).

Kochia, due to its C_4 metabolism, is not too shade tolerant and shading reduces its competitive ability. Fischer *et al.* (2000) reported that photosynthesis rate of Kochia increases in PAR values more than 400 mmol.m^{-2} and as PAR reduces below this limit, CO_2 loss due to dark respiration will increase and its NAR and WUE will be decreased.

Because of its deep rooting system, Kochia is exceptionally drought tolerant (FEIS-Fire Effects Information System, 1996; Madrid *et al.*, 1996; Mullenix, 1998) and can rapidly extend under dry condition (Unknown, 2001). It is in particular suited as a forage crop for drylands (Hager and Sprague, 2001) that produce proper forage at the annual rainfall less than 150 mm (Undersander *et al.*, 1990). Kochia also has a good tolerance to moisture stress during germination (Mullenix, 1998). Kochia is spring lodging sensitive (FEIS-Fire Effects Information System, 1996) and does not tolerate exposure to relatively humid weather. It seems the dry matter yield and protein content of Kochia is not affected by applied water volume (Fuehring *et al.*, 1985; Mullenix, 1998).

The possibility of domestication and sustainable utilization of Kochia as a feed source for ruminants needs a multilateral study of its different characteristics under different climates and conditions. In order to achieve this we performed a comprehensive study about *K. scoparia*, and its responses to changes in growth medium, including in the field as well as controlled environment. Part of our result is presented in this chapter.

Materials and Methods

The field studies consisted of two experiments at two different locations. In the first one, response of plants to different salinity levels of irrigation water (1.5, 8.6 and 28.2 dS.m^{-1}) was done at Birjand, a dry area with limited water resources for irrigation, particularly fresh water. In the second one, we cultivated Kochia at different combinations of plant densities (5, 10, 15, and 20 pl.m^{-2}) and irrigation intervals (7 and 14 days) at Birjand (13°59' N, 53°32'E) and Mashhad (36°20'N, 59°35'E). Based on longtime meteorological information, the average annual precipitation in Birjand and Mashhad is 176 and 260 mm respectively. The chemical characteristics of applied waters and soil, before and after experiments, are represented in Tables 11.1 and 11.2.

Table 11.1: Chemical Analysis of Soil of Experimental Fields in Mashhad and Birjand

Location		*EC (dSm^{-1})*	*pH*	*Absorptable Elements (ppm)*			*Soluble Ions ($meq.l^{-1}$)*						
				N	*P*	*K*	*HCO_3^-*	*Cl^-*	*SO_4^-*	*Na^+*	*K^+*	*Mg^{2+}*	*Ca^{2+}*
Before experiments													
Birjand		10.33	7.36	9.33	1.25	150	3.6	90	20.8	43.48	1.03	37.2	22.2
Mashhad		12.4	7.66	24.5	28	420	4.5	75	44.5	67	1.3	28	29
After experiment													
Birjand	S1	1.65	8.3	30.33	0.25	140	3.6	8.8	3	8.7	0.18	2.6	1.4
	S2	13.24	7.73	11.66	0.25	145	1.6	94	46	76.1	0.9	48.2	27.6
	S3	22.7	7.93	21	0.25	160	2.2	18.4	64	186.96	1.28	54.4	27

1. S1, S2, and S3 refer to salinity of irrigation water equal to 1.5, 8.6 and 28.2 dS m^{-1} respectively
2. Soil texture in Birjand and Mashhad were sandy loam and silty clay loam, respectively

Table 11.2: Chemical Analysis of Waters Used During Different Experiments at Birjand and Mashhad

Location		*EC (dSm^{-1})*	*pH*	*Soluble Ions ($meq.l^{-1}$)*						
				HCO_3^-	*Cl^-*	*SO_4^-*	*Na^+*	*K^+*	*Mg^{2+}*	*Ca^{2+}*
Birjand		5.6	7.57	3.7	32.5	19.8	38	0.18	11.6	7.0
Mashhad		5.5	6.81	–	39.2	–	24.7	0.19	15.4	5.4
Salinity Experiment	S1	1.5	6.92	–	8	–	10.8	0.15	5.8	1.4
	S2	8.6	6.3	–	74.4	–	48.7	0.40	26.6	12.4
	S3	28.2	6.68	–	204	–	250	1.4	–	9.2

1. S1, S2, and S3 refer to salinity of irrigation water equal to 1.5, 8.6 and 28.2 dS m^{-1} respectively

Germination at Different Temperatures

K. scoparia has a high initial germination rate. It is clear that Kochia germinates quickly under optimum conditions and usually penetrates its seed coat 24 hours after imbibitions (Everitt *et al.*, 1983; Steppuhn and Wall, 1993). Kochia often grows in very dry environments; therefore, its rapid germination has a special importance for using limited moisture in spring (Dyer *et al.*, 1993; Eberlin and Fore, 1996; Mullenix, 1998). Weiner and Fishman (1994) showed that almost half of seeds germinated in three days after planting. Seed viability of *K. prostrata* under room temperature has been reported to be 53 per cent (Lotfi, 1996), but about *K. scoparia* there are some reports declaring its germination percentage more than 88 per cent (Everitt *et al.*, 1983) and even equal to or more than 96 per cent (Dyer *et al.*, 1993).

Kochia scoparia has a good germination in a wide thermal range between 4 to 41°C (Eberlin and Fore, 1996; Mullenix, 1998). Young *et al.* (1981) studied germination of different accessions of *K. prostrata* under different temperatures and found the maximum germination of all accessions occurred in 20 °C. Khan *et al.* (2000a) observed germination rate was the highest at higher temperatures. However, there is a disagreement about its optimum temperature of germination. For example, its optimal germination has been reported at constant temperatures of 16 °C (Eberlin and Fore, 1996), 17 °C (Everitt *et al.*, 1983), 10 and 15 °C (Everitt *et al.*, 1983), inconstant 15/30 or 20/30 °C (Eberlin and Fore, 1996), and inconstant night/day temperatures of 25-3°C (Khan *et al.*, 2001). Also, it has been reported that germination rate of its newly harvested seeds at temperatures near to freezing (2 °C) is not simultaneous between plant samples and vary highly between different ecotypes (Harrison *et al.*, 2000).

Concerning importance of these temperatures in researches and agronomic planning, we studied *K. scoparia* germination response to different temperatures, to determine its cardinal germination temperatures. We evaluated the Kochia seed response to temperature at a thermal range from 10 to 40 °C, with 5 °C intervals, with an additional temperature equal to 8 °C in growth chambers at Instituto De Agricultura Sostenible (CSIC), Cordoba, Spain. Germination was recorded when tip of radicle with a 5 mm length was uncoiled (Young *et al.*, 1981). Germination counting was done at 12 h intervals, continuing until cumulative germinated seeds became stable in two consecutive records or when 100 per cent germination was done.

Germination Percentage

The trend of cumulative Kochia germination in response to temperature showed different patterns in different temperatures, especially during first 96 h (Figure 11.1). During this period, seeds at 20 and 25 °C germinated rapidly and reached nearly their peak germination. The lowest temperatures (8, 10, and 15 °C) caused less seeds to germinate during the same period and final germination percentages in these temperatures were also lower, when compared with 25-35 °C.

On the average, 85 per cent of seeds germinated in a range from 8 to 35 °C, when at 40 °C, germination percentage showed a significant reduction ($p<0.01$) and dropped to 31.5 per cent. Our results are also in agreement with those of Everitt *et al.* (1983), however they report that the significant reduction in germination of *K. scoparia* began from 30 °C, and there were only 11 per cent germination at 40 °C, that is lower than our findings; though our findings support their results and those of Romo and Haferkamp (1987) that cold to moderate temperatures improve Kochia's germination while constant high temperatures constraints germination severely.

Germination Rate

As daily germination trend indicates, the highest germination percentage equal to 86 per cent at 25 °C occurred during the first 48 h after start of the test, and about half of seeds finished their germination in the first 24 h of test at this temperature (Figure 11.1). The inverse of time span until 50 per cent germination (T_{50}) was

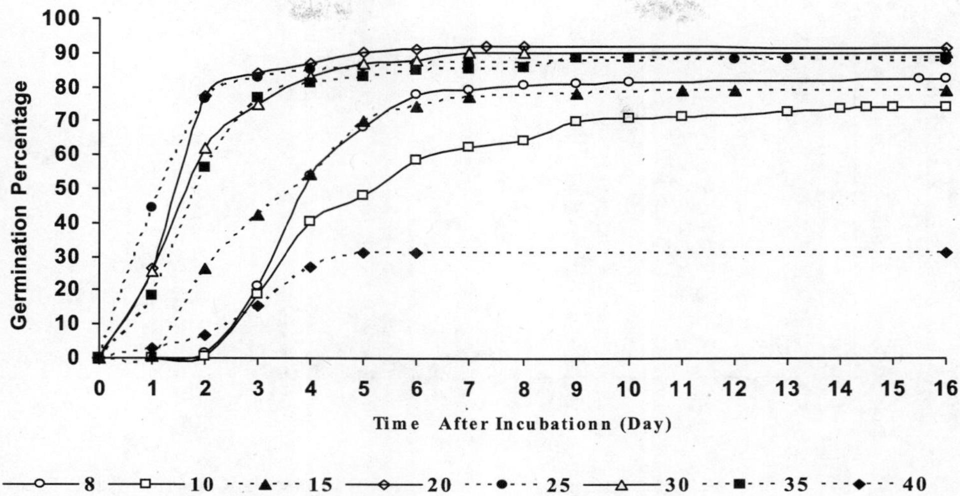

Figure 11.1: Cumulative Germination Percentage of Kochia Seed at 24 h Intervals in Different Temperatures as Shown in Figure. Each data point represents the average of 4 replications (together 160 seeds).

determined as an index of germination rate (Aflakpui *et al.*, 1998; Tobe *et al.*, 2000; Flores and Briones, 2001).

Analysis of variance showed a significant difference ($p<0.01$) in required time for 50 per cent germination (T_{50}; Figure 11.2). As mentioned before, the minimum time span for 50 per cent germination to be reached was 24 h at 25 °C, which indicates this temperature induced the highest germination rate, however, it was not significantly different between 20 and 30 °C, but the general trend in Figure 11.2 is germination rate reduction (increase of half time of germination) with changing temperature above or below 25 °C. This causes the response curve to temperature of time taken to T_{50} to possess an anticline like shape ('U' shape), that is typical of the temperature reaction of many developmental processes.

Flores and Briones (2001) in a study of six desert species at three temperatures (12, 20 and 26 °C) also found that as temperature increased, the beginning of germination and time span up to 50 per cent germination decreased. Temperatures below 15 °C delayed germination of *Salicornia rubra* (Khan *et al.*, 2000b). In winterfat (*Kraschenninkovia lanata*), a dry tolerant shrub, when temperature increased, germination rate raised and time to peak of germination decreased.

Cardinal Temperatures

Calculation method of cardinal temperatures is based on dependence of developmental rate to temperature. In order to describe cardinal temperatures we used an intersected lines model that has been used in many literatures, in which

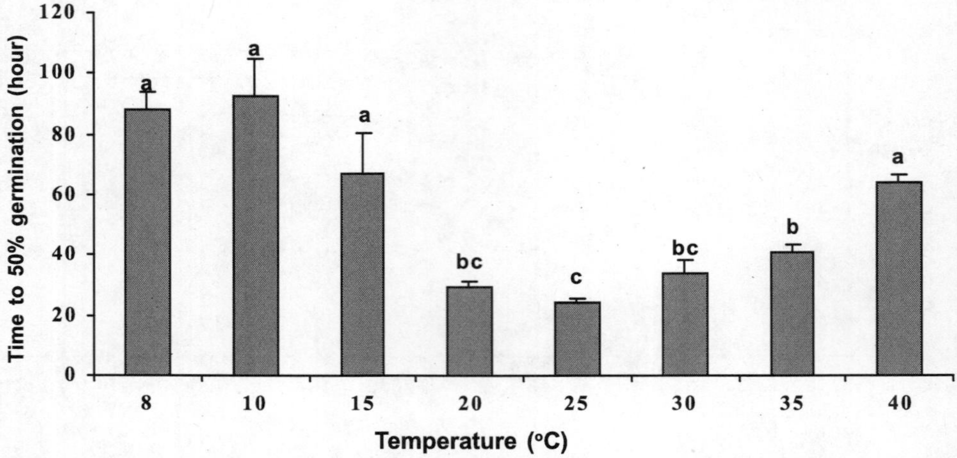

Figure 11.2: Temperature Response of Time Taken to 50 per cent Germination. Error bars represent standard error of means for each temperature. Non-similar letters above the bars indicate significant different between temperatures at p<0.01 level. The coefficient of variance (CV) is 4.6 per cent.

germination rate index and temperatures were dependent and independent variables respectively:

$Y=a_1+b_1x$ if $T_b< Td \leq T_o$, $b>0$

$Y=a_2+b_2x$ if $T_o< Td \leq T_c$, $b<0$

In which Y is germination rate and T, T_b, T_o, and T_x are average, base, optimal and maximum temperatures respectively. In this model T_b and T_x are derived from the interception of each regression line with the abscissa and T_o is the intersection point of two linear regression lines that describe response of germination rate to temperature at sub-optimal and super-optimal temperatures.

We used the inverse of time to T_{50} as dependent variable in the model. The model could estimate plant germination response to temperature satisfactorily ($p<0.01$, $r^2=0.946$, with a standard error of estimate equal to 0.0029; Figure 11.3). Based on model estimation T_b, T_o, and T_x of Kochia are 3.4, 24, and 49.7 °C, respectively. There is no any report about cardinal temperatures of Kochia yet and it seems these results are the first ones in this field. It is evident that more experiments are required for these results to be generalized. In interpretation of cardinal temperatures, it should be considered that from these relations one can obtain T_b and T_o, however, extrapolated values of T_x are not constant for all germination percentages and then should be accepted carefully.

The estimation of model from T_x reflects the high tolerance of Kochia to high temperatures, especially in germination stage. It should be noted that in the nature, seeds are subject to thermal fluctuation when germinating (Flores and Briones, 2001), thus these results under constant temperatures may differ slightly from germination in the field. The wide thermal range of Kochia germination from 3.5 °C (T_b) to maximum

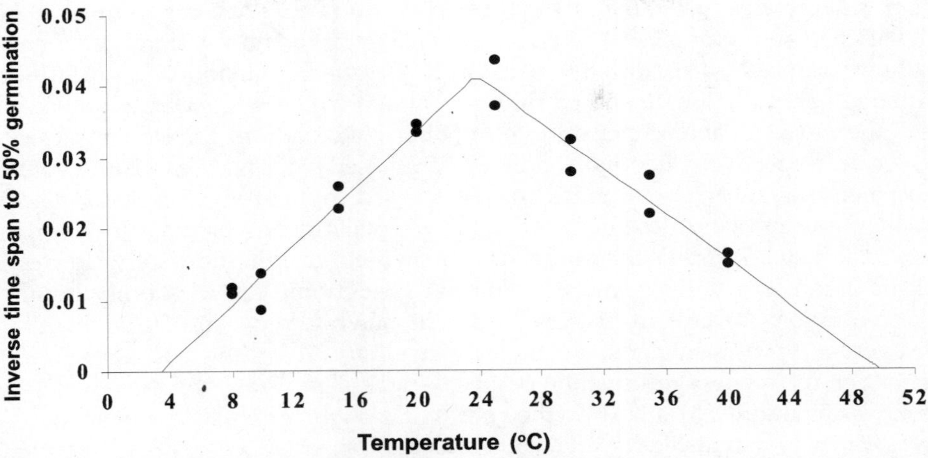

Figure 11.3: Kochia Germination Rate Response to Temperature Based on Inverse Time Taken to 50 per cent Germination Index. Each data point is mean of 2 replications of each series of experiment. Fitted line between points has been obtained using nonlinear triangular regression, and is the basis of determination of cardinal temperatures.

50 °C, and its rapid germination ability indicate the high potential of Kochia to establish in most regions of the world. Kochia can be cultivated when soil temperature is above 5 °C and any time during the season thereafter. Since Kochia can germinate at low soil temperatures and is high cold tolerant in early growth stages, it can be sown at late winter; thus Kochia germinates with first moisture in spring. With this fast start at growth season, competition with other annual plants would not be complicated. The ability of Kochia to germinate at low temperatures and its fast germination in high temperatures allows us to use Kochia as a second crop or to cultivate it twice a year (2 plantations), thus allowing maximum exploitation of resources, especially in arid and semi arid regions.

Germination at Different Levels of Salinity

Initial establishment of plant species in saline habitats depends on their seed germination response to thermal and saline regimes and it is the level of this response that usually determines whether a population can persists until reproductive maturity or not (Aflakpui *et al.*, 1998; Khan and Ungar, 2001). Indeed, the seed ability to germinate under high salinity levels is an important factor for successful halophytes establishment through direct sowing (Miyamoto *et al.*, 1992). Most of halophytes germinate better under non saline situations (Huang *et al.*, 2003), but their restriction to saline environment demonstrates their requirement to relatively high salt concentration, tolerance to excessive salt levels, or competition reduction from other plants, compared to less stressful environments. Halophytes are also capable to germinate in higher soil salinity level than glycophytes (Katerji *et al.*, 2000; Huang *et al.*, 2003).

Kochia scoparia is a halophyte that germinates well under moderate and even high salinity level (Clarke and West, 1969; Everitt *et al.*, 1983; Steppuhn and Wall, 1993; Khan and Ungar, 2001). Everitt *et al.* (1983) reported no any significant effect of salinity induced by six different salts on *K. scoparia* germination up to 20 dS m^{-1}, although germination decreased significantly with increase in salinity, yet at 40 dS m^{-1} there was 40 and 45 per cent germination with sodium chloride and calcium chloride, respectively. Steppuhn and Wall (1993) used equal NaCl and $CaCl_2$ volumes to generate salinity concentrations from zero to 30 dS m^{-1} and a revise relation between maximum cumulative Kochia germination with salinity of substrate. However, they could not determine a threshold limit in which germination starts to decline dramatically. In their study, germination reductions due to salinity begun at concentrations lower than 20 dS m^{-1} and reduction rate between 12 to 30 dS m^{-1} was in average -3.3 per cent per dS m^{-1} increase in salinity. According to Khan *et al.* (2001), although the highest germination occurred in distilled water, still there was 14 per cent germination at 31 dS m^{-1}. In their study, although germination percentage (GP) was 95 per cent in distilled water, it decreased to 14 per cent in 30 dS m^{-1}. Reduction of GP and rate had been reported in another species of Kochia (*Kochia americana*) and also other halophytes (Clarke and West, 1969; Katembe *et al.*, 1998; Khan *et al.*, 2001; Huang *et al.*, 2003).

We studied seed germination characteristics of *Kochia scoparia* under unequal combination of NaCl and $CaCl_2$ salts, and threshold of its germination tolerance to salinity. For this the saline solutions were prepared by adding proper amounts of NaCl and $CaCl_2$ salts in a 2:1 molar ratio to distilled water, obtaining an electrical conductivity range from 0 to 20 dS m^{-1}, with 5 dS m^{-1} intervals. After adding saline solutions, Petri dishes were put in the second plastic container and to avoid evaporation this second container was encased with a plastic film.

Seeds germination counting was done like above mentioned temperature experiment. Germination rate (GR) was determined with dividing cumulative germinated seed number in each counting by total germinated seeds, and also the inverse of time taken to 50 per cent germination (T_{50}) was calculated as an index of germination rate (Tobe *et al.*, 2000; Flores and Briones, 2001). As the lowest germinated seed number was equal to 8 seeds in one of 20 dS m^{-1} replications, the time taken to germination of 8 seeds was calculated as an arbitrary index of germination rate.

Germination Percentage (GP)

There was a reverse relation between the peak of cumulative germination with salinity of the medium (Figure 11.4). GP reduction with increase in salinity had been reported in Kochia and other halophytes (Clarke and West, 1969; Everitt *et al.*, 1983; Steppuhn and Wall, 1993; Katembe *et al.*, 1998; Khan *et al.*, 2001; Huang *et al.*, 2003). In addition to the reduction of total germinated seeds, enhancement of salinity caused the time taken to germination stability to be increased. Though peak germination occurred at first 100 h of experiment, but a few seeds germinated even 14 days after the start of tests (Figure 11.4). This indicates high genetical diversity of Kochia seed lots, which maybe a reason of its success to grow in harsh environments and also shifting to a successful weed in agroecosystems.

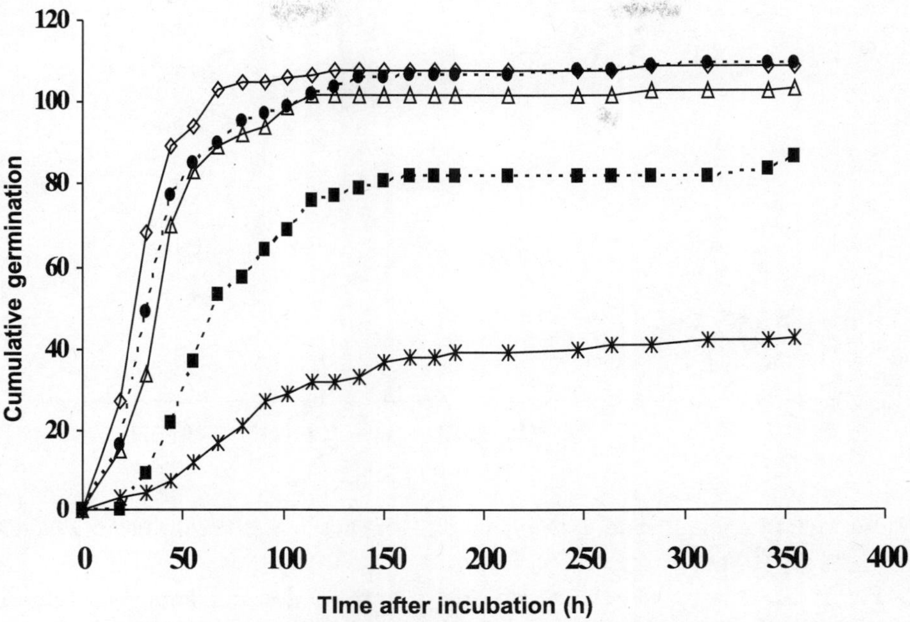

Figure 11.4: The Trend of Cumulative Kochia Seed Germination at 0(◇), 5(●), 10(△), 15(■), and 20(✳) dS m^{-1} Salinity Levels

Salinity up to 10 dSm^{-1} did not have a considerable effect on seed germination, but passing this threshold, germination percentage and rate decreased significantly ($p<0.01$, Figures 11.4 and 11.5; Table 11.3). The extreme reduction occurred at 20 dS m^{-1}, but still more than 35 per cent of seeds germinated at this level of salinity. In the study of Steppuhn and Wall (1993), there was 14 per cent germination for Kochia at 30 dS m^{-1}. Regardless of inequality of units used by researchers to explain salinity concentration, these reports clearly showed that each species has very particular germination requirements and response to salinity and also each species displays a distinct germination, based on the location and dominated climate conditions during growth and seed production.

Table 11.3: Effects of Salinity on Germination Properties of *Kochia scoparia*

Salinity (dSm^{-1})	*Average Germinated Seeds in Each Petridishes*	*Germination Percentage*	*Time Taken to Germination of 8 Seeds (Hour)*	*Time Taken to 50% Germination (Hour)*
0	27.25 a**	90.84 a**	20 b*	27.8 b**
5	27.5 a	91.67 a	25.3 b	34.25 b
10	26 ab	86.67 a	30.25 b	37.8 b
15	21.75 b	72.5 ab	54.5 b	60.30 a
20	10.75 c	35.84 b	143.3 a	77.8 a

In each column, non similar letters indicate significant difference at 0.05 (*), or 0.01(**) probability level.

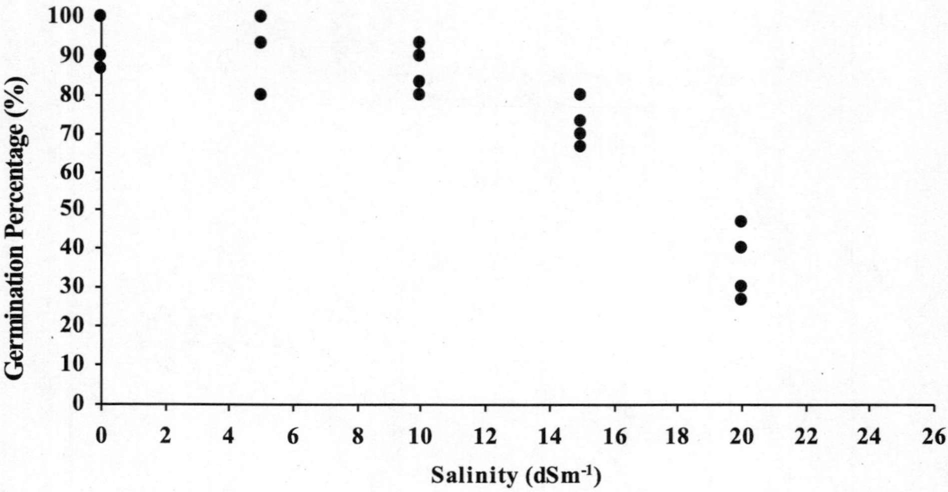

Figure 11.5: Changing Trend of Germination Percentage in Different Salinity Levels

Figure 11.5 clearly shows the decline of germination percentage in salinities above 10 dS m^{-1}. Of course, with this data it is very difficult to obtain a strict threshold of Kochia seed tolerance to salinity, which above that, seed germination severely affected by salinity and would be reduced, and need further experiments with narrower salinity levels. Also with germination percentage it is very difficult to attain ultimate limit of germination tolerance to salinity, in which germination of Kochia seeds will cease, maybe using of germination rate will give us better result.

Decrease of germination percentage with increase in salinity may be due to osmotic effects or specific ion toxicity. As salinity prevents seed germination with reduction of water availability or interference with some aspects of metabolism like changing equilibrium of growth regulators (Khan *et al.,* 2001), possibly in this experiment mixing of NaCl and $CaCl_2$ caused more disruption in critical process of seed than applying NaCl alone.

Germination Rate (GR)

Increase in salinity caused reduction of seed GR (Figure 11.6) and seeds required a longer period to complete their germination in high salinity levels (Table 11.3). Figure 11.7 indicates as salinity increased, seed GR (the inverse of time taken to 50 per cent germination) decreased linearly, as time required to this stage reached from 28 hours in distilled water to 78 hours at 20 dS m^{-1}. On the other hand, when about half of seeds completed their germination at the first day of test in control, it prolonged to 3 days in 20 dSm^{-1}, that clearly shows a delay in germination due to salinity enhancement; however there was not any significant difference between salinity levels for GR, up to 10 dS m^{-1}. In the study of Khan *et al.* (2001) also as salinity increased, average required time to reach germination stability (the end of germination) increased, from 4 days in distilled water reached to 9 days in 24 and 30 dS m^{-1}. Romo

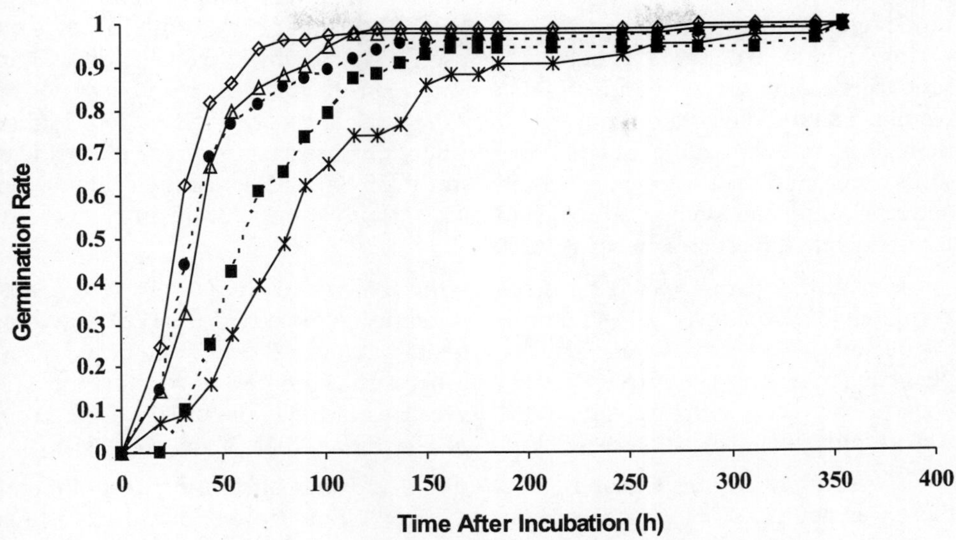

Figure 11.6: The Trend of Germination Rate of Kochia Seeds at 0(◇), 5(●), 10(△), 15(■), and 20(✳) dSm^{-1} Salinity Levels

and Haferkamp (1987) showed in *K. prostrata*, there was a linear relation between osmotic potential and days to 50 per cent germination.

Fitting a line between data points of GR and extrapolating it until crossing horizontal axis, the threshold of germination tolerance appears to be slightly more

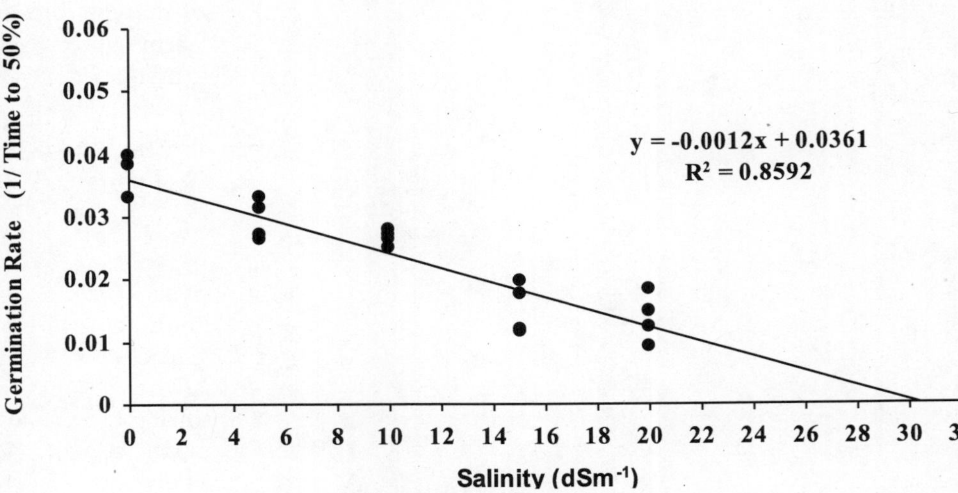

Figure 11.7: Germination Rate (Inverse of time taken to 50 per cent germination) of *K. scoparia* Under Different Salinity Levels. Each series of point belong to 4 replications of each treatment.

than 30 dS m^{-1} (Figure 11.7). It seems after this salinity level, GR approaches to zero; however, according to Steppuhn and Wall (1993), there was still 14 per cent germination at 30 dS m^{-1}. Of course, they used equal volume of NaCl and $CaCl_2$ in preparing saline solution and this may be the reason for difference between the results. It should be noted in present study that the effects of concentrations higher than 20 dS m^{-1} were not evaluated. In addition, our used temperature (20 °C) was different from temperature used in their study (15 °C) and concerning interaction between salinity and temperature (Khan *et al.*, 2001); this temperature difference may contribute in difference between results.

Germination rate based on 8 seeds (the inverse of time taken to 8 seeds germination) also showed a linear decrease as salinity increased (Figure 11.8), however, the declining slope of GR was slightly more than that of GR calculated based on 50 per cent germination (Figure 11.7). It maybe because at the start of germination, GR was very high and with elapsing time, it decreased gradually (Figures 11.4 and 11.6). Accordingly, it is reasonable that GR of 8 seeds is higher than 50 per cent of seeds.

Referring to Figures 11.5 and 11.6 and Table 11.3 and considering the point that there was not any significant reduction in GR and GP up to 10 dSm^{-1}, it seems it is possible to cultivate *K. scoparia* at farms using saline water with same salt concentration, provided soil is wetted until establishment of seedlings so that soil water potential remains at a level that does not interfere with seeds. Furthermore, it should be considered that several factors (including water, temperature, light, and salinity), which interfere at soil surface, regulate seed germination and may, besides

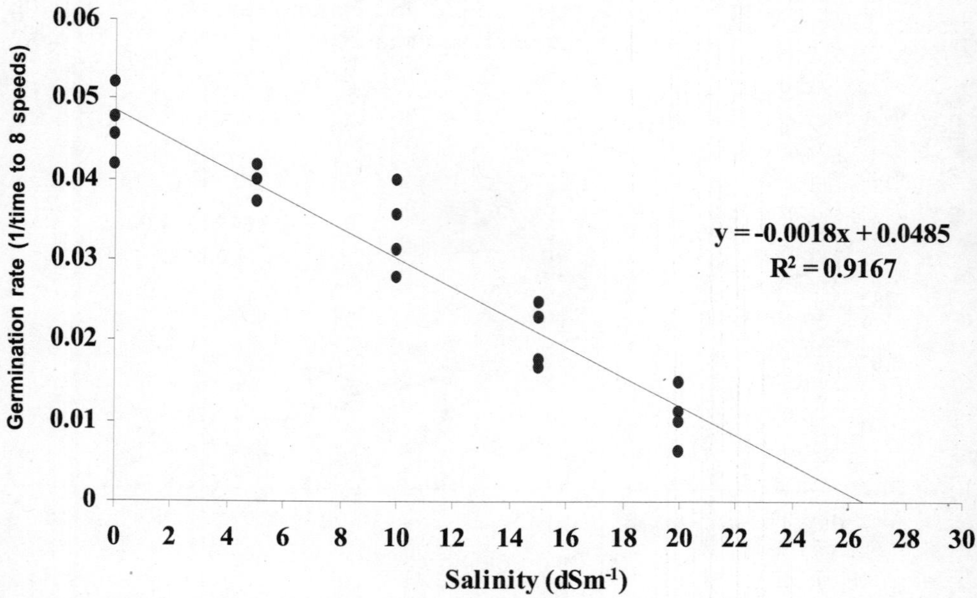

Figure 11.8: Germination Rate (Inverse of time taken to germination of 8 seeds) of *K. scoparia* Under Different Salinity Levels. Each series of point belong to 4 replications of each treatment.

seasonal variation in temperature, affect temporal pattern of germination. Osmotic and matric potentials restrict effective thermal range for germination of temperate halophytes (Khan and Ungar, 2001), and interaction between salinity and temperatures also influenced Kochia germination, somehow in higher salinity levels, increase of temperature improves its germination percentage (Khan *et al.*, 2001).

Therefore, at planting time in the early spring, two issues should be consulted: if air temperature (prior, soil temperature) is high, it is evident that GP is more important than GR. In this situation, one can sow more seeds to ensure a good crop stand. If temperature is low or water is salty, GR would be reduced and other factors such as microorganisms may affect the seed; thus in this situation irrigation should be done, which should not be saline enough to reduces GR severely. Considering these points, it could be expected a desirable plant establishment using saline water or soil.

Growth and Development

Vegetative Growth

In Birjand plant emergence was very rapid, only at 5DAP and accumulation of 64TU (thermal unit), but in Mashhad plant emergence was delayed to 11DAP during which only 19TU was accumulated, due to lower temperatures and snowing.

Karachi and Pieper (1987) stated rapid Kochia establishment in new areas is mainly due to its high seed reservoirs and rapid initial seedling growth. Its cotyledons are highly tenuous, essentially linear in outer shape, with faint green in colour and covered with hairs.

It has been shown that low to moderate salinity level stimulate halophytes growth. Noaman and El-Haddad (2000) demonstrated halophytic plants response differently to increase of salt concentration in irrigation water. Salinity increase to 40 gr liter^{-1} caused growth reduction of five halophytes, except *Salicornia bigelevii*, which showed growth stimulation even in that concentration. In one experiment in Saudi Arabia, with increase of soil salinity from 12 up to 20 mmho cm^{-1} after 11 weeks irrigation with saline water, no negative effects was observed in growth of two Kochia species (*K. scoparia* and *K. indica*) (Zahran, 1993).

In our experiments also there were no significant differences in respect to LAI between different salinity levels, particularly during later growth stages (Figure 11.9a), and the maximum difference was observed in maximum of LAI, which was reached about 100 DAP. From Figure 11.9a, it can be concluded that salinity caused the increasing trend of plant leaf area to be slowed, however it seems Kochia can tolerate it interestingly, approaching to maximum LAI comparable to non-saline condition.

Considering high correlation between produced dry matter with leaf area (Bhardwaj *et al.*, 1988), which was also observed here (r^2=0.88, p>0.01), it can be concluded that variation in total dry matter (TDM) largely follows changing trend of LAI (Figures 11.9a and b). The moderate salinity increased dry matter accumulation (Figure 11.9b). Stimulation of growth with moderate salinities also has been reported in *K. indica* (Zahran, 1993). These results also coincide with Fowler *et al.* (1988) who observed at 36 DAP dry weight of *Salsola iberica* in 10.5 dS m^{-1} exceeded from 1.4 dS m^{-1}, however, higher salinity caused reduction of dry matter accumulation.

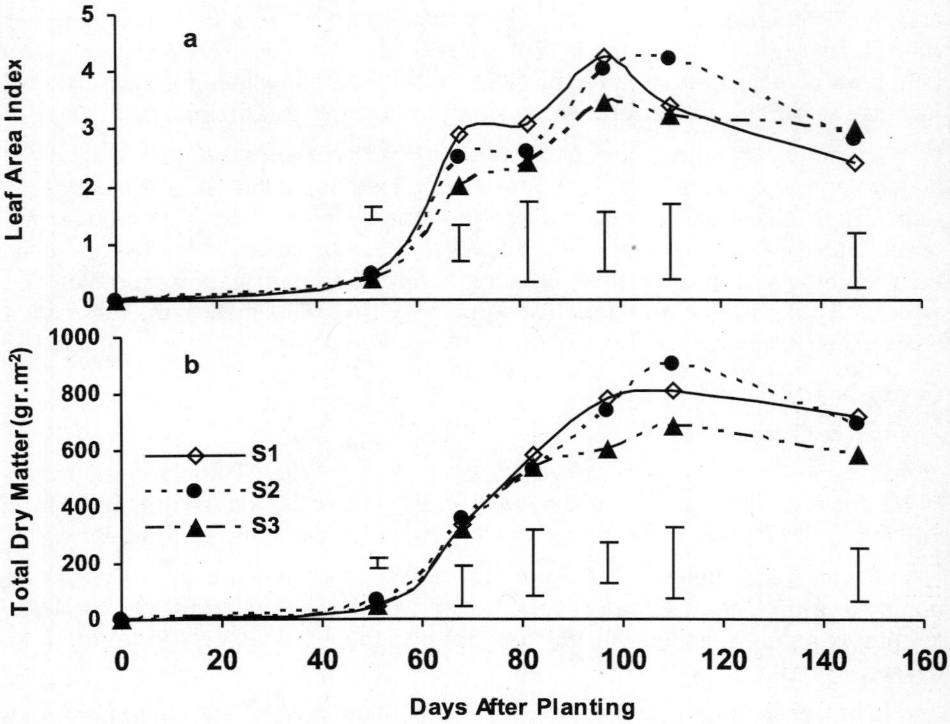

Figure 11.9: Changing Trend of LAI (a) and Total Dry Matter (b) in Response to Different Salinities of Irrigation Water. S1, S2, and S3 represent a salinity of 1.5, 8.6, and 28.2 dS m^{-1}, respectively. The vertical bars indicate the least significant difference (LSD) at 0.05 probability level.

According to these results, it appears that *K. scoparia* needs salts for its optimum growth and low to moderate salinity levels stimulate its growth. Thus, contrary to pervious assumption, testifying Kochia is a voluntary halophyte (Brevedan *et al.*, 1992), it seems it is rather an obligate halophyte species.

The maximum LAI (6.4) was observed in highest density in Mashhad. This treatment also had the highest maximum dry matter accumulation (1545.8 g m^{-2}). As growth progressed, the amounts of leaves (as a percent of total dry matter) decreased and stem dry matter increased, and this resulted in reduction of leaf to stem ratio (Figure 11.10, Table 11.4). This trend has also been observed in other plants, like as alfalfa (Zahran, 1993). Sherrod (1971) harvested Kochia forage from a natural green vegetation at pre-, mid- and late-flowering stages and observed the percentage of leaf-weight decreased with progress in plant growth, leading to reduction of leaf to stem ratio from 1.2 at pre-flowering to 0.4 at late-flowering. Increase of salinity to 28.2 dS m^{-1} caused a noticeable alteration in partitioning coefficients, in favour to continuity of photosynthetate allocation to leaf production. Other reports (ex. Waller *et al.*, 1983; Noaman and El-Haddad, 2000) also emphasized increase of leaf to stem ratio in response to increase of salinity, especially in halophytes. These and others

(Glenn *et al.*, 1992; Khan *et al.*, 2000 b,c), all coincide on this point that salinity exerts more restriction for stem growth than leaves.

Table 11.4: Equations and Coefficients of Dry Matter Partitioning to Stem and Leaves at Different Salinity Levels. The equations were obtained with fitting linear regression through data points illustrated in Figure 11.2, separately for each salinity treatments.

Salinity (dSm^{-1})	*Leaf*		*Stem*	
	Equation	*R^2*	*Equation*	*R^2*
1.5	LW=0.3773TDW	0.8981	SW=0.6227TDW	0.9713
8.6	LW=0.3548TDW	0.7772	SW=0.6452TDW	0.9555
28.2	LW=0.4617TDW	0.9449	SW=0.5383TDW	0.9131

LW, SW, and TDW are leaves, stem and total dry matter, respectively, as $gr.m^{-2}.day^{-1}$.

Concerning halophytic nature of Kochia, it may be more realistic that increase of leaf thickness be a factor contributing to tolerance mechanism of the plant to higher salinities. As maximum leaf to stem weight ratio showed a direct correlation with electrolytes leakage from leaves (r^2=0.48, p<0.01), it seems continuity of leaf production in highest salinity (Figures 11.9a, and 11.10c) is a response to more salt accumulation in leaves that can damage plant cells, leading to defoliation of aged leaves. In this situation, plant continues to produce new leaves for keeping high phosynthesizing area and control intensive salt accumulation at growth points (Kafi and Damghani, 2000; Smart *et al.*, 2001). This will cause reduction of salt concentration inside tissues and evasion of salinity damage, but for photosynthesis to continue at the cost of some reduction in dry matter accumulation.

Reproductive Growth

Kochia is a short-day plant, which produces flowers and seeds in summer (Eberlin and Fore, 1996; Unknown, 2001). Reproduction of Kochia species can be done with self- or out-crossing (Mullenix, 1998). The day lengths that cause flowering induction vary from 13 to 15 h between different Kochia populations (Eberlin and Fore, 1996). In our trials, plants in each location flowered and matured rather simultaneously between different treatments (± 7 days). In Birjand, flowering started only after a lapse of 68 DAP (1083TU accumulation), however peak of pollination occurred about 14 days later when 1348 thermal units were accumulated. Flowering was terminated at 110 DAP (1824TU), thus duration of flowering was about 42d in Birjand. In Mashhad, flowering started at 75 DAP, with only 737.4TU accumulation, which is much lower than Birjand. The maximum pollination occurred at 85 DAP (916.8TU), and at 112 DAP (1373.6TU), flowering was finished. Thus, flowering period in Mashhad was 37 days. It seems extended flowering period of Kochia mostly is due to high diversity between individuals, which is affected by environmental condition *per se*. This is confirmed by observing high variation between individuals in growth indices measurements.

The final harvest for seed (growth termination) was done at 174 DAP (2986TU) and 168 DAP (2404TU) in Birjand and Mashhad, respectively. Although it has been

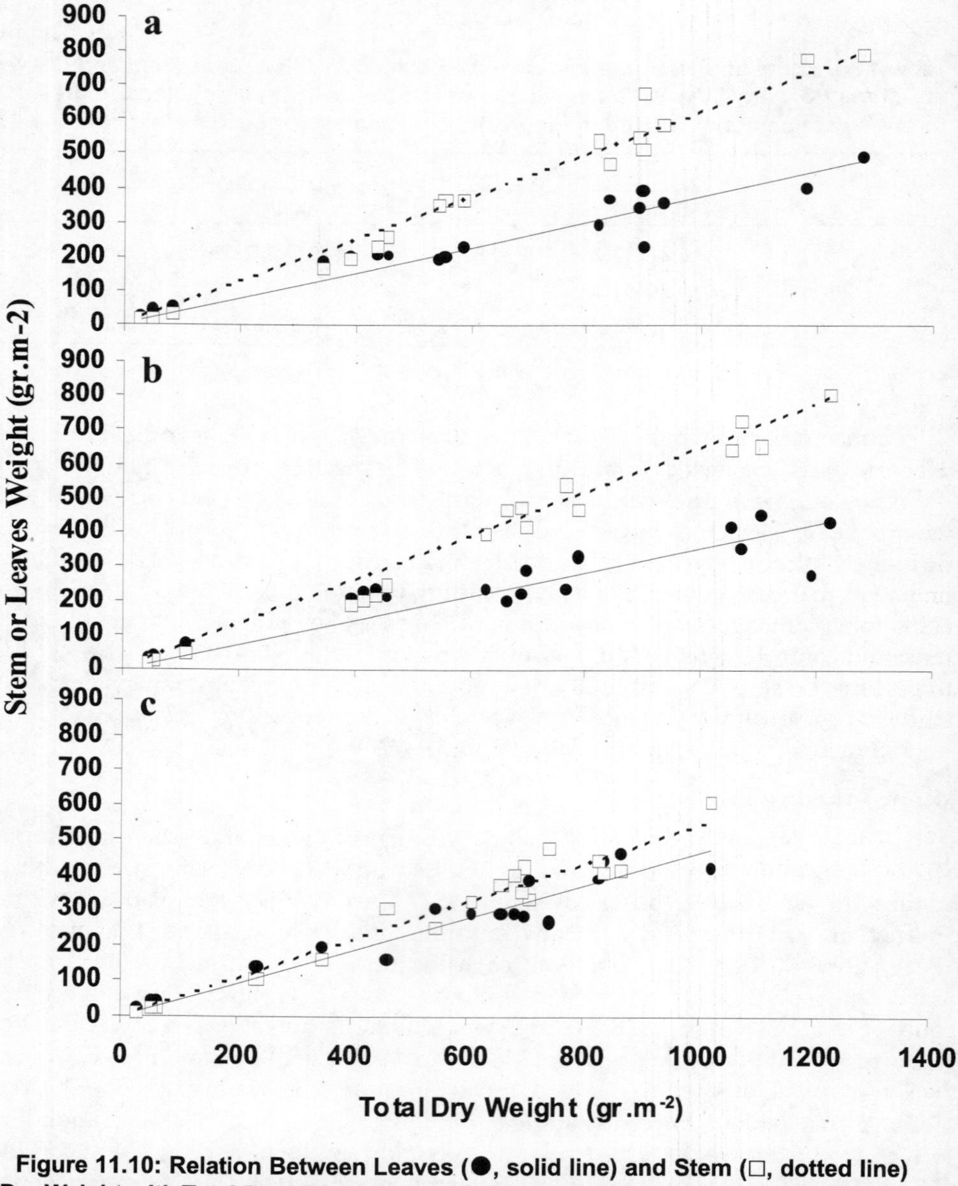

Figure 11.10: Relation Between Leaves (●, solid line) and Stem (□, dotted line) Dry Weight with Total Dry Weight at Three Different Salinities as 1.5 (a), 8.6 (b), and 28.2 (c) dS m^{-1}. Data points were calculated from measurements made during growth season. Equations are presented in Table 11.1.

stated that any plant needs a constant thermal unit to reach a given developmental stage (Kader and Jutzi, 2004), but the growth period differences between Birjand and Mashhad was notable. It seems a precise classification of Kochia growth stages based on plant phonology can help one to describe Kochia responses to environment conditions.

Salinity and Light Utilization Characteristics

Kochia benefits from C_4 photosynthesis pathway (Fischer *et al.*, 2000) and, therefore, is not a shade tolerant plant. The only available information about its light characteristics had been presented by Fischer *et al.* (2000). After performing replacement series experiments in order to study competition of Kochia with wheat and barley, they found Kochia was more effective in interception of radiation than wheat and barley due to its higher light extinction coefficient. They reported K values of 0.65, 0.36, and 0.33 for Kochia, wheat, and barley respectively. In their study radiation absorption by canopy of Kochia raised up linearly as LAI and plant height increased. Kochia maintained its high rate of photosynthesis at PAR values above 400 mM m^{-2} s^{-1} but at PAR values below this threshold, its CO_2 loss due to dark respiration was more and its net assimilation rate was lower than cereals plants. This result shows clearly that wheat and barley tolerate shading better than Kochia.

To calculate canopy light extinction coefficient, we measured incident radiation using a linear solarimeter (Sunscan, ΔT) at hours from 11 to 14 of solar noon. Since diffuse component of solar radiation has a better uniform distribution than direct one, all measurements were done with artificial shading. Charles-Edwards and Lawn (1984) stated that if the measured radiation at solar noon were used to determine radiation interception, the ratio of absorbed radiation would be underestimated up to 10 per cent. Thus, F_{abs} was corrected for whole day using equation represented by Charles-Edwards *et al.* (1986, Equation 1):

$$Q_{day} = 2.\ Q_{noon} / [1+Q_{noon}] \quad [1]$$

Daily radiation was estimated using the method described by Goudriaan and Van Laar (1993). These values were corrected based on sunny hours and Angstrom coefficients estimated by Khalili and Rezaei Sadr (1998) for Birjand.

$$I_{abs}=I_{tot}.\ F_{abs} \quad [2]$$

The amounts of absorbed radiation by canopy were calculated using a constant K value between salinity treatments, assuming 45 per cent of intercepted radiation is photosynthetically active (Goudriaan and Van Laar, 1993).

Considering that all radiation measurements were done at solar noon, Figure 11.11a indicates the relation between natural logarithm (Ln) of the F_{abs} by canopy at solar noon with LAI. The slope of this relation is light extinction coefficient (K) value of Kochia at solar noon that is 0.59. This value is lower than reported K of 0.65 by Fischer *et al.* (2000) for Kochia. However, in their experiment K was calculated under competitive condition with wheat and barley. It should be noted that measurements of radiation at above and bottom of canopy in our experiment were done with elimination of direct radiation by artificial shading, because according to Sinclair and Horie (1989), ignoring this component may generate a considerable error in calculations.

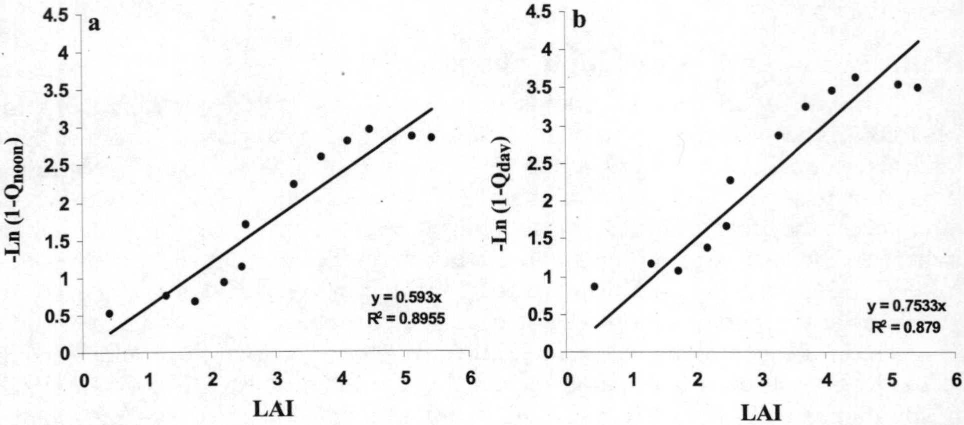

Figure 11.11: Linear Regression of Natural Logarithm of Transmitted Radiation Through Canopy Against LAI for Solar Noon (a) and Whole Day (b). The slope of fitted line between data points indicate light extinction coefficient (K) of *K. scoparia.*

The amount of radiation intercepted by canopy and its transmitted fraction change with time and both of them reach to peak around solar noon (Charles-Edwards *et al.*, 1986). Thus, the daily variation in flux intensity of incident radiation above the canopy (I_0) and the fraction transmitted through canopy (I/I_0) should be considered in calculating K. After correcting the amount of radiation absorption by canopy using equation 1, the K of Kochia for whole day increased to 0.75 (Figure 11.11b), which was higher than the value reported by Fischer *et al.* (2000). This value was used in further calculations and was assumed constant in all treatments. Figure 11.12 shows that the fraction of canopy-absorbed radiation in whole day is higher than its value at noon. Begue *et al.* (1991) also stated that F_{abs} at noon is lower than its value at morning or afternoon and to obtain real K values, the daily variation of K with change in solar angle should be considered in calculations.

The fraction of radiation intercepted by canopy increased with LAI and 95 per cent of incident radiation was absorbed at a LAI of 4.5 to 5 (Figure 11.13). Figure 11.9a indicates that plants reached to this LAI in about 100 DAP, *i.e.* in mid to late of flowering period. This time almost coincides with the early summer and after that total daily radiation that reach upon the plant canopy starts to decrease (Figure 11.14). Lack of synchrony between maximum incoming seasonal radiation and maximum LAI will lead to inefficient absorption and utilization of light by canopy.

Radiation Absorption

Figure 11.15 indicates estimated absorbed radiation during growth season in different salinities. It seems likely that plants at lower salinities closed their canopies earlier and the extension rate of leaf area at early season reduced as salinity increased. Around 90 DAP, the fraction of radiation absorption in moderate salinity (8.6 dS m^{-1}) was the same as lower salinity (1.5 dS m^{-1}) and after 100 DAP, *i.e.* just at peak of

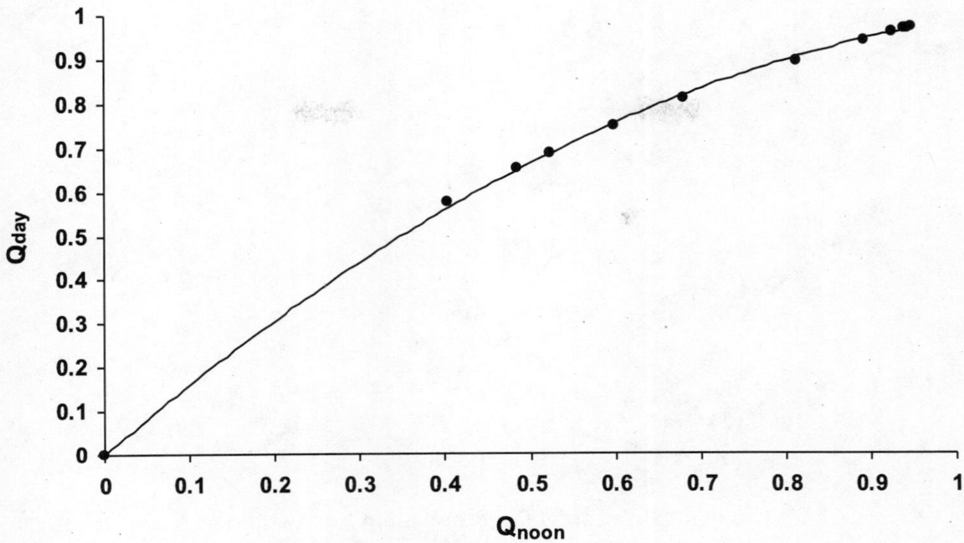

Figure 11.12: The Predicted Relation Between Fractions of Absorbed Radiation at Noon (Q_{noon}) with its Value for Whole Day (Q_{day}). The Q_{day} values were calculated with recommended algorithm by Charles-Edwards *et al.* (1986).

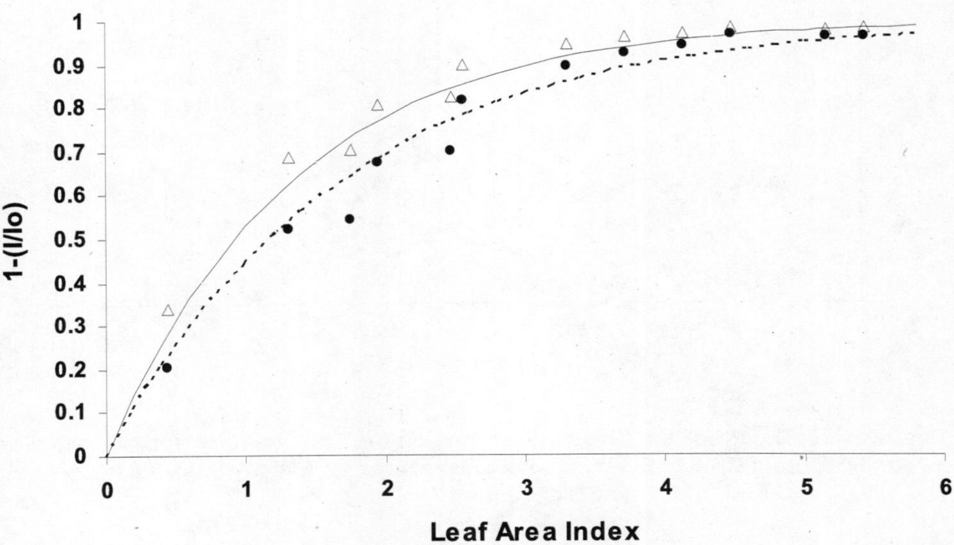

Figure 11.13: The Observed and Predicted Relations Between the Fraction of Absorbed Radiation at Solar Noon (Solid circles, dotted line) and Whole Day (Empty triangular, solid line) with Leaf Area Index

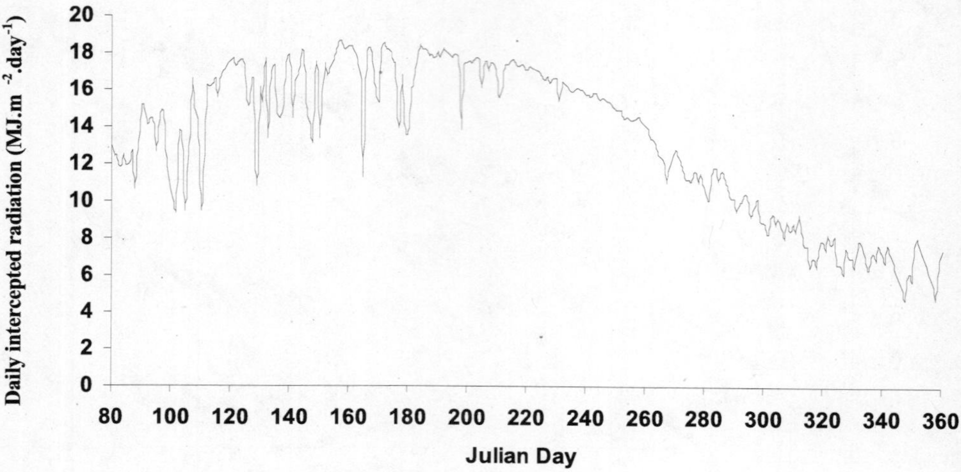

Figure 11.14: The Variation of Total Daily Radiation Intercepted by Plant Canopy During Year at Birjand, South Khorasan

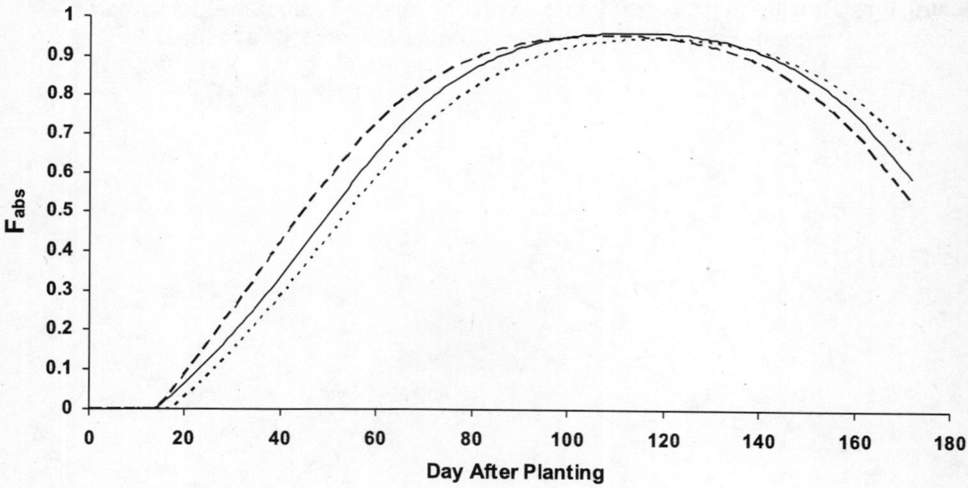

Figure 11.15: The Predicted Effects of Salinity on the Fraction of Absorbed Radiation of Kochia During Growth Season. Salinity levels are 1.5 (patchy line), 8.6 (solid line), and 28.2 (dotted line) dS m^{-1}.

radiation absorption, the peak of F_{abs} at moderate salinity exceeded that of lower salinity. At high salinity level (28.2 dS m^{-1}) plants approached their maximum F_{abs} later in growth season. The important point is that the peak of F_{abs} at this salinity level coincided with the time that incident radiation to canopy started to decrease (Figures

11.14 and 11.15). Therefore, lower absorbed radiation is expected in this treatment. The higher F_{abs} in higher salinity at late season were due to its higher LAI in this time (Figure 11.9a). Reduction of F_{abs} at late season is consequent to normal leaf desiccation.

Radiation Use Efficiency (RUE)

Figure 11.16 illustrates the relation between total dry matter produced and cumulative PAR in three different salinities. Fitting a linear regression between data point for each salinity level showed that increase in salinity level, from 1.5 to 8.6 dS m^{-1}, caused a slight increase in RUE that decreased again with more salinity increase; however, there was not much difference between different salinity levels in respect to RUE (Figure 11.16, Table 11.5). The moderate salinity level, which had the highest RUE, is the treatment which had the highest above ground biomass accumulation (Figure 11.16), that is a consequence of stimulating effect of moderate salinity on growth of halophytes such as Kochia (Zahran, 1993; Khan *et al.*, 2000b; Noaman and El-Haddad, 2000,). Therefore, it seems that moderate salinity caused Kochia plants use intercepted radiation more efficiently. It had been observed that plants exposed to salinity assimilate carbon with higher efficiency and this is a result of carbon effects on osmotic adjustment so that increase in efficiency of carbon fixation and assimilation lead to improved osmotic adjustment process (Wyn Jones *et al.*, 1983). Although it is evident that this is true until the salinity level does not exceed plant tolerance threshold and its capacity to osmotic adjustment. In addition, it is clear that activity of PEP- carboxylase and C_4 carbon metabolism would be stimulated

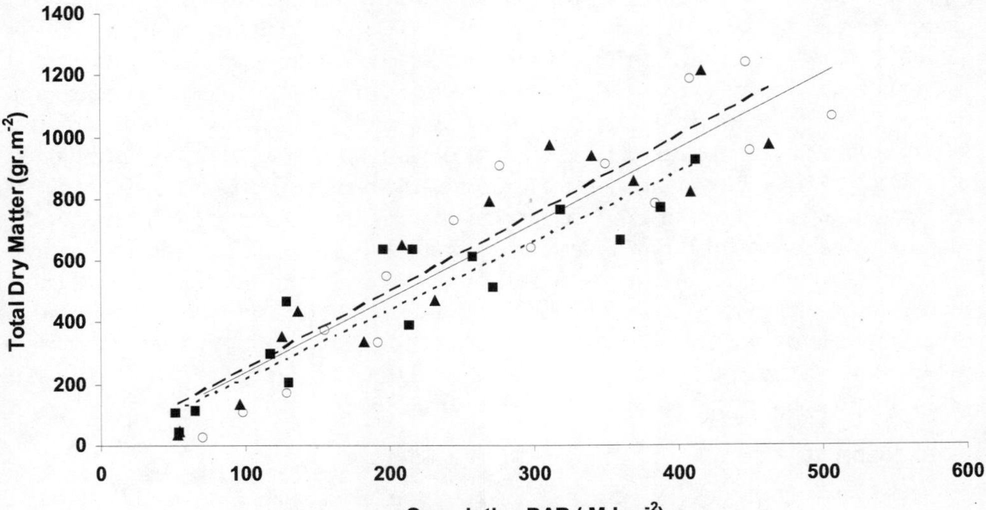

Figure 11.16: The Observed Relation Between Total Dry Matter with Cumulative PAR in Three Salinities as 1.5 (Empty circle), 8.6 (Solid triangular), and 28.2 (Solid square) dS m^{-1}. The linear relation between dry matter and cumulative PAR has been obtained with fitting a linear regression between data points for 1.5 (solid line), 8.6 (patchy line), and 28.2 (dotted line) dS m^{-1}. RUEs for different salinity which is the slope of regression line are presented in Table 11.5.

in presence of salinity (Wyn Jones *et al.*, 1983). Hence, it seems the positive feedback of osmotic adjustment on carbon metabolism and assimilation in response to salinity at C_4 plants like Kochia causes to improve their RUE in response to salinity.

Table 11.5: Radiation Use Efficiencies of Kochia in Different Salinities Level

EC (dSm⁻¹)	*RUE (gr.MJ⁻¹ PAR)*	R^2
1.5	2.41	0.86
8.6	2.48	0.87
28.2	2.20	0.85

Our results which are the first report about this species, indicated that Kochia has good tolerance to salinity, so that salinity increase of irrigation water from 1.5 to 28.2 dS m^{-1}, that cause plant death in many species, led to a negligible decrease in RUE. Moreover, the moderate salinity level of 8.6 dS m^{-1} induced stimulation in plant growth and improvement of RUE. Therefore, *K. scoparia* not only is a saline tolerant species, but also shows a good response to moderate levels of accumulated salts in soils and can be used successfully in saline and desert environments to produce forage for domestic animals in these harsh areas, which already have been faced with forage shortage.

Relative Water Content (RWC)

The leaves RWC were not influenced by salinity, but it reduced with a lapse of time significantly ($p<0.01$) from 59.8 per cent at 82 DAP (mid of flowering) to 50.8 per cent at 111 DAP (end of flowering). Unlike these results, Khan *et al.* (2000a) stated that water content of *Atriplex* plants decreased severely with increase of salt concentration, however, in their experiment, increase of salinity did not influence significantly leaf water content of plants after 30 and 60 days of salinity exertion. It seems unaffected leaf RWC of Kochia by salinity is an approach for resistance to salinity that makes possible the persistence of growth. Increase of irrigation interval from seven to 14 days in Mashhad and Birjand also reduced slightly leaf RWC at first measuring. Reduction of leaves RWC in response to increase of irrigation intervals indicates water supply from roots had not been matched with water loss from leaves, due to reduction in soil water content. Under these circumstances inevitable water loss, when stomata are open for photosynthesis, lead to reduction in leaf RWC, which it self will affect the photosynthesis (Lambers *et al.*, 1998).

Ion Accumulation

Most of halophytes (Francois, 1976; Wyn Jones *et al.*, 1983; Nerd and Pasternak, 1992; Khan *et al.*, 2000b,c), and some non-halophytes (Wang *et al.*, 2001) accumulate Na^+ rather than K^+, and Na^+ is responsible for a considerable part of tissues osmotic pressure in these plants. This trend particularly is evident in dicotyledon halophytes, when monocotyledon halophytes have lower Na^+ levels (Wyn Jones *et al.*, 1983). There is a genetical diversity between halophytes species in respect to degree of osmotic adjustment and sodium to potassium ratio (Fowler *et al.*, 1988; Choukr-Allah *et al.*, 1992; O'Leary *et al.*, 1992; Katerji *et al.*, 2003; Tomar *et al.*, 2003), thus, it has been stated that selective absorption mechanisms of ions which provide possibility of chemical discrimination between similar ions (like as Na^+ and K^+), possess an adaptive value (Shannon, 1997).

Salinity of irrigation water did not affect ion accumulation, but leaf potassium concentration was changed during time significantly. Table 11.6 indicates although potassium concentration of leaves was increased significantly at mid-flowering but after that it reduced substantially toward end of flowering and approached to concentrations lower than pre-flowering values. Sherrod (1971) also reported that leaf potassium percentage reduces with increase of plant age. End of flowering is time of termination of plant vegetative growth. Because K^+ is a major osmoticum in retention of low water potential of plant tissues (Wang *et al.*, 2004), it is possible that increase in its absorption during vegetative growth stage may be due to its demand for induction cellular expansion for keeping turgor potential in plant tissues (Khan *et al.*, 2000c). With terminating vegetative growth, potassium requirement may be reduced and due to high-energy requirements of solute absorption and conservation, and due to nature of Kochia in preference for Na^+ rather than K^+ (Sherrod, 1971; Steppuhn and Wall, 1993), sodium replaces potassium and excess excreted ions.

Table 11.6: K^+ Concentration and Ionic Ratio of Na^+ to K^+ in Kochia Leaves at Different Growth Stages

	Stage of Flowering		
	Pre-	*Mid-*	*Post-*
K^+	101.34 (±2.36)	115.69 (±5.25)	82.36 (±3.83)
Na^+ / K^+	1.45 (±0.067)	1.11 (±0.092)	1.70 (±1.17)

1. The numbers inside parenthesis present standard error of means (SEM).

2. Concentration of K^+ as ppm.

Table 11.6 also indicates that as growth progressed, Na^+ to K^+ ratio decreased at first and then displayed a very significant increase ($p<0.01$). Another point is that with progress of growth season, accumulation of Na^+ and K^+ in soil increases because of irrigation with saline water. Under high external salt condition, membranes become more permeable for Na^+ and absorption of K^+ may be inhibited, especially due to this fact that sodium inhibit K^+ absorption *per se* (Kafi and Damghani, 2000). However, Tables 11.1 and 11.2 indicate that Na^+/K^+ in water and soil were higher than that in plant tissues (Table 11.6, Figure 11.17). Thus, it seems plants have created some constraints for excessive Na^+ absorption to avoid severe damage by sodium, particularly to plasma membranes (Kafi and Damghani, 2000) and to keep equilibrium between these ions inside its tissues.

Although non-significant, but trends indicate reduction in amounts of K^+ and increase of Na^+ in leaves (increase of Na^+/K^+) with increasing salinity (Figure 11.17). It have been observed that most of halophytes, particularly dicotyledons (Francois, 1976; Wyn Jones *et al.*, 1983; Fowler *et al.*, 1988; Nerd and Pasternak, 1992; Khan *et al.*, 2000b,c;), and some non-halophytic species (Ferreyra *et al.*, 1997; Wang *et al.*, 2001)

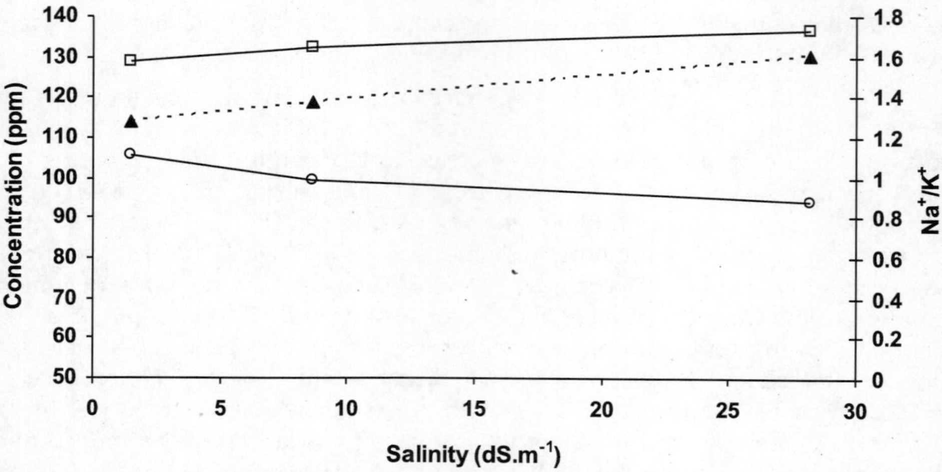

Figure 11.17: Changing Trends of Average Concentrations of K^+ (O) and Na^+ (□), and Na^+/K^+ (▲) in Kochia Leaves with Increase of Salinity Level of Irrigation Water

accumulate Na^+ rather than K^+, and Na^+ is responsible for osmotic pressure of their tissues. Considering raising of Na^+ and K^+ concentrations in post-experimental soil samples with increase of irrigation water salinity (Table 11.1), it seems insignificant increase of these ions inside plants leaves, as Short and Colmer (1999) stated, is an indicative of self-controlling of Na^+ and K^+ concentrations in plant tissues. In this case, despite their increase in root medium, their concentration in plant tissues did not increase accordingly. Another reason may refer to high salinity tolerance of Kochia. It means tolerance mechanisms of Kochia act actively up to high salinity level as 28.2 dS m^{-1} and restrict excessive salt accumulation in plant. In general, concerning this self-controlling of ion concentration in leaves and non-significant increase of Na^+ and K^+ concentrations in leaves with raising salinity, it can be confirmed that Kochia is an excretive halophyte (Pessarakli, 1994).

Electrolytes Leakage

The electrolytes percolation was measured using leaf discs at 82 and 111 DAP (about middle to late flowering). In average between two measurements, electrical conductivity of water, after floating leaf discs for 24 h, at 28.2 dS m^{-1} was 30.3 µS m^{-1}, higher than ($p<0.01$) two other salinity levels, which did not show any significant difference (23.8 and 25.6 µS m^{-1} in 1.5 and 8.6 dS m^{-1} treatments, respectively). As Ghoulam *et al.* (2002) stated, it appeared that at the highest salinity level, salts were not compartmented in cells effectively. Thus, the bulk of ions in leaves was not involved in cell osmotic adjustment and in fact caused disruption in membrane permeability.

The amount of electrolyte leakage was not influenced by the increase of irrigation intervals. It can be concluded that despite observed growth depression, drought stress resulted by increase of irrigation interval has not been too severe to induce some changes in membrane permeability. Furthermore, considering that cell membrane stability has been assigned as main constituent of drought tolerance (Kader and Jutzi, 2004), non-significant change in membrane permeability indicates high tolerance of Kochia to drought stress, which has been emphasized in many reports (Zahran, 1993; FEIS-Fire Effects Information System, 1996; Madrid *et al.*, 1996; Mullenix, 1998).

Irrigation Intervals and Dry Matter Production

Kochia yield is influenced by environmental parameters. Zahran (1993) subjected *K. indica* under two different irrigations with 20 and 30 days intervals and did not observe any significant differences. He also reported that there is a genetical variation between Kochia species in respect to yield production and yield of *K. indica* is more than *K. scoparia*.

We also studied yield of Kochia with 7 and 14 days irrigation intervals. There were not any significant differences between irrigation intervals in plant dry weight and dry matter content (per cent) of produced forage; however, some interactions with harvest time was observed. The plant dry weight showed a negative correlation with LWR (r^2=0.53, p<0.01), and plant density. We repeated our experiment in another location with more rainfall and higher relative humidity (Mashhad). The total average of single plant in Birjand and Mashhad was 215.3 and 134.7 gram for fresh weight and 75.1 and 38.9 gram for dry weight respectively, showing lower difference in fresh weight compared to dry weigh (Figure 11.18).

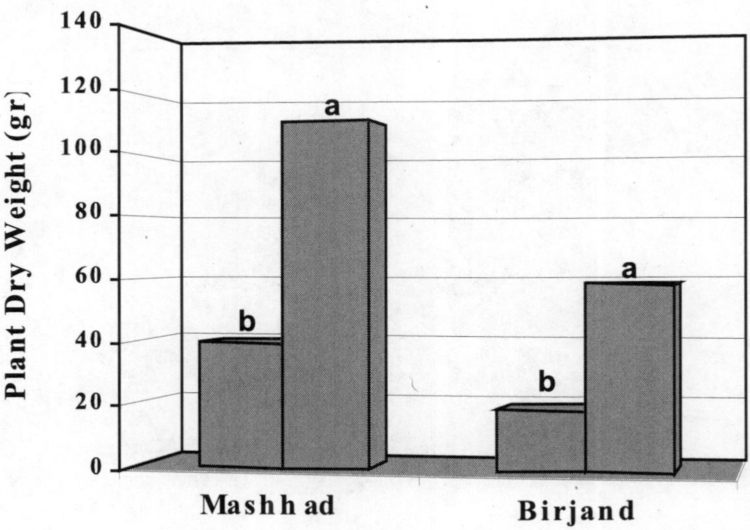

Figure 11.18: Plant Dry Weight at Two Different Cuttings. The first cut was done at 81 and 76 DAP (dotted), and second cut was done at 145 and 141 DAP (dashed) in Birjand and Mashhad, respectively. For each location, non-similar letters above bars indicate significant difference at p>0.01.

Different climatic condition also caused irrigation interval influence dry matter content of forage only in Birjand, with lower rainfall and higher temperatures, which accordingly would increase evapotranspiration requirement. In general plants in Birjand had lower dry matter content (and higher water content) than Mashhad 27.8 per cent compared 30.1 per cent, in Birjand and Mashhad respectively (Figure 11.18). This, and also higher RWC, indicates that plants in Birjand retained more water in their tissues and shifted to succulence as a strategy in order to withstand more unfavorable conditions in this location.

Effect of irrigation intervals on dry forage yield was also significant only in Birjand ($p<0.01$), so dry forage yield with only 25 per cent reduction dropped from 12044 kg ha^{-1} in 7 days interval to 9004 kg ha^{-1} in 14 days interval (Figure 11.19). Zahran (1993) did not observe any yield reduction up to 30 days irrigation interval in *K. indica*, so one may conclude that *K. indica* is more tolerant than *K. scoparia*. Positive correlations between dry yield with LAI (r^2=0.50, $p<0.01$), and SPAD index (r^2= 0.44, $p<0.05$) indicate that lower irrigation affected photosynthesis through reduction of LAI and chlorophyll content. Indeed, as irrigation interval increase, stomata would be more closed and, thus, after each irrigation amount of fixed carbon will decrease gradually, coinciding with soil drying till next irrigation (Khan *et al.*, 2000c), which causes dry matter accumulation to decrease.

There was also an interaction between irrigation intervals and time of cutting; as at first cutting (81 DAP) yield was not influenced by irrigation, whereas in second cutting (145 DAP) plants which grow in warmer and drier conditions during growth

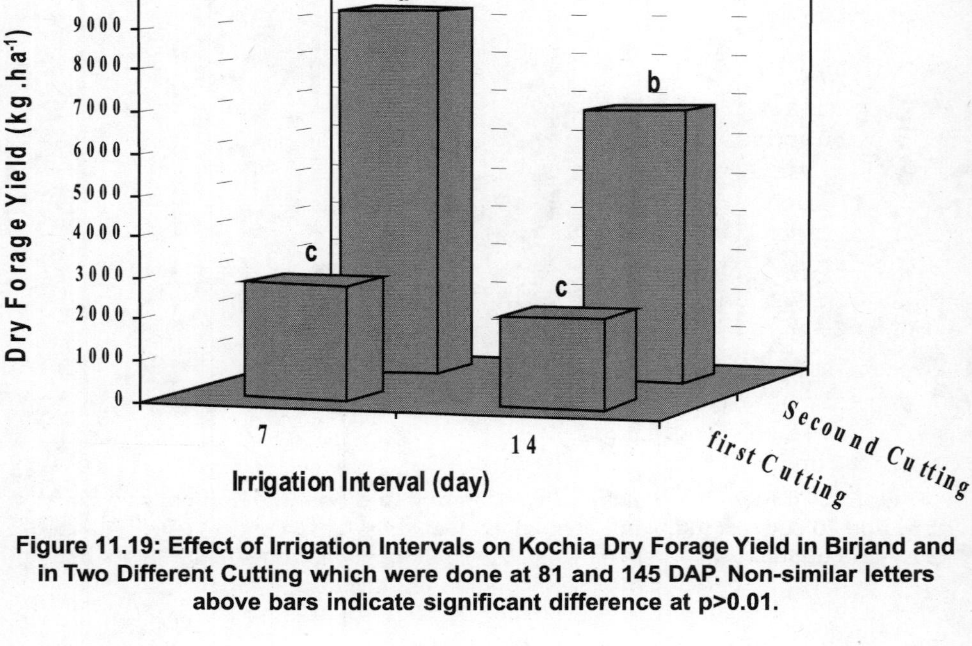

Figure 11.19: Effect of Irrigation Intervals on Kochia Dry Forage Yield in Birjand and in Two Different Cutting which were done at 81 and 145 DAP. Non-similar letters above bars indicate significant difference at p>0.01.

season (from re-growth in the onset of summer until late of August) responded to increasing irrigation interval (Figure 11.19). Moreover, only yield of second cutting showed direct correlation with leaves RWC (r^2=0.51, p<0.01). Concerning good drought tolerance of Kochia, which was observed in our study and also other reports (Nemati, 1977; Zahran, 1993; FEIS-Fire Effects Information System, 1996; Madrid *et al.*, 1996; Mullenix, 1998), and high growth potential of its roots (Eberlin and Fore, 1996; FEIS-Fire Effects Information System, 1996; Mullenix, 1998; Zahran, 1993) it seems that it is possible to increase irrigation intervals at initial growth stages, even more than 14 days, without any considerable yield losses. It will also cause plant roots to be more extended that will help in drought tolerance at next growth stages.

Total forage yield (sum of two cuttings) in Mashhad and Birjand were 19697.3 and 10524.2 kg ha^{-1}, respectively. Undersander *et al.* (1990) reported a forage yield between 2 to 8 ton.ha^{-1} dry matters, which are lower than the obtained yield in our studies. Our harvested yields are similar to those of Sherrod (1971) who harvested Kochia from a wild green stand at 3 different times (pre-, mid-, and post-flowering) and observed a linear dry matter yield increment with progress in maturity and 11326.7 kg ha^{-1} dry matter was obtained at final harvest. Whereas Foster (1980, cited in Everitt *et al.*, 1983) in New Mexico could harvest 26 ton ha^{-1} on dry basis from a well irrigated and fertilized stand of *K. Scoparia* with 4 harvests during growth season; however, this report seems superlative. These reports indicate a great variation between obtained yields of *K. scoparia*, which is probably a result of its high genetic variation (Smith *et al.*, 1975; Mullenix, 1998; Undersander *et al.*, 1990) and existence difference between landraces or wild collections. Another effective factor on these yield fluctuations is acclimation of Kochia in different environmental conditions, which also was observed in our study and caused different growth behaviour and yield of Kochia in two different locations (Mashhad and Birjand).

Salinity and Dry Matter Production

Kochia is a high salt tolerant species (Fuehring *et al.*, 1985; Steppuhn and Wall, 1993; Eberlin and Fore, 1996), which may supply a good source of livestock forage by using saline irrigation water (Fuehring *et al.*, 1985). Francois (1976) observed high significant difference between two accessions. Miyamoto *et al.* (1992) in a review of salinity tolerance of different halophytes, indicated two Kochia species (*K. prostrata* and *K. brevifolia*) showed a 25 per cent yield reduction in electrical conductivity of saturated soil extract about 12 dS m^{-1} and there was 50 per cent yield reduction with *K. brevifolia* at 27 dS m^{-1}.

In our study moderate salinity (8.6 dS m^{-1}) caused slight stimulation of growth and increase of plant dry weight. At 28.2 dS m^{-1} plant dry weight decreased significantly (Table 11.7). These findings are in agreement with those of Ashour *et al.* (1997), who underlined stimulating effect of moderate salinity on growth and yield of halophytic grasses. More plant dry weight reduction at higher salinities can be attributed to several factors such as leaf area loss, increase of respiration, turgor reduction, raising osmotic adjustment costs and salt uptake and excretion (Khan and Ungar, 2001).

Table 11.7: Effects of Different Salinities of Irrigation Water on Yield and Yield Attributes of *K. scoparia*

Irrigation Water Salinity (dS m^{-1})	*Plant Dry Weight (gr)*	*Forage Dry Matter (per cent)*	*Forage Dry Yield (kg ha^{-1})*
1.5	37.12 (± 4.02) *	34.20 (± 1.87)	9670.6 (± 564.8)
8.6	47.64 (± 5.75)	33.58 (± 2.68)	9907.0 (± 1221.8)
28.2	26.7 (± 3.00)	26.11 (± 1.00)	6182.1 (± 687.5)

* Numbers inside parenthesis indicate Standard Error of Means (SEM)

Many halophytes tolerate temporal arise of salts in apoplast by increasing water content of mesophyll cells, thus dilute salts and herewith enhance their capacity to extract salts from apoplast solution (Kafi and Damghani, 2000). Leave succulence, which is a way to dilute salts, was observed in our study because higher salinity resulted in reduction of specific leaf area (SLA) and increase of specific leaf weight (SLW), *i.e.* increase of leaf thickness (data not shown).

Average forage dry matter percentage at first cutting (26.7 per cent) was different from second one (35.9 per cent) and effects of salinity was obvious only in second harvest. Whereas dry matter content increased with the progress of growth season. Plants in the highest salinity conserved more water content to withstand the negative effects of salinity. Thus, it may be concluded that control of plant water is part of adaptive process in *K. scoparia* for osmotic adjustment, especially because it has been demonstrated that plant ion concentration and water content together determine total osmotic pressure, which have a physical relation with salinity of external medium of roots (Kafi and Damghani, 2000).

Increase in salinity from 1.5 to 8.6 dS m^{-1} increased the dry forage yield slightly, but with more salinity increase to 28.2 dS m^{-1}, yield reduced significantly (Table 11.7); however, plant under this high salinity level produced a forage yield equal to 64 per cent of plants at the lowest salinity level (a 36 per cent reduction). These results confirm higher salinity tolerance of *K. scoparia* than *K. brevifolia*, which had a 50 per cent yield reduction at 27 dS m^{-1}. The reduction of forage yields with salinity was mainly a result of reduction in stem dry weight due to a reduction of plant height. Evaluation of partitioning coefficients (Figure 11.10, Table 11.4) also indicates reduction of dry matter allocation to stems in the highest salinity level. These results are in agreement with the findings of Noaman and El-Haddad (2000) with 6 and Steppuhn *et al.* (2001) with 4 halophyte species. The more negative effects of salinity on stem growth and increase of leaf to stem ratio in response to salinity enhancement have been emphasized at *Atriplex griffithi* var. *stocksii* and *Suaeda fruticosa* (Khan *et al.*, 2000a,b).

It seems *K. scoparia* has a very high yield potential and in this respect is competitive with most of conventional forage crops. However, it should be kept in mind, that these yields in our study have been produced by plants that resulted from seeds that were not subjected to any breeding program. Yet, produced yield in Mashhad was comparable with alfalfa yield and even it is not expected that under Birjand condition obtained yields be lower than most of forage crops. We believe that under these conditions, particularly after breeding and domestication, *K. scoparia* could compete with any conventional forage crop.

Seed Yield

Seed yield is a complex trait, which is a function of some distinct yield components that are influenced greatly by weather conditions. Kochia plants show a drastic reproductive flexibility, depending upon the environment and competition conditions as seed production may vary greatly between single plants from only 5 seeds per year under stressful conditions to more than 50000-100000 seeds per year under favourable conditions (FEIS-Fire Effects Information System, 1996; Hager and Sprague, 2001; British Columbia Government, 2004). Nevertheless, a single Kochia plant usually produces about 14,600 seeds per year (Eberlin and Fore, 1996; FEIS-Fire Effects Information System, 1996; Mullenix, 1998; British Columbia Government, 2004). Kochia seed yield has been reported about 1500 to 2000 kg ha^{-1} (Undersander *et al.*, 1990). Cox *et al.* (1969, cited in Zahran, 1993) estimated that average seed yield of *K. scoparia* is equal to 2200 kg ha^{-1}.

In our study, increase of water salinity from 1.5 to 28.2 dS m^{-1} caused a seed yield reduction from 2701 to 2198 kg ha^{-1}. The seed yield reduction in response to salinity increase has been reported in some plant species (Noaman and El-Haddad, 2000; Katerji *et al.*, 2003) and it seems plants behave differently in respect of yield response to salinity (Katerji *et al.*, 2003).

Nevertheless, non-significance effect of salinity on seed yield in our study is against results of Noaman and El-Haddad (2000) who reported seed yields of six halophytic grasses at high salinity levels generally decreased more than vegetative growth. Thus, it seems under saline stress, Kochia plants, with reducing portion of vegetative growth, always allocate a constant photosynthetate volume to seed production and this trait is not influenced by external factors.

Feeding Quality Properties

Kochia produces rich hay (Zahran, 1993) that its nutrient level, especially if harvested at initial growth stages, is sufficient to meet requirements of most of domestic animals (FEIS-Fire Effects Information System, 1996). Numerous references (Finley and Sherrod, 1971; Sherrod, 1971; Everitt *et al.*, 1983; Shroyer and Erickson, 1987; Undersander *et al.*, 1990; Le Houerou, 1992; Zahran, 1993; FEIS-Fire Effects Information System, 1996; Mullenix, 1998; Saskatchewan Agriculture, 2000; Khan *et al.*, 2001; Smart *et al.*, 2001; Koch and Asay, 2004) emphasize to high nutritional value of Kochia as forage, particularly at first growth stages. Kochia forage, when harvested at proper time, contains up to 60 per cent leaves (Undersander *et al.*, 1990) and its nutritional value is equal to, or slightly lower than, alfalfa (Everitt *et al.*, 1983; Mir *et al.*, 1987; Zahran, 1993; Pessarakli, 1994; FEIS-Fire Effects Information System, 1996; Saskatchewan Agriculture, 2000).

In study of several halophytes of Chenopodiaceae, it was proven whereas ash and salts content of all halophytes were high, Kochia ash content was low and near to alfalfa (Miyamoto *et al.*, 1992). Zahran (1993) in a comparison between forages obtained from *K. indica* (harvested at mid-flowering) and alfalfa, observed that crude protein content, ash, dry matter and organic matter digestibility and nitrogen-free extracts of Kochia forage were higher than in alfalfa, while protein and crude fiber digestion coefficients were comparable. However, total digestible nutrients of Kochia (46.9 per cent) were slightly lower than alfalfa (50.3 per cent).

Forage quality sometimes is enhanced by application of saline water (Gihad and El-Shaer, 1992). This situation also has been observed in *Salsola iberica* (Fowler *et al.*, 1992). It has been observed that oxalates levels in Kochia reduce significantly as salinity increases (Fuehring *et al.*, 1985; Fowler *et al.*, 1992). As salinity increased, alkaloids of Kochia increased at first cutting but were reduced at second and third cuttings, indicating reduction of toxicity levels by increase in salinity level (Fuehring *et al.*, 1985). Reduction of oxalates is a result of reduction in water soluble oxalates, which are problematic in domestic feed (not insoluble oxalate), that is a result of insoluble calcium oxalate formation induced by salinity (Fowler *et al.*, 1992). Chemical analysis of forage quality attributes has also shown that plant growth reduction in response to water stress can lay a positive effect on quality of produced forage (Wilson and Ng, 1975).

Crude protein percentage reduces alongside plant maturity and mature forage crops have low crude protein content and energy (Buxton, 1996). This has been observed in *K. prostrata* (ZoBell *et al.*, 2003) and *K. scoparia* (Undersander *et al.*, 1990).

It has been reported that protein content of *K. scoparia* varies between 11-22 per cent (Undersander *et al.*, 1990). In addition, maturity progress caused increase in organic matter, reduction of ash content, reduction or rigidity of micro-and macronutrients concentration and increase of crude fiber.

Finley and Sherrod (1971) also evaluated quality properties of *K. scoparia* forages that were harvested from natural vegetation at pre, mid and post flowering stages. They observed that forage of first cutting (pre-flowering) was intake completely by domestics and digestibility of this forage was comparable with that of alfalfa; however, with progress of maturity, dry matter intake was reduced, which was because of increase of crude fiber content during maturity and also reduction of leaf to stem ratio.

In vitro Dry Matter Digestibility (IVDMD)

IVDMD is an indication of available energy for livestock and provide a reliable estimation of forage digestibility by livestock (Galitzer and Oehme, 1979). There is generally a positive relation between forage digestibility and intake by livestock (Mir *et al.*, 1987; Buxton, 1996). Wilson (1992) observed that dry matter digestibility of some species from Chenopodiaceae varied from 55 per cent to more than 77 per cent. In another study digestibility of several halophytes of Chenopodiaceae, including *K. scoparia* were near to conventional forage plants (Miyamoto *et al.*, 1992). Koocheki and Mahallati (1992) evaluated 12 predominant halophytic species in Southern Khorasan and declared that digestibility of *K. prostrata* was 64.6 per cent, resting among four species, which had the greatest IVDMD.

We used a two-step pepsin-cellulose digestion technique (Jones and Hayward, 1973), for predicting Kochia dry matter digestibility. Increase of irrigation water salinity resulted in significant increase of IVDMD ($p<0.01$; Table 11.8) as difference between the lowest and highest salinity levels reached to more than 10 per cent in each cutting. It seems that forage quality improvement by increase of water salinity is a common phenomenon (Fowler *et al.*, 1992; Gihad and El-Shaer, 1992; Glenn *et al.*, 1992; Miyamoto *et al.*, 1992), and one of its reasons is a delay in plant maturity. The more important reason for digestibility enhancement by increase in salinity is by increase in leaf to stem ratio. Considering leaf blades have more digestibility, increase of leaf to stem ratio can increase forage digestibility (Albrecht *et al.*, 1987; Smart *et al.*, 2001; Knight and Walter, 2002; Huang *et al.*, 2003). In our study also, the highest salinity level, which had the highest dry matter partitioning to leaves (Figure 11.10, Table 11.4) and least diminishing slope of LWR during season, had also the greatest forage digestibility (Table 11.8).

Table 11.8: Effects of Irrigation Water Salinity on Quality Properties of *K. scoparia*, Averaged Between Two Cuttings

Factor	*Irrigation Water Salinity (dS m^{-1})*		
	1.5	*8.6*	*28.2*
IVDMD (%)	68.92 (± 1.18)	73.78 (± 1.11)	79.46 (±1.09)
Leaf Crude Protein (%)	21.20 (±0.52)	21.75 (±0.63)	15.32 (±0.73)

IVDMD was increased in the second cutting Forage (75.01 per cent versus 73.1 per cent in the first cutting). Though this increase was not significant but indicates forage quality enhancement from increase of harvest frequency, which is a result of increase of leaf to stem ratio (Knight and Walter, 2002) and more increase of stem digestibility (Wilson and Ng, 1975; Buxton, 1996) due to more frequent forage cutting.

Delayed irrigation at Birjand caused significant increase ($p<0.05$) of IVDMD in second harvest from 66.03 to 72.6 percent. This trend, however non-significant, was also observed in Mashhad, as irrigation interval increased from 7 to 14 days, IVDMD increased from 51.3 to 54.9 per cent.

The average forage IVDMD at second harvest in Birjand (71.7 per cent) was significantly ($p<0.05$) higher than the first one (68.1 per cent). In some tropical rangeland species, it has been observed that older leaves in less irrigation treatments usually had a higher digestibility than same leaves under more irrigated treatments (Wyn Jones *et al.*, 1983), which is a result of their lower structural materials and higher soluble nitrogen and carbohydrates (Wilson and Ng, 1975).

In average overall treatments and cuttings, IVDMD in Birjand (69.9 per cent) was much higher than its value in Mashhad (53.4 per cent). It seems higher IVDMD values in Birjand have been caused by lower precipitation and higher temperatures, especially because it has been accepted as a rule that forage quality and palatability of a given species increase with dryness of the location (Wilson and Ng, 1975; Le Houerou, 1992; Zahran, 1993; Buxton, 1996; Knight and Walter, 2002) and even it has been observed that the produced forage in dry years has improved nutritional quality (Wilson and Ng, 1975).

Madrid *et al.* (1996) stated *K. scoparia* IVDMD as 57.3 per cent. In Saskatchewan also, *in vitro* organic matter digestibility (not dry matter) of Kochia has been reported to be 57.7 per cent at full flowering (Saskatchewan Agriculture, 2000). These values are slightly higher than IVDMDs of forage produced in Mashhad but lower than Birjand values. Glenn *et al.* (1992) also pointed out that many halophytic species have a higher dry matter digestibility (DMD) than alfalfa. Obtained values in these experiments lay in range of DMD values of some species of Chenopodiaceae (from 55 per cent to more than 77 per cent) reported by Wilson (1992).

Crude Protein Content

Kochia forage is rich in protein, sometimes comparable with alfalfa; however, there are different, and sometimes inconsistent, reports about its CP content (Brevedan *et al.*, 1992; Gihad and EL Shaer, 1992; O'Leary *et al.*, 1992). *K. prostrata* CP content is between 7.7 to 9.5 per cent, which is more than some grass and forage legume species (Koch and Asay, 2004). ZoBell *et al.* (2003) stated that in rangelands of USA the average CP percentage in *K. prostrata* and Wheat grass were 7.2 and 5.9 per cent respectively.

It has been observed that *K. scoparia*, alongside with five *Atriplex* species, had the highest CP content over 97 species from 29 plant families at Mediterranean region of Spain (Paton, 2003). In addition, analysis of natural halophytic vegetations indicated that CP content of some species of Chenopodiaceae did not vary noticeably with those of alfalfa and clover and among them, *K. scoparia* had maximum CP content (16 g per 100 g free-ash dry mater) (Miyamoto *et al.*, 1992). Erickson and Moxon (1947, cited in Steppuhn and Wall, 1993) reported that Kochia CP between 11.3 to 22.6 per cent. In Saskatchewan CP per cent of Kochia was reported as 10.5 per cent at full flowering and 6.4 per cent in early seed set (FEIS-Fire Effects Information System, 1996) and in another report CP content of *K. scoparia* has been assessed about 10 per cent (Kefu *et al.*, 2002). Madrid *et al.* (1996) in comparison of CP content of Kochia forage harvested at early- to mid-flowering with ammoniated barley straw observed while ammonium treatment increased CP content of barley straw up to three-fold (9.2 per cent), it was still very lower than Kochia, which had a CP percentage equal to 20.6 per cent.

Salinity did not influence leaf CP content in each sampling significantly. Francois (1976) pointed out CP per cent of *K. prostrata* showed a slight increase as salinity level increased. Our findings also support this because leaf CP content which was 18.9 per cent at 1.5 dS m^{-1}, with increase in salinity to 8.6 and 28.2 dS m^{-1}, raised to 19.1 and 20.3 per cent respectively, indicating a narrow increase with increase in salinity.

It also has been observed in *K. prostrata* (ZoBell *et al.*, 2003) and *Salsola iberica* (Fowler *et al.*, 1992) in which protein percentage at full flowering is lower than early flowering. This results also coincide with those of Sherrod (1971) who stated that leaf CP content of *K. scoparia*, which was 29 per cent before flowering, fell to 24.4 per cent at end of flowering; however, in present study, obtained leaf CP contents are lower than his reported values (21.2 per cent and 15.3 per cent at pre and post flowering, respectively; Figure 11.21).

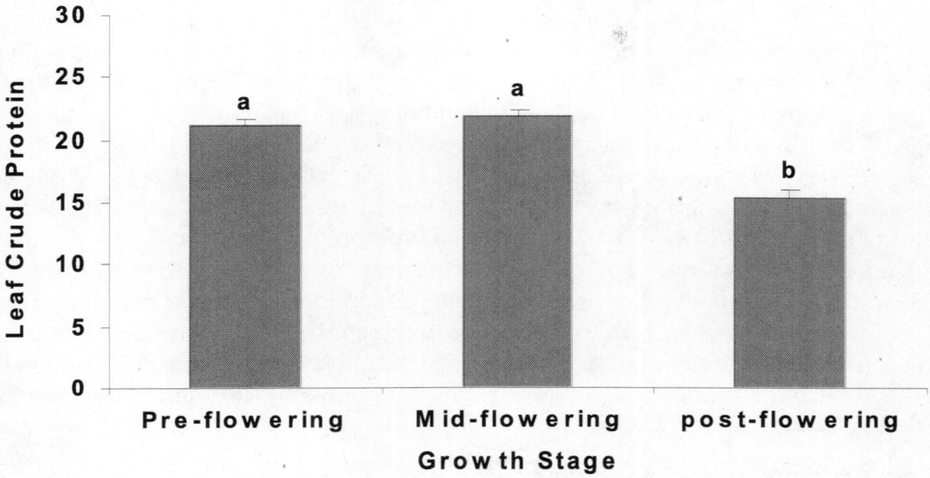

Figure 11.21: Crude Protein Content (per cent) of Kochia Leaves, Measured at Three Stages as Pre, Mid, and Post-flowering. Vertical lines above each column represent standard error of means (SEM). Non-similar letters above columns indicate significant differences at 0.01 probability level.

The general CP average of Kochia leaves between treatments and samplings was 19.4 per cent, which is higher than the minimum required crude protein for livestock (7 per cent, ZoBell *et al.*, 2003).

Several studies with different Kochia species (Finley and Sherrod, 1971; Sherrod, 1971; Miyamoto *et al.*, 1992; Steppuhn and Wall, 1993; Madrid *et al.*, 1996; ZoBell *et al.*, 2003) have reported total N of Kochia forage between 7.2 to 2.5 per cent in different growth stages. These reports and the measured CP content of Kochia in our study altogether confirm the high crude protein content of Kochia in comparison with other plants and are even comparable with alfalfa crude protein, which has been reported in different references to be among 12 to 22 per cent (ex. Buxton, 1996; Choukr-Allah *et al.*, 1992; Miyamoto *et al.*, 1992).

In general, it can be concluded that whereas forage of Kochia is not completely equal to alfalfa in respect to nutritional value (Undersander *et al.*, 1990), but it can be considered as a valuable, if not say incomparable, forage crop in terms of water efficiency and unfavourable conditions of soil (salinity, alkalinity, etc.). The point should be considered that in these experiments volunteer populations of Kochia, which were not subjected to any breeding program, were compared with forage crops, which have been cultivated for long time and have been under severe breeding and selection pressure. It is obvious that due to extensive genetical variation in wild Kochia it is possible to overcome problems related to this plant as a forage crop by plant breeding programs. In all probability with domestication and breeding of Kochia, we could harvest forage close to alfalfa from adverse regions.

Conclusion

The extended temperature range for Kochia germination, from a base temperature of 3.5 °C to maximum of 50 °C and its ability to germinate at salinities of up to 30 dS m^{-1}, is an indicative of its high establishment potential in most of dry and saline areas. Capability of Kochia for quick germination, emergence and growth will favor its establishment inside saline seedbeds. Field trials indicated that Kochia display a good tolerance to salinity after establishment so that increase of irrigation water salinity from 1.5 to 28.2 dS m^{-1} accompanied with only 36 per cent yield reduction. Indeed, moderate salinity level of 8.6 dS m^{-1} stimulated plant growth and improved RUE and quality properties of produced forage. Therefore, Kochia is not only a saline tolerant species but also displays a positive response to moderate levels of accumulated salts in soil and seems to be a salt excretive halophyte, which has a self-control mechanism that keeps plant intra-tissue ion concentration at low levels, independent to their concentration in the soil.

Kochia also showed good tolerance to reduction of irrigation amount, so that increase of irrigation intervals from 7 to 14 days caused 25 per cent yield reduction only in arid climate of Birjand. This reduction was mostly a consequent of LAI and RUE reductions, but on the contrary increase of irrigation intervals improved forage digestibility by restricting stem growth and increasing partitioning to leaves. According to diminishing returns law, increase of applied irrigation water in arid and semi-arid regions did not cause proportional yield increase, due to increasing water loss by infiltration, runoff and especially surface evapouration. Thus, it is possible to allow Kochia to experience some water shortage and yield loss and instead of that, valuable saved water can be applied to irrigate more lands. This approach will decrease cost of production and maximize farmer's gain per field area unit.

References

Aflakpui, G. K. S., P. J. Gregory and R. J. Froud-Williams. 1998. Effects of temperature on seed germination rate of *Striga hermonthica* (Del.) Benth. Crop Protection 17: 129-133.

Albrecht, K. A., W. F. Wedin and D. R. Buxton. 1987. Cell-wall composition and digestibility of alfalfa stems and leaves. Crop Science 27:735-741.

Begue, A., J. F. Desprat, J. Imbernon and F. Baret. 1991. Radiation use efficiency of pearl millet in the Sahelian zone. Agric. For. Meteo. 56:93-110.

Bhardwaj, S. N., A. D. Saini Munshising and K. D. Singh. 1988. Contribution of area, thickness and conductance of leaf in biomass production in upland cotton (*Gossypium hirsutum*). Indian Journal of Agricultural Science 58: 100-103.

Biswas, A. K. 1991. Land and water management for sustainable development and constraint. Report to Economic and Social Policy Division. FAO, Rome.

Bouwer, H. 1994. Irrigation and global water outlook. Agric Water Manage. 25: 221-231

Brevedan, R. E., O. A. Fernandez and C. B. Villamil. 1992. Halophytes as a resource for livestock husbandry in South America. *In* Squires, V. R., and A. T., Ayoub (eds.) Halophytes as a resource for livestock and for rehabilitation of degraded lands. Kluwer Academic Publishers. Netherlands. pp. 175-199.

British Columbia Government. A Guide to weeds in British Columbia.: Kochia. [On-line]. http://www.weedsbc.ca/pdf/kochia.pdf. [24 Dec 2004]

Buxton, D. R. 1996. Quality-related characteristics of forage as influenced by plant environment and agronomic factors. Animal Feed Science and Technology 59:37-49.

Charles-Edwards, D. A., and Lawn, R. J. 1984. Light interception by grain legume row crops. Plant, Cell, and Environment. 7:247-251.

Charles-Edwards, D. A., D. Doley and G. M. Rimmington. 1986. Modelling plant growth and development. Academic Press, Australia.

Choukr-Allah, R., G. W. Jones and L. Kenny. 1992. The potential economic use of halophytes for agricultural development of southern Morocco. *In* Squires, V. R., and A. T., Ayoub (eds.) Halophytes as a resource for livestock and for rehabilitation of degraded lands. Kluwer Academic Publishers. Netherlands. pp.259-261.

Clarke, L.D. and N. E. West. 1969. Germination of *Kochia americana* in relation to salinity. Journal of Range Management 22:286-287.

Dyer, W. E., P.W. Chee and P. K. Fay. 1993. Rapid germination of sulfonylurea-resistant *Kochia scoparia* L. accessions is associated with elevated seed levels of branched chain amino acids. Weed Science 41:18-22.

Eberlin, C.V. and Z.Q. Fore. 1996. Kochia biology. [On-line]. http://mandakzerotill.org/book11. kochia.html. [12 Dec 2002]

Everitt, J. H., M. A. Alaniz and J. B. Lee.1983. Seed germination characteristics of *Kochia scoparia*. Journal of Range Management 36: 646-648.

FEIS-Fire Effects Information System. 1996. Wildlife, animals, and plants: *Kochia scoparia*. [On-line] http://www.1upinfo.com/wildlife-plants-animals/plants/forb/kocsco /all.html. [12 Mar 2003].

Ferreyra, R. E., A. U. Aljaro, R. S. Ruiz, L. P. Rojas and J. D. Oster. 1997. Behaviour of 42 crop species grown in saline soils with high boron concentrations. Agric. Water Management 34: 111-124.

Finley, L. G. and L. B. Sherrod. 1971. Nutritive value of *Kochia scoparia*. II. Intake and digestibility of forage harvested at different maturity stages. Journal Dairy Science 54: 231-234.

Fischer, A. J., C. G. Messersmith, J. D. Nalewaja and M. E. Duysen. 2000. Interference between spring cereals and *Kochia scoparia* related to environment and photosynthetic pathways. Agronomy Journal 92:173-181.

Flores, J. and O. Briones. 2001. Plant life-form and germination in a Mexican inter-tropical desert: effects of soil water potential and temperature. Journal of Arid Environment 47:485-497.

Fowler, J. L., J. H. Hageman, K. J. Moore, M. Suzukida, H. Assadian and M. Valenzuela. 1992. Salinity effects on forage quality of Russian thistle. Journal of Range Management 45:559-563.

Fowler, J. L., J. H. Hageman, M. Suzukida and H. Assadian. 1988. Evaluation of salinity tolerance of Russian-thistle, a potential forage crop. Agronomy Journal 80:250-258.

Francois, L. E. 1976. Salt tolerance of prostrate summer cypress (*Kochia prostrata*). Agronomy Journal 68: 455-457.

Fuehring, H. D., R. E. Finkner and C. W. Oty. 1985. Yield and composition of kochia forage as affected by salinity of water and percent leaching. [On-line]. http://wrri.nmsu.edu/publish/techrpt/abstracts/abs199.html. [4 Jun 2005].

Galitzer, S. J. and F. W. Oehme. 1979. Studies of the comparative toxicity of *Kochia scoparia* (L.) schrad (fireweed). Toxicology letters 3: 43-49.

Gelin, Z., S. L. Mosyakir and S. E. Clemants. 2003. Chenopodiaceae. Flora of China. 5:351-414.

Gihad, E. A. and H. M. El-Shaer. 1992. Utilization of halophytes by livestock on rangelands: Problems and prospects. *In* Squires, V. R., and A. T., Ayoub (eds.) Halophytes as a resource for livestock and for rehabilitation of degraded lands. Kluwer Academic Publishers. Netherlands. pp. 77-96.

Glenn, E. P., R. S. Swingle, J. J. Riley, C. U. Mota, M. C. Watson and V. R. Squires. 1992. North American halophytes: Potential use in animal husbandry. *In* Squires, V. R., and A. T., Ayoub (eds.) Halophytes as a resource for livestock and for rehabilitation of degraded lands. Kluwer Academic Publishers. Netherlands. pp. 165-174.

Hager, A. and C. Sprague. 2001. Kochia biology and management. The pest management and crop development bulletin. K. Steffey (ed.). University of Illinois. 2001. [On-line]. http://www.ag.uiuc.edu/ cespubs/ pest/ articles/200/ 03j.html. [7Aug 2002].

Harrison, R. D., N. J. Chatterton, B. L. Waldron, B. W. Davenport, A. J. Palazzo, W. H. Horton and K. H. Asay. 2000. Forage Kochia: its compatibility and potential aggressiveness on intermountain rangelands. [On-line]. http://www.agx.usu.edu/Research Reports/KOCHIA/kochia.html. [12 Dec 2002]

Huang, Z., X. Zhang, G. zheng and Y. Gutterman. 2003. Influence of light, temperature, salinity, and storage on seed germination of *Haloxylon ammodendron*. Journal of Arid Environment 55: 453-464.

Iranian Ministry of Agriculture. 1998. Agricultural information package. Vol. 1. (in Persian)

Jones, D. I. H., and M. V. Hayward. 1973. A cellulose digestion technique for predicting the dry matter digestibility of grasses. J.Sci. Fd Agric. 24:1419-1426.

Kader, M. A. and S. C. Jutzi. 2004. Effects of thermal and salt treatments during imbibition on germination and seedling growth of sorghum at 42/19°C. Journal of Agronomy and Crop Science 190: 35-38.

Kafi, M. and A. Mahdavi Damghani. 2000. Mechanisms of environmental stress resistance in plants (in Persian). Ferdowsi University Press.

Karachi, M. and R. D. Pieper. 1987. Allelopathic effects of Kochia on blue grama. Journal of Range Management 40: 380-381.

Katembe, W. J., I. A. Ungar and J. P. Mitchell. 1998. Effect of salinity on germination and seedling growth of two Atriplex species (Chenopodiaceae). Annals of Botany 82:167-175.

Katerji, N., J. W. van Hoorn, A. Hamdy and M. Mastrorilli. 2000. Salt tolerance classification of crops according to soil salinity and to water stress day index. Agricultural Water Management 43:99-109

Katerji, N., J. W. van Hoorn, A. Hamdy and M. Mastrorilli. 2003. Salinity effect on crop development and yield, analysis of salt tolerance according to several classification methods. Agricultural Water Management 62:37-66.

Kefu, Z., F. Hai and I.A. Ungar. 2002. Survey of halophyte species in China. Plant Science 163:491-498.

Khalili, A. and H. Rezaei Sadr. 1997. Estimation of overall solar radiation in Iran based on climatic data (in Persian). Periodical of Geographical researches. 84:15-35.

Khan, M. A., B. Gul and D. J. Weber. 2000. Germination response of *Salicornia rubra* to temperature and salinity. Journal of Arid Environment 45:207-214

Khan, M. A., I. A. Ungar and A. M. Showalter. 2000. Effects of salinity on growth, water relations and ion accumulation of the subtropical perennial halophyte, *Atriplex griffithi* var. *stocksii*. Annals of Botany 85: 225-232.

Khan, M. A., I. A. Ungar and A. M. Showalter. 2000. The effect of salinity on the growth, water status, and ion content of a leaf succulent perennial halophyte, *Suaeda fruticosa* (L.) Forssk. Journal of Arid Environment 45:73-84.

Khan, M. A. and I. A. Ungar. 2001. Seed germination of *Triglochin maritima* as influenced by salinity and dormancy relieving compounds. Biologia Plantarum 44:301-303.

Khan, M. A., B. Gul and D. J. Weber. 2001. Influence of salinity and temperature on germination of *Kochia scoparia*. Wetlands Ecological Management 9:483-489.

Knight, A. P. and R. G. Walter. 2002. Plants causing sudden death [On-line]. http://www.iris.org/special_books/ knight/chap1/chapter_frm.asp?la=1[13 Nov. 2003].

Koch, D. W. and W. Asay. Forage Kochia–a forage with fall and winter grazing potential. [On-line]. http://www.uwyo.edu/CES/PUBS/B1122-5.pdf. [15 Dec 2004]

Koocheki, A. and M. N. Mahalati. 1992. Feed value of some halophytic range plants of arid regions of Iran. *In* Squires, V. R., and A. T., Ayoub (eds.) Halophytes as a resource for livestock and for rehabilitation of degraded lands. Kluwer Academic Publishers. Netherlands. pp. 249-253.

Lambers, H., F. S. Chapin III and T. L. Pons. 1998. Plant physiological ecology. Springer. New York. 540p

Le Houerou, H. N. 1992. Forage halophytes and salt-tolerant fodder crops in the Mediterranean basin. *In* Squires, V. R., and A. T., Ayoub (eds.) Halophytes as a resource for livestock and for rehabilitation of degraded lands. Kluwer Academic Publishers. Netherlands. pp. 123-137.

Lieth, H. and M. Lohmann. 2000. Cash crop halophytes for future halophyte growers. Institute of Environmental Systems Research, University of Osnabrück.

Lotfi, M. 1996. Study of ecological characteristics (autecology) of *Kochia prostrata* in Gorgan and Gonbad (in Persian). MSc thesis of forestry and range. Tarbiat-Modarres University, Noor, Iran

Madrid, J., F. Hernandez, M. A. Pulgar and J. M. Cid. 1996. Nutritive value of *Kochia scoparia* L. and ammoniated barley straw for goats. Small Ruminant Research 19: 213-218.

Mengistu, L. W. and C. G. Messersmith. 2002. Genetic diversity of Kochia. Weed Science 50:498-503.

Mir, Z., S. Bittman and L. Townley-Smith. 1987. Voluntary intake and digestibility of complete diets comprising varying levels of *Kochia scoparia*, alfalfa hay and barley. Canadian Journal of Animal Science 67: 1206 (abstract)

Miyamoto, S., E. P. Glenn and N. T. Singh. 1992. Utilization of halophytic plants for fodder production with brackish water in subtropic deserts. *In* Squires, V. R., and A. T., Ayoub (eds.) Halophytes as a resource for livestock and for rehabilitation of degraded lands. Kluwer Academic Publishers. Netherlands. pp.43-75.

Mullenix, W. 1998. Kochia (*Kochia* spp.) biology outline and bibliography. [On-line]. http://www.agron.iastate. edu/~weeds/WeedBiolLibrary/kochiabiblio.html. [3 Dec 2000].

Nemati, N. 1977. Comparative palatability of *Atriplex canescens*. Journal of Range Management 30:368-369.

Nerd, A. and D. Pasternak. 1992. Growth, ion accumulation, and nitrogen fractioning in *Atriplex barclayana* grown at various salinities. Journal of Range Management 45:164-166.

Noaman, M. N. and E. El-Haddad. 2000. Effects of irrigation water salinity and leaching fraction on the growth of six halophyte species. Journal of Agricultural Science (Cambridge). 135: 279-285.

O'Leary, J. W. and E. P. Glenn. 1992. Global distribution and potential for halophytes. *In* Squires, V. R., and A. T., Ayoub (eds.) Halophytes as a resource for livestock and for rehabilitation of degraded lands. Kluwer Academic Publishers. Netherlands. pp. 7-17.

Paton, D. 2003. Elaboration of a multi-variate model for determination of the metabolizable energy of Mediterranean bushes based on chemical parameters. J. Arid Environ. 53:271-280.

Pessarrakli, M. 1994. Handbook of plant and crop stress. Marcel Dekker, Inc. New York. 680p.

Phelps, S. D. and J. W. Bauder. "The role of plants in the bioremediation of coal bed ethane product water". Montana State University [On-line]. http://water quality.montana.edu/docs/methane/halophytes.shtml [28 Jan 2002].

Romo, J. T. and M. R. Haferkamp. 1987. Forage Kochia germination response to temperature, water stress, and specific ions. Agronomy Journal 79:27-30.

Saskatchewan Agriculture, Food and Rural Revitalization. 2000. Feeding Kochia. [On-line]. http://www.agr.gov. sk. ca/docs/crops/forage_pasture/ forage_management/ Feedingkochia.as. [3 Dec 2000]

Shannon, M. C. 1997. Adaptation of plants to salinity. Advance Agronomy 60:75-120.

Sherrod, L. B. 1971. Nutritive value of *Kochia scoparia*. I. yield and chemical composition at three stages of maturity. Agronomy Journal 63: 343-344.

Shroyer, J. P. and D. B. Erickson. 1987. Specialty and non-traditional crops. Cooperative extension service, Kansas State University, Manhattan.

Siadat, H., M. Bybordi and M. J. Malakouti. 1997. Salt-affected soils of Iran: a country report. Proceeding of International symposium on "Sustainable Management of Salt Affected Soils in the Arid Ecosystems". Cario. Egypt.

Smart, A. J., W. H. Schacht and L. E. Moser. 2001. Predicting leaf/stem ratio and nutritive value in grazed and nongrazed big bluestem. Agronomy Journal 93:1243-1249.

Smith, D. T., A. F. Wiese and A. W. Cooley. 1975. Post-emergence control of Kochia and Russian thistle in early spring. Agronomy Journal 67: 752-754.

Steppuhn, H. and K. Wall. 1993. *Kochia scoparia* emergence from saline soil under various water regimes. Journal of Range Management 46:533-538.

Tobe, K., X. Li and K. Omasa. 2000. Seed germination and radicle growth of a halophyte, *Kalidium capsicum* (Chenopodiaceae). Annals of Botany 85:391-396.

Tomar, O. S., P. S. Minhas, V. K. Sharma and R. K. Gupta. 2003. Response of nine forage grasses to saline irrigation and its schedules in a semi-arid climate of north-west India. Journal of Arid Environment 55: 533-544.

Undersander, D. J., B. R. Durgan, A. R. Kaminski, J. D. Doll, G. L. Worf and E. E. Schulte. 1990. Alternative Field Crops Manual: Kochia. [On-line]. http://www.hort.purdue.edu/newcrop/afcm/kochia.html. [3 Dec 2000].

Unknown. 2001. Kochia [On-line]. http://www.wa.gov/agr/weedboared/weed-info/kochia.html [13 Nov. 2003].

Waller, S. S., C. M. Britton, D. K. Schmidt, J. Stubbendieck and F.A. Sneva. 1983. Germination characteristics of two varieties of *Kochia prosterata* (L.) Schrad. Journal of Range Management 36:242-245.

Wang, D., M. C. Shannon and C. M. Grieve. 2001. Salinity reduces radiation absorption and use efficiency in soybean. Field Crops Research 69: 267-277.

Wang, S., C. Wan, Y. Wang, H. Chen, Z. Zhou, H. Fu and R. E. Sossebee. 2004. The characteristics of Na^+, K^+ and free praline distribution in several drought-resistant plants of the alxa Desert, China. Journal of Arid Environment 56:525-539.

Weiner, J. and L. Fishman. 1994. Competition and allometry in *Kochia scoparia*. Annals of Botany 73:263-271.

Wilson, A. D. 1992. Halophytic shrubs in semi-arid regions of Australia. *In* Squires, V. R., and A. T., Ayoub (eds.) Halophytes as a resource for livestock and for rehabilitation of degraded lands. Kluwer Academic Publishers. Netherlands. pp.101.113.

Wilson, J. R. and T. T. Ng. 1975. Influence of water stress on parameters associated with herbage quality of *Panicum maximum* var. *trichoglume*. Australian Journal of Agricultural Researh 26:127-136.

Wyn Jones, R. G. and J. Gorham. 1983. Osmoregulation. pp. 35-58. *In* Lange, O. L., P. S. Nobel, C. B. Osmond and H. Ziehler (eds.). Encyclopedia of plant physiology. New Series, Vol. 12C. Physiological Plant Ecology. Springer-Verlag. Berlin.S

Young, J. A., R. A. Evans, R. Stevens and R. L. Everett. 1981. Germination of *Kochia prostrata* seed. Agronomy Journal 73:957-961.

Zahran, M. A. 1993. Juncus and kochia: fiber- and fodder-producing halophytes under salinity and aridity stress. In M. Pessarakli (Ed.). Handbook of plant and crop stress. Marcel Dekker Inc., N.Y., USA. pp: 505-503.

ZoBell, D. R., B. L. Waldron, K. C. Olson, R. D. Harrison and H. Jensen. 2003. Forage Kochia for fall/winter grazing. [On-line]. http://extension.usu.edu/files/ publications/ zpbell7.pdf. [7 Aug 2003].

12

Effect of Nitrogen and Phosphorous Fertilizers on the Growth of Some Halophytic Forage

M. Karimi[1], S.A.M. Cheraghi[1*], M.H. Banakar[1] and S.H. Ismail[2]

[1]*National Salinity Research Centre, Iran*

[2]*International Centre for Biosaline Agriculture, UAE*

**E-mail: cheraghi@insrc.org*

ABSTRACT

In order to study the effect of nitrogen and phosphorous fertilizers on yield and yield components of five halophytic species, namely *Atriplex nummularia, A. lentiformis, A. halimus, A. canescens* and *Kochia indica*, an outdoor experiment was conducted in National Salinity Research Centre, Yazd, Iran. The experiment was arranged in triplicate with a split-plot factorial design with the five halophyte species as main plots and three levels of Nitrogen (0, 25, 50 kg N/ha) and three levels of phosphorous (0, 25, 50 kg P_2O_5/ha) fertilizers as sub-plots, using saline water of 14 dS m^{-1} electrical conductivity. Nitrogen and phosphorous were applied in the form of ammonium nitrate and triple super phosphate, respectively. Application of nitrogen fertilizer showed a significant effect on shoot fresh weight, stem, leaf and total dry weight of the tested halophytes. The highest yield and yield components of the tested halophytes were obtained for maximum nitrogen application rate. On the other hand, application of phosphorous fertilizer did not increase yield and yield components of the tested halophytes, except for *Kochia indica*. The results for interactive effect of nitrogen and phosphorous on yield and its components indicate that for all levels of phosphorous, nitrogen fertilizer increased yield and yield components of all the halophytic species, while there was no consistent effect over all three nitrogen levels of applying of phosphorous on yield and yield components of the halophytic species studied. It should be noted that *Kochia indica* produced maximum shoot fresh weight (40.2 tons/ha.year) followed by *A. canascens* (23.6 tons/ha.year), *A. halimus* (10.16 tons/ ha.year), *A. nummularia* (8.97 tons/ ha.year) and *A. lentiformis* (7.76 tons/ ha. year), respectively.

Introduction

Halophytes are plant species that grow naturally in saline habitats. They can produce forage for feeding the animals and can make a significant contribution to mankind and to human food security (Watson and O'Leary, 1993). The efforts for irrigated halophytic forage production have been successful in USA (Watson and O'Leary, 1993), Mexico (Glenn *et al.*, 1991) and in the Gulf states (Glenn *et al.*, 1991; Riley and Abdul, 1993).

Malcolm (1996) inferred that fertilization can influence halophyte production, in Australia. His review shows that, until last few years little work had been done on the use of fertilizers on halophytic forage species. It was reasoned that minimum cost methods of salt land revegetation were required to gain acceptance by farmers. However, it has been shown by farmers that nitrogen fertilization had a positive effect on pastures of *Puccinellia ciliate* Bor. and *Thinopyrum elongatum*. It has also been shown that forage shrubs respond to N fertilizer. An application of 60 kg/ha of N to a stand of *A. amnicola* caused a 40 per cent increase in bush volume over the first three months. Differences due to the initial application of di-ammonium phosphate applied down a rip mark at planting time appeared to increase the size of *A. amnicola* bushes on a sandy loam site but the differences were not significant. By contrast, no response had been obtained to P fertilizer applications on *Atriplex Spp*. planting in the field in Australia.

Although results of some experiments indicate that halophyte production is positively affected by fertilizer application, fertility aspects of halophyte species in Iran have not been studied in detail. Thus there is a need to determine the optimum fertilizer rates to increase the forage production and quality of halophytic species.

The objective of this study was to determine the influence of nitrogen and phosphorous fertilizers on forage production and quality of five halophytic species, growing with saline water in Iran.

Materials and Methods

The experiment was conducted in outdoor conditions of National Salinity Research Centre (NSRC) in Yazd, Iran. The weather in the area is very hot and dry in the summer and very cold in the winter with an annual average temperature of 20.2 degrees Centigrade. The total annual rainfall is less than 60 mm. Seeds of five halophytic species namely: *Atriplex nummularia, A. lentiformis, A. halimus, A. canescens* and *Kochia indica* were propagated in 20 cm plastic bags in the greenhouse in January 2003. They were irrigated twice daily with tap water. Before transplanting, during April, the soil was leached with irrigation water of 3.5 dS m^{-1}. This was carried out twice to lower the soil salinity. At this time the irrigation system (water tank, pipes and bubblers) was installed and tested. In April 2003, all halophytic seedlings were transplanted into 110-liters drums in an outdoor condition. The experiment was arranged in triplicate with a split-plot factorial design with the five halophyte species as main plots and three levels of nitrogen (0, 25, 50 kg N/ha) and three levels of phosphorous (0, 25, 50 kg P_2O_5/ha) fertilizers as sub-plots. Nitrogen and phosphorous were applied in the form of ammonium nitrate and triple super phosphate, respectively. Nitrogen fertilizer was applied after each cut while phosphorous fertilizer was applied once a year. Seedlings were irrigated with irrigation water of 3.5 dS m^{-1} until the first of June and then gradually with saline water of 14 dS m^{-1} electrical conductivity to avoid osmotic shock, via a bubbler irrigation system once a week. Throughout the experiment seedlings were thinned gradually and the process was completed on the last of April so that each drum had one plant. Saline water was prepared by adding NaCl and $CaCl_2$ with a ratio of 2:1. All species were cut at the height of 30 cm from the soil surface then they were dried at 70 °C for 48 h and stem and leaf dry weights were measured separately. It should be noted that *Atriplex species* are perennial but *Kochia* is an annual plant and had to be re-transplanted each year. *Atriplex species* were harvested 2 times (in August and December 2003) while *Kochia indica* was harvested one time (in August 2003). Soil samples were also taken from different depths (0-20, 20-40 and 40-60 cm increments) to determine some soil properties such as EC and SAR (in 1:1 soil: water extract) and it was done after each cut.

Results and Discussion

In order to evaluate the response of halophytic species to different nitrogen and phosphorous application rates, analysis of variance was performed on yield and yield components using SAS statistical software. Table 12.1 presents some simple and interactive effects of treatments on yield and yield components, including shoot fresh weight (SHFW), leaf dry weight (LDW), stem dry weight (SDW) and total dry weight (TDW).

Table 12.1: Analysis of Variance for Shoot Fresh Weight (SHFW), Leaf Dry Weight (LDW), Stem Dry Weight (SDW) and Total Dry Weight (TDW)

Source	*df*	*Mean Square*			
		SHFW	*LDW*	*SDW*	*TDW*
Rep	2	184.35**	2.52 **	12.7**	26.334**
Var	4	39.66**	83.47**	278.22**	602.08**
Rep × Var	8	89.92**	1.48*	6.99**	12.38**
N	2	304.88 **	11.17 **	29.33*	76.73 **
P	2	5.07 ns	0.15 ns	0.073ns	0.15 ns
Var × N	8	38.16**	1.46*	3.16*	8.26**
Var × P	8	9.72 ns	0101 ns	0.801 ns	1.25 ns
N × p	14	4.76 ns	0.29 ns	0.51 ns	1.22ns
Var × N × P	16	9.66 ns	0.388 ns	0.71 ns	1.76 ns

*, **, ns = significant at $p<0.05$, $p<0.01$ and non-significant, respectively.

Results of ANOVA indicate that the simple effect of nitrogen fertilizer and halophyte species, on all aspects of growth of halophyte species, were significant ($P<0.01$), while phosphorous fertilizer effect was not significant. Also, all the variables, except P-fertilizer, showed a significant ($P<0.01$) interactive effect for all the characteristics studied.

Figure 12.1 indicates that nitrogen fertilizer application caused a positive effect on shoot fresh weight of the tested halophytes. The highest nitrogen treatment increased shoot fresh weight of *A. canescens, A. halimus, A. lentiformis, Kochia indica* and *A. nummularia* by 63.3 (9.18 tons/ha), 47.2 (3.26 tons/ha), 27 (1.66 tons/ha), 33.6 (10.12 tons/ha) and 27 (1.91 tons/ha) per cent, respectively.

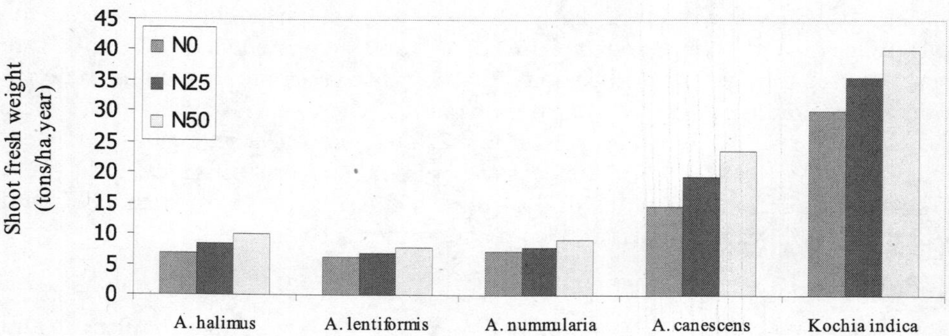

Figure 12.1: Shoot Fresh Weight of Five Halophytic Species at Three Nitrogen Levels

The same trend was found for the other characteristics of the halophyte species (data not presented). This positive reaction of halophytes growth to nitrogen was due to low soil total nitrogen (0.01 per cent) and organic matter (0.09 per cent) content. The highest nitrogen treatment increased stem, leaf and total dry weight of *A. canescens* by 103 (2.54 tons/ha), 60.3 (2.21 tons/ha) and 76.3 (4.74 tons/ha) per cent. The similar trend was found for the other halophytic species. This results are in agreement with Noaman (2004) findings, reporting that halophyte species (such as *Spartina sp., Distichlis palmerii., Paspalum vaginatum, Juncus roemerioanus* and *Batis maritima*) can produce high yield and biomass if they are fertilized with K and N at moderate levels such as 120 kg N/ha and 48 kg K_2O/ha and can grow satisfactory using brackish water of 20 g/l salinity.

Donald and Irwin (1989) also studied growth response of *A. triangularis* to salinity, nitrogen fertilizer and population density in laboratory and field conditions. For laboratory plants, nitrogen fertilization was the most important variable, with biomass and reproductive performance being greatest at the high nitrogen level. For field plants, density was the most important variable, with biomass per plant and survival both decreasing as density increased. Nitrogen fertilization slightly enhanced survival, but did not affect biomass production. Malcolm (1996) reported that an application of 60 Kg/ha of N to a stand of *A. amnicola* caused a 40 per cent increase in bush volume over the first three months of growth. Differences due to the initial application were reduced to 10 per cent after two years. Preliminary studies in India indicate a possible growth response to farmyard manure applied to *A. lentiformis* planted in salt-affected soils in Rajasthan (Malcolm, 1996). Studies on seed production by *A. amnicola* have so far failed to show any response to either N or P fertilizer in terms of fruit fill or germination of the seeds (Malcolm, 1996). Experiments conducted under controlled conditions have shown that as salinity of solution culture increased, nitrogen requirement of *A. amcicola* increased. At 50 mM NaCl in solution culture, maximum growth was obtained with 7.14 mM nitrogen but at 400 mM NaCl, the best growth was obtained by 14.3 mM nitrogen. It should be noted that *A. amnicola* responded to additions of NaCl to the growth medium up to 25-50 mM but then as salt was increased growth declined to 50 per cent of the maximum by 420 mM (Malcolm, 1996).

It must be noted that application of nitrogen fertilizer showed the greatest effect on dry matter yield and yield components of *A. canescens* and it was followed by *A. halimus, Kochia indica, A. lentiformis* as well as *A. nummularia*, respectively. *A. nummularia* showed the minimum response to nitrogen fertilizer application and its shoot fresh weight, stem, leaf and total dry weight increased by 27 (1.91 tons/ha), 34.6 (0.69 tons/ha), 21.7 (0.33 tons/ha) and 28.5 (1 tons/ha) per cent, respectively.

Each of the tested halophyte species responded differently in terms of shoot fresh weight, with increasing phosphorous fertilizer application rate (Figure 12.2). While, phosphorous fertilizer did not increase stem fresh weight (SFW) of *A. canescens, A. halimus, A. lentiformis* and *A. nummularia*, application of 25Kg P_2O_5/ha showed a 25 per cent increase in shoot fresh weight of *Kochia indica*. The same trend was found for the other characteristics of the halophytic species. This may be due to relatively high phosphorous content of the soil (soil available phosphorous was about 10 ppm) and genetical differences between halophyte species in terms of phosphorous requirement. Phosphorous requirement of *Kochia indica* and *Atriplex Spp.* also were

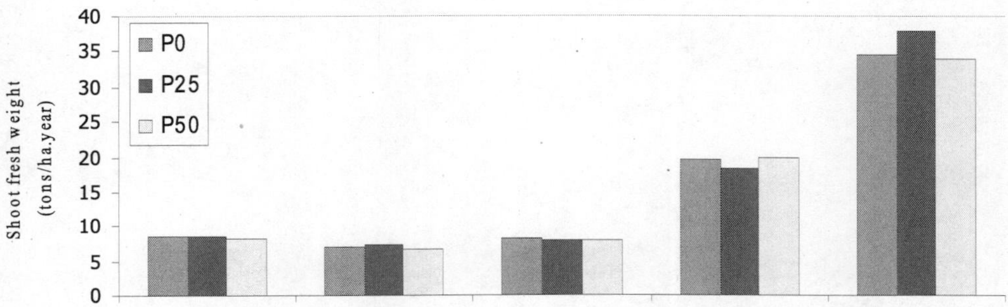

Figure 12.2: Shoot Fresh Weight of Five Halophytic Species at Three Phosphorous Levels

studied by the other researchers. Results of a field experiment (Zahran, 1994) indicated that there was a highly significant increase in productivity with the increase in the doses of both N and P fertilizers. The highest values of dry weights of shoots were obtained in pots treated with 480 Kg/ha urea and 120 Kg/ha super phosphate, being about 6, 15 and 22.5 tons/ha for three successive harvests, respectively, using fresh water. Results of another experiment, done by Zahran (1994), showed that phosphorous fertilizer seems to be the most important element for the vegetative growth of *Kochia indica*. This was obvious when the dry weights of plants growing in pots treated with complete nutrient solution containing P are compared with other pots treated with nutrient solution lacking P. Contrary to the previous reports, the presence of N, under experimental condition, seemed to hinder the uptake of P by the plants from the soil. Results of another experiment (Malcolm, 1996) indicated that no response was obtained with P fertilizer applications on *Atriplex spp*.

Interactive effect of nitrogen and phosphorous fertilizers on shoot fresh weight of *A. canascens* is shown in Figure 12.3. For all levels of phosphorous, nitrogen fertilizer increased shoot fresh weight of *A. canescens*, while there was not any consistent effect over all three nitrogen levels of applying of phosphorous on yield and yield components of *A. canescens*. The similar trend was found for the other halophyte species as well as yield components (data not presented). Although maximum shoot fresh weight for *A. canescens* (24.92 tons/ha), *A. halimus* (10.59 tons/ha), *A. lentiformis* (8.69 tons/ha), *Kochia indica* (40.47 tons/ha) and *A. nummularia* (9.7 tons/ha) were obtained from application of 50 kg N/ha and 25 Kg P_2O_5/ha, 50 kg N and P_2O_5/ha, 50 kg N/ha, 50 kg N and P_2O_5/ha and 50 kg N and P_2O_5/ha, respectively (Table 12.2), avoiding phosphorous application dose did not greatly change maximum shoot fresh weight of the tested halophytes (data not presented). Application of 50 Kg N/ha produced 23.49, 9.67, 8.69, 40.11 and 9.06 tones shoot fresh weight per hectare for *A. canescens*, *A. halimus*, *A. lentiformis*, *Kochia indica* and *A. nummularia*, respectively.

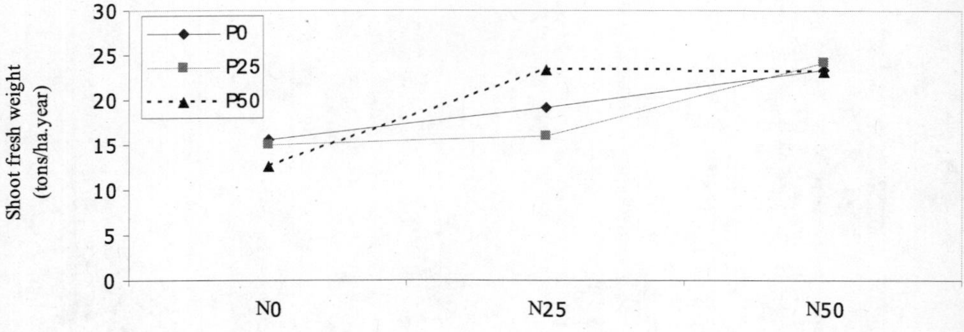

Figure 12.3: Interactive Effect of N and P-fertilizers on *A. canescens* SHFW

Table 12.2: Nitrogen and Phosphorous Fertilizer Requirement of the Five Halophyte Species Producing Maximum Yield

Halophyte species	*Maximum Shoot Fresh Weight (tons/ha.y)*	*Fertilizer Requirement*	
		N (kg/ha.y)	*P_2O_5 (kg/ha.y)*
A. canescens	24.29	50	25
Kochia indica	40.47	50	50
A. halimus	10.59	50	50
A. nummularia	9.7	50	50
A. lentiformis	8.69	50	0

The effect of fertilizer on quality of different halophytic forages, also, studied. Results of the first cutting indicate that the quality of the tested halophytic species were significantly different, as given in Table 12.3. *A. lentiformis* contained higher protein (14.02 per cent) followed by *A. nummularia* (13.2 per cent), *Kochia indica* (11.12 per cent), *A. canescens* (10.47 per cent) and *A. halimus* (9.66 per cent), respectively. This trend was not found in the second cutting and the protein content of the tested halophytic species was not significantly different (Table 12.4). O'Leary *et al.* (1985) reported that the crude protein levels of halophytes range from about 11 to 25 per cent of the dry weight, depending on the age of the plant tissue at harvest and the salinity at which the plant was grown. A few halophytes, such as *A. lentiformis*, grown on sea water showed higher values (more than 16 per cent) of protein content.

Table 12.3: Ash, Fiber and Protein Content of Different Halophytic Species in the First Cutting

Halophytic Species	*Plant Quality*		
	Ash (per cent)	*Fiber (per cent)*	*Protein (per cent)*
A. canescens	24.85(c) [1]	10.41(b)	10.47(bc)
A. halimus	33.66(b)	6.99(c)	9.66(c)
A. nummularia	36.74(a)	7.5(c)	13.2(a)
A. lentiformis	33.81(b)	7.2(c)	14.02(a)
Kochia indica	23.07(c)	12.45(a)	11.12(b)

1. Numbers followed by the same letter are not significantly different at the 5 per cent level.

The ash content of the halophytes reduces the potential forage values of these plants. *A. nummularia* had maximum ash content (36.74 per cent) followed by *A. lentiformis* (33.81 per cent), *A. halimus* (33.66 per cent), *A. canescens* (24.85 per cent) and *Kochia indica* (23.07 per cent), respectively (Table 12.3). The same trend was found for the second cut, as given in Table 12.4. Similar finding was also reported by O'Leary *et al.* (1985). He noted that *A. lentiformis* and *A. barclayana* grown on seawater had ash contents of 27 per cent and 33 per cent, respectively.

Table 12.4: Ash, Fiber and Protein Content of Different Halophytic Species in the Second Cutting

Halophytic Species	Plant Quality		
	Ash (per cent)	Fiber (per cent)	Protein (per cent)
A. canescens	24.07(c) [1]	11.35(a)	11.62(a)
A. halimus	32.66(b)	8.01(c)	10.29(a)
A. nummularia	35.81(a)	8.87(c)	9.78(a)
A. lentiformis	34.7(a)	9.85(b)	11.62(a)

1. Means followed by a letter are significantly different at the 5 per cent level from means not followed by that letter.

In general, nitrogen and phosphorous fertilizer treatments did not significantly change forage quality of the halophytic species, in terms of protein, fibre and ash content, especially for the first cut. On the other hand, protein content of the halophytes significantly increased as nitrogen fertilizer increased, in the second cut, as given in Table 12.5.

Table 12.5: Effect of Nitrogen Fertilizer on Quality Parameters of the Halophytic Species in the Second Cutting

Nitrogen Application Rate (Kg/ha.year)	Plant Quality		
	Ash (per cent)	Fiber (per cent)	Protein (per cent)
0	32.27(a) [1]	9.06(b)	10.12(b)
25	31.8(a)	9.8(a)	11.05(a)
50	31.3(a)	9.61(ab)	11.31(a)

1. Means followed by a letter are significantly different at the 5 per cent level from means not followed by that letter.

Nearly ten months after halophyte cultivation, soil organic matter, Zn, Mn as well as Fe contents increased (Data not presented) but available K and pH decreased. Positive effect of halophyte cultivation on chemical properties of the soil is because of addition of root organic residuals and CO_2 to soil by expanding root system as well as leaf falling, during the course of the experiment. Decreasing available soil potassium causes the necessity of using potassium fertilizer in the future.

Conclusion

Controlling soil fertility especially N under saline condition is considered to be one of the most important factors for successful production of halophytes. Proper fertilization of the tested halophyte species clearly (compared to non-fertilized halophytes) increased yield and yield components. This is mainly because of nitrogen fertilizer application while phosphorous fertilizer showed little effect on yield and yield components, indicating that nitrogen is the main plant growth limiting nutrient.

Among the tested halophytes, *Kochia indica* produced maximum shoot fresh weight (40.47 tons/ha. year) followed by *A. canescens* (24.29 tons/ha. year), *A. halimus* (10.59 tons/ha. year), *A. nummularia* (9.7 tons/ha. year) and *A. lentiformis* (8.69 tons/ha. year), respectively.

References

Donald, R.D. and Irwin, A.U. 1989. Effects of salinity, nitrogen and population density on the survival, growth and reproduction of Atriplex triangularis. American Journal of Botany 76: 1125-1135.

Glenn, E.P., J.W. O'Leary, M.C. Watson, T.L. Thompson and R.O. Kuehl. 1991. Salicornia bigelovii Torr.: An oilseed halophyte for seawater irrigation. Science 251: 1065-1067.

Malcolm, C.M. 1994. Use of halophyte forage for rehabilitation of degraded lands. Ch. 4, In: V.R. Squires and A.T. Ayoub, (eds.), Halophytes as a resource for rehabilitation of degraded lands. Kluwer Academic Publishers, Dordrecht, The Netherlands.

Malcolm, C.V. 1996. Ways to improve the productivity of halophyte forages. In: Choukr-allah, R., Malcolm, C.V. and Hamdy, A. (eds.), Halophytes and Biosaline agriculture. New York, Marcel Dekker, Inc.

Noaman, M.N. 2004. Effect of potassium and nitrogen fertilizers on the growth and biomass of some halophytes grown under high levels of salinity. Journal of Agronomy 3: 25-30.

O,leary, J.W., E.P. Glenn and M.C. Watson. 1985. Agricultural production of halophytes irrigated with sea water. Plant and Soil 89: 311-321.

Riley, J.J. and M. Abdul. 1993. Preliminary evaluation of Salicornia production and utilization in Kuwait. In: H. Leith and A. Al Masoom, eds., Towards the rational use of high salinity tolerant plants. Tasks in Vegetation Science, 28:319-329, Kluwer Academic Publishers, Dordrecht, The Netherlands.

Watson, C.M. and J. O'Leary. 1993. Performance of Atriplex species in the San Joaquin Valley, California, under irrigation and with mechanical harvest. Agriculture Ecosystems and Environment 43: 255-266.

Zahran, M.A. 1994. Juncus and Kochia: Fiber and fodder producing halophytes under salinity and aridity stress.. In: Pessarakli, M. (ed.), Handbook of Plant and Crop Stress. New York; Marcel Dekker.

13

Evaluation of Promising Bread Wheat (*Triticum aestivum* L.) Lines/ Varieties in Saline Condition

M.H. Saberi*, A. Azari Naserabad and H. Tajalli
Agricultural and Natural Resources Research Centre, Iran
**E-mail: saberil338@yahoo.com*

ABSTRACT

This two year experiment was conducted from 2003 to 2005 as randomized complete block design (RCBD) with three replications in the Amirabad (field Research station of University of Birjand). Eighteen lines/varieties that were selected from advanced regional wheat yield trial in saline areas and two checks (Kavir and Rowshan) were used. Electrical conductivity (EC) of irrigation water and soil extract saturation was 11 and 14.1 dSm^{-1} respectively. Traits such as days to heading (DHE), days to maturity (DMA), numbers of spike per m^{-2}, kernel/spike, plant height, spike length, peduncle length, and thousand kernel weight (TKW) and grain yield were measured. Results of combined analysis showed that lines / varieties showed a significant difference at 1 per cent level probability on grain yield, TKW, numbers of kernel per spike, spike length, peduncle length, plant height, DHE and DMA. Interaction between year and variety was significantly different for number of seed per spike, plant height, DHE, DMA at 1 per cent probability level. Finally lines 3, 12, 7, 9, 19 and 13 with grain yield of 6.6, 6.5, 6.4, 6.4, 6.3 and 6.3 ton/ha shown better performance than controls.

Introduction

Increasing salinity of soil and irrigation water is one of the most important limiting factors for plant growth and production in arid and semi-arid areas. In arid environments water resources with high quality are limited; therefore, application of saline water is essential for crop production. In Iran about 12 per cent of total country area about 19 million hectares are under cultivation for agricultural production and about 50 per cent of this area is encountered with salinity, sodidicity, and water logging problem in different degrees. One of the best methods for application of saline water for irrigation is use of salt tolerant crop varieties. Wheat (*Triticum aestivum* L.) crop is the major cereal grown in Iran. Improving the salt tolerant genotypes of wheat is one of the practical solutions for this problem.

Materials and Methods

This experiment was conducted in research field of Birjand University, South Khorasan, Iran, with Latitude of 32°, 53' and longitude 59°, 13' and 1480 m Altitude. This experiment included 18 selected lines and varieties from advanced trial of last year with two checks (Kavir and Rowshan) and was done as randomized complete block design (RCBD) with 3 replications in years 2003-2005. Precipitation during growing season in the first and second years was 75 mm and 238 mm respectively. Mean of minimum and maximum absolute temperature was 5 and 22 for first and second year were 5 and 19 centigrade degree, respectively.

Each lines and varieties was planted on 6 rows in 2.5 m length and 20 cm width. Plant density was based on thousand kernel weight and 450 kernels per m^2 from each genotype were sown. Planting date in first year was 23 Nov. 2003 and 1 Nov. 2004 for second year. Chemical fertilizers were applied based on soil test of each year. Number of irrigation in the first year was 5 times in autumn and winter and 6 times in spring and 3 times in autumn and winter and 10 times in spring for second year. Harvesting date was 11 June 2004 in first and 12 June 2005 for second year.

Electrical conductivity (EC) of irrigation water and saturated soil extract was 11 and 14 dS m^{-1} respectively. Characteristics such as days to heading, days to physiological maturity, plant height, spike length, peduncle length, number of fertile spike, number of kernel per spike, thousand kernel weight, and grain yield were measured in each years as simple analysis of variance and then combined over two years by using of MSTAT-C program and mean comparison was done, using least significant difference (LSD) method.

Results and Discussion

Results of combined analysis showed that lines/varieties were significantly different in grain yield, thousand kernel weight, number of kernel per spike, spike length, peduncle length, plant height, days to heading, and days to maturity at 1 per cent probability level (Table 13.1). Means comparison of lines/varieties showed that lines number 3, 12, 7, 9, 19 with 6.6, 6.5, 6.4, 6.3 and 6.3 ton ha^{-1} grain yield were better than superior check (Rowshan) with grain yield of 6 ton ha^{-1}.

Table 13.1: Combined Analysis of Variance on Some Agronomic Traits, Grain Yield and Yield Components in Elite Lines of Bread Wheat in Saline Condition (MS)

Days to Maturity	*Days to Heading*	*Plant Height*	*Peduncle Length*	*Spike Length*	*Number of Fertil Spike/m²*	*Kernels/ Spike*	*1000 Kernel Weight*	*Grain Yield*	*d.f.*	*S.O.V*
ns 7.0	ns 31.0	ns 519.5	ns 106.4	ns 0.003	ns 126100.8	ns 437.8	ns 6.5	ns 2.7	1	year
ns 8.3	ns 18.3	ns 212.1	ns 56.5	ns 4.3	ns 88003.3	ns 134.5	ns 61.7	ns 5.7	4	replication × year
** 35.7	** 70.1	** 438.4	** 93.4	** 2.55	ns 54569.2	** 307.5	** 96.1	** 3.7	19	lines/varieties
** 29.9	** 103.1	** 84.1	ns 37.9	ns 1.4	313318. 3 ns	** 193.2	ns 34.8	ns 1.93	19	varieties × year
4.6	15.2	27.9	24.3	0.9	34570	73.2	36.3	1.65	76	Error
1.4	3.6	6.2	12.8	8.8	26.4	14.5	14.9	23		CV%

ns, * and **: Non significant, significant at 5 per cent and 1 per cent probability levels respectively.

Table 13.2: Means Comparison of Some Agronomic Traits, Grain Yield and Yield Components in Elite Lines of Bread Wheat in Saline Condition

Days to Maturity	Days to Heading	Plant Height (cm)	Peduncle Length (cm)	Spike Length (cm)	Kernels/ Spike	Number of Spike per (m^2)	1000 Kernel Weight	Grain Yieldt h^{1}	Lines/ Varieties
ef 149.7	cde 105	ef 77.2	bc 39.2	abc 10.5	abc 59.5	c 523.3	b 36.9	d 4.4	Kavir (Check)
abc 154.2	a 116.3	a 115.5	a 50.8	ab 11.5	d 35.2	a 890	a 52.8	abcd 6.0	Roshan (Check)
bcdef 151.3	bcde 107.7	bcdef 84.7	bc 38.6	bc 10.0	bc 53.8	abc 663.3	ab 45.3	a 6.6	Dh4-209-15571 (Check)
cdef 150.7	bcde 105.8	bcdef 82.2	c 36.7	abc 10.1	abc 55.8	abc 746.7	b 42.8	cd4.6	Emu'S'......-.........
ef 149.8	cde 104.3	cdef 81.2	c 37.1	abc 10.9	abc 56.8	abc 698.3	b 42.6	abcd 5.3	Emu'S'......
cdef 150.5	bcde 107.2	bcde 84.2	c 36.8	abc 10.4	abc 59.5	abc 721.7	b 39.2	cd 4.7	Emu'S'.......
abcde 153.7	bcde 106.5	bcdef 82.7	c 35.5	abc 10.5	abc 64.8	abc 640	b 37.1	abc 6.4	Nad.Ww..Lee.Fn.......
abcde 153.3	bcde 106.7	b 90.6	bc 39.8	abc 11	abc 59.2	abc 778.3	b 38.4	cd 4.6	Nad.Ww..Lee.Fn.........
f 148.8	de 103.7	bcde 86.4	bc 39	bc 9.9	abc 58.2	ab 795	b 40.8	abc 6.4	Nad.Ww..Lee.Fn........
f 149.2	cde 104.5	b 91.4	c 37	abc 11.0	ab 69.0	abc 693.3	b 38.2	abcd 5.9	Sakha8.Darab#2
abcdef 152.7	bcd 108.5	bcde 85.8	ab 46.5	ab 11.3	abc 60.8	abc 680	b 34.7	abcd 5.5	Sakha8.Darab#2
f 149.2	cde 105.7	bcde 85.7	c 37.6	abc 11.9	abc 61.3	abc 646.7	b 40.8	ab 6.5	Mahooti.Shiroodi
cdef 151.0	bed 108.3	bcdef 82.9	c 35.7	abc 11.2	abc 63.8	a 826.7	b 39.9	abc 6.3	Mahooti.Shiroodi
ef 149.7	bcde 106.8	bcd 86.8	c 36.1	abc 11.1	a 69.9	abc 773.3	b 37.7	abcd 5.9	Mahooti.Shiroodi
g 145.5	e 101.0	cdef 79.2	c 37.9	abc 10.5	abc 61.5	bc 563.3	b 38.2	bcd 4.8	Mahooti.Shiroodi
def 151	ab 112.7	ef 77	c 36.2	abc 10.8	abc 60.2	abc 646.7	b 37.5	abcd 5.9	1-66-22..Bow'S'Crow'S'
abcd 153.7	bcde 107.2	f 76.3	bc 38.1	c 9.4	c 51.8	abc 718.3	b 36.2	abcd 5.3	1-66-22..Bow'S'Crow'S'.
def 150	de 103.5	def 78	c 34.7	abc 10.1	abc 58.4	bc 7536.7	b 42.3	d 4.3	1-66-22..Bow'S'Crow'S'
a 155.2	abc 111.0	bc 88.2	bc 41.3	a 11.7	abc 54.8	abc 773.3	ab 43.4	abc 6.3	1-66-22..Bow'S'Crow'S'.
ab 155	cde 105.5	def 77.8	c 24.6	abc 10.8	abc 61.9	abc 766.7	b 40.8	abcd 5.0	Mv17.Mahooti.4. EVWYT2.Azd

Lines 3 and 19 with 45.3 and 43.4 g thousand kernel weight were the same as check (Rowshan) with 52.8 gram and other lines were the same as check (Kavir).

Table 13.3: Chemical and Physical Characteristics of Soil Field Research Station of University of Birjand

Depth (cm)	*Clay%*	*Silt%*	*Sand %*	*Kmg/ kg*	*Pmg/ kg*	*N%*	*O.M %*	*Gypsum meq/ 100gr*	*pH*	*×EC 10^3*	*Tex–ture*
0–30	9	34	57	168	8	0.007	–	–	7.5	14.1	Silty loam
					SAR	*CO_3^{2-}*	*HCO_3^-*	*Cl^-*	*Sum of cation*	*Ca^{2+}*	*Na^+*
					9.68	–	3	66.3	145.7	21	82.7

Table 13.4: Chemical Characteristics of Irrigation Water Field Research Station of University of Birjand

pH	*EC (dSm^{-1})*	*S.A.R*	*Sum of Cations*	*K^+*	*Na^+*	*Mg^{2+}*	*Sum of Anions*	*Ca^{2+}*	*SO_4^{2-}*	*Cl^-*	*HCO_3^-*	*CO_3^{-2}*
7.7	10/5	11.2	85	0/8	47.8	27.6	86	8.8	21	61	3	–

S.P: Saturation percentage; S.A.R: Sodium Absorption ratio; Ec: Electrical conductivity; O.M: organic mater; N: Total nitrogen.

Lines number 19 and 11 with 11.7 and 11.3 cm plant height were the same with check (Rowshan) and the rest lines/varieties were the same to Kavir. Peduncle length in lines number (13, 9, 19, 3, 12 and 7) with 826.7, 795, 773.3, 663.3, 646.7 and 640 spike/m^2 respectively were the better than the superior check 7 and 153.7 days to physiological maturity were the late mature lines (Table 13.2). Finally lines 3 and 12 with highest grain yield number of fertile spike per m^2, number of kernels per spike, spike length and peduncle length were selected for on-farm experiments.

References

Chauhan, C. and S. Singh. 1993. Wheat cultivation under saline irrigation. Wheat Inf. Ser.77: 33-38.

Cullu, M.A. 2003. Estimation of the effect of soil salinity on crop yield using remote sensing and geographic information system. Turk Agric. 27: 23-28.

Looss, S.P. and K.H.M. Siddiaue. 1994. Morphological traits associated with wheat yield increased in Mediterranean environments. Adv. Agron. 25: 229-276.

Poustini, K. 1994. Physiological response of tow cultivars wheat to salt stress. Journal of Agricultural Science 26: 57-63.

Poustini, K. and Shoehtab Salmasi. 1997. Effect of salinity on production and transformation dry maters in tow cultivars of wheat. Journal of Agricultural Science 29: 11-16.

Winter, S.R., J. Musick and K.B. Porter. 1988. Evaluation of screening techniques for breeding drought–resistance winter wheat. Crop Science 28: 512-516.

14

Panicum antidotale: A Potential Grass for Salt Affected Soils

Mansoor Hameed and Muhammad Ashraf
Department of Botany, University of Agriculture, Faisalabad, Pakistan

ABSTRACT

Adaptation of some wild populations of blue panic grass (*Panicum antidotale* Retz.) to salinity stress was assessed in a hydroponic system under glasshouse conditions. Two independent experiments were conducted in a glasshouse in the Botanic Garden of the Department of Botany, University of Agriculture, Faisalabad. For experiment I, two populations from distinct habitats, normal non-saline (ECe 3.73 dS m^{-1}, pH 7.47) and saline soils (ECe 17.34 dS m^{-1}, pH 6.79) were subjected to two salt treatments (0, Hoagland nutrient solution and 200 mol m^{-3} NaCl in Hoagland nutrient solution) for 46 days. For experiment II, three populations collected from different habitats, normal non-saline (ECe 2.92 dS m^{-1}; pH 7.38), moderately saline (ECe 12.41 dS m^{-1}; pH 7.13) and highly saline (ECe 19.33 dS m^{-1}; pH 6.62) were grown for 60 days under three salt treatments (0, Hoagland nutrient solution 100 m*M* NaCl and 200 m*M* NaCl both in Hoagland nutrient solution). In both experiments the population collected from the saline habitats excelled their normal counterparts in growth in terms of plant biomass production, in physiological attributes such as chlorophyll contents, net CO_2 assimilation, transpiration rate and stomatal conductance, and anatomical characteristics such as increased metaxylem area in roots, phloem area, stem area, and sclerenchyma thickness etc. Such natural populations with high adaptability to saline conditions could be potential colonizers of salt affected soils. Since *P. antidotale* inherently produces a large quantity of biomass, it could be used as a potential fodder and forage crop for livestock and small ruminants in most of the salinity stricken areas.

Introduction

Panicum antidotale Retz. (Blue panic grass, Blue panic or Giant panic grass; locally called Bansi, Ghemor or Gramma), a native of Indo-Pakistan, is found growing in a variety soils throughout the country. It is distributed throughout Pakistan, but predominantly in desert regions (Cope, 1982). This tall grass is a coarse perennial with creeping woody rootstock that can attain a height of 2 m or more. The leaves have a distinct midrib and wavy margins, bluish-green in colour. The grass begins growing in late spring and blooms from July to October (Cope, 1982).

Blue panic can withstand multiple stresses, mainly salinity, drought, fire etc. Its suitability to light saline soils had been reported by Whyte *et al.* (1959), Ryan *et al.* (1975) and Cope (1982), but Ryan *et al.* (1975) reported its more tolerance to alkalinity caused by sodium and magnesium than to chloride. However, Chaudhary (1989) and Bokhari *et al.* (1987) reported tolerance to very high levels of salinity, up to 15000 ppm. Moreover, blue panic is highly tolerant to drought (Cope, 1982; Bokhari *et al.*, 1987; Chaudhary, 1989), and water logging (Ashraf, 2004).

Blue panic is a highly productive grass with high nutritional value that can produce 150 to 180 tons of fresh biomass per hectare per year with 15 to 18 per cent protein (Bokhari *et al.*, 1987). Sen and Ray (1964) and Göhl (1975) recorded fresh early pasture with 18.8 per cent crude protein in the dry matter, 40.5 per cent crude fibre, and 7.9 per cent ash.

Blue panic has proven to be a valuable cultivated fodder grass (Chaudhary, 1989). Comparison with other forage crops, blue panic is more productive and nutritionally more acceptable (Bokhari *et al.*, 1987), particularly for cattle (Sarwar *et al.*, 2006). A report by the Tucson Plant Materials Centre (http://www.az.nrcs.usda.gov) indicated that *P. antidotale* has been used by several wildlife species, such as antelope and California jackrabbits. Blue panic has a good economic value for wildlife and livestock, but it can cause nitrate poisoning when grazed where it is richly fertilized.

Salt tolerance is a complex trait involving multiple mechanisms, and is defined as the ability of the plant to thrive under salinity and complete the growth cycle with an acceptable growth and yield (Wilson and Read, 2006). The factors affecting plant growth under salinity may be evaluated in 3 groups: i) water stress, ii) ion toxicity, iii) problems in nutrient uptake and translocation to green plant parts, and, as a result, disorders in cells due to disruption of ionic balances such as in the case of K^+ and Ca^{2+} (Qiu *et al.*, 2006). Under salt stress, physiological drought may play an important role by limiting water uptake from the soil. On the other hand, excess salt uptake by plants disrupts cellular functions and damages physiological processes such as photosynthesis and respiration (Leopold and Willing, 1984; Marschner, 1995).

There is a continuous spectrum of plant tolerance to saline conditions ranging from glycophytes that are sensitive to salt, to halophytes which survive in very high concentrations of salt (Volkmar *et al.*, 1998). Unfortunately, most crops are not halophytic in nature. Studies in crops suggest that salt tolerance is a multigenic trait (Niknam and McComb, 2000), which makes it more difficult to study and improve. It has also been noted that in some species the acquired salt tolerance can be passed along to offspring (Niknam and McComb, 2000). Tolerant species use more than one strategy to tolerate or avoid stress. It is important to keep the levels of ions low in the leaves, particularly in the young ones. This can be done by excluding the ions at the point of uptake and reducing the translocation of ions to the shoot (Niknam and McComb, 2000).

Salt tolerant plants have the ability to tolerate high levels of salt due to developing a series of anatomical, morphological, and physiological adaptations (Poljakoff-Mayber, 1975, 1988), such as an extensive root system and salt secreting glands on the leaf surface (Liphschitz and Waisel, 1974; Oross and Thomson, 1982; Gould and Shaw, 1983; Oross *et al.*, 1985; Gorham *et al.*, 1985; Sinha *et al.*, 1986; Marcum and Murdoch, 1990; Marcum *et al.*, 1998).

The possible mechanisms in the salt tolerance of plants may involve: i) no or less uptake of salt into the plant, ii) tissue tolerance, iii) accumulation of salt in vacuoles without any physiological interference, iv) ion discrimination, such as K^+, Na^+, Cl^- and SO_4^{2-} during uptake by roots and translocation to shoots, v) different biochemical processes such as the production of some enzymes, hormones, antioxidants, etc. (Pitman, 1984; Wise and Naylor, 1987; Gorham and Wyn Jones, 1990; Spychalla and Desborough, 1990; Begum and Karmoker, 1999; Ashraf, 2004; Qiu *et al.*, 2006; Wilson and Read, 2006). One or more of these mechanisms may be responsible for variations in the salt tolerance of plant genotypes and species.

The salinity problem can be overcome through either soil reclamation or growing tolerant crops. However, soil reclamation is a very expensive process, and hence the cultivation of tolerant species and varieties is the most practical solution when the salinity is low. It is well known that there are significant genotypic differences with respect to salt tolerance between and within plant species (Rana, 1986; Bozcuk, 1991; Suhayda *et al.*, 1992; Açýkgöz and Gevrek, 1994; Zhong *et al.*, 1995).

As an excellent forage species, *P. antidotale* has been introduced into many countries of the world, *e.g.* to the United States from India via Australia (Magness *et al.*, 1971; Ruyle and Young, 1997), and thereafter several accessions and selections have been released (examples are A-130 from Australia, P-15630 from Afghanistan, and a Seedling Drought Tolerant-SDT; selected from type A-130 (Wright and Dobrenz, 1970; Tucson Plant Materials Center-http://www.az.nrcs.usda.gov). Strain 341 is recommended for use under arid conditions in western Rajasthan, India (Prasad and Singh, 1973). Since *P. antidotale* has been found to be grown on a variety of habitat types (Ashraf, 2003), there would be a fair chance of hunting high salt tolerant populations from such diverse habitats in the surroundings of the Faisalabad region, as well as from other salt and drought-prone areas. Such selected highly salt tolerant populations of *P. antidotale* could be used for economic utilization of salt affected soils.

Materials and Methods

Populations of blue panic grass (*Panicum antidotae* Retz.) were collected from the Faisalabad region to evaluate their adaptation to salt affected soils. Two independent experiments were hydroponically conducted in glass house in Botanical Garden Research Area, Department of Botany, University of Agriculture, Faisalabad.

Experiment I

For the experiment, two habitats were selected, normal non-saline soil at Botanical Garden Research Area (ECe 3.73 dSm^{-1}, pH 7.47, coordinates 31°25′ 43.96" N, 73°04′ 17.40 E) and saline soil Rajawala near University of Agriculture, Faisalabad (ECe 17.34 dSm^{-1}, pH 6.79, coordinates 31° 25′ 30.61" N, 73° 03′ 53.84" E).

Twenty plants of blue panic grass from each habitat were collected. The experiment was conducted in a naturally-lit glasshouse in the Botanic Garden of the University of Agriculture, Faisalabad during July–October 2001. The average day and night temperatures were 37±3 and 24±3 °C, respectively, and photoperiod from 11 to 12 h. The relative humidity ranged from 45.9 to 58.6 per cent.

A small ramet of each plant (comprising five tillers) was transplanted into normal soil in 30 cm diameter earthen pots. The plants were grown for 30 days in a glasshouse with natural sunlight. Two tillers of almost uniform size were separated from each plant, and all the tillers were mixed randomly within each population. Five tillers per replicate of each population were transplanted equidistant into a plastic pot (30 cm diameter and 25 cm deep) that contained 8.0 l of aerated full strength Hoagland's nutrient solution.

The experiment was conducted in a completely randomized design with two populations, four replicates, and two NaCl treatments (0, Hoagland nutrient solution; 200 mol m^{-3} NaCl in Hoagland nutrient solution). The plants of each treatment were aerated continuously for 46 days. All nutrient solutions were changed after every week. The NaCl treatments were increased stepwise until the appropriate treatment concentration was attained.

After 46 days of growth in different salt treatments, data for plant biomass, root, stem, and leaf anatomical parameters for experiment 1 and for leaf gas exchange parameters, chlorophyll contents, and water relations were recorded and the third leaf (fully expanded youngest) from the top of each plant tiller was used for these measurements.

Experiment II

For experiment II, 20 plants were collected from normal non-saline soil at University Campus agricultural fields, Lalazar Colony (electrical conductivity of soil saturation extract (ECe 2.92 dSm^{-1}; pH 7.38, coordinates 31° 25′ 40.21" N, 73 ° 04′ 25.04" E), moderately saline at University Campus near Khadeja Hall along roadside (ECe 12.41 dSm^{-1}; pH 7.13, coordinates 31° 25′ 45.86" N, 73° 03′ 53.84" E) and highly saline soil at Punjab Forest Research Institute, Gatwala, Faisalabad (ECe 19.33 dS m^{-1}; pH 6.62, coordinates 31° 28′ 25.25" N, 73° 12′ 42.87" E). The plant material was grown in the Botanical Garden research Area for about two months till the complete establishment.

Ramet of each plant of equal size (comprising five tillers) was transplanted into normal soil (mixture of loam and sand in 1:1 ratio) in earthen pots. The plants were grown for 30 days in a glasshouse with natural sunlight. Three tillers of uniform size were separated from each plant and transplanted into into a plastic pot (30.0 cm diameter and 25 cm deep) that contained 8 L of aerated full strength Hoagland's nutrient solution. The experiment was arranged in a completely randomized design during June-September 2005 with three populations, four replicates, and three NaCl treatments (0, Hoagland nutrient solution; 100 mM and 200 mM NaCl in Hoagland nutrient solution).

After two months of growth, growth attributes like plant height, root length, number of sminal roots per plant and root and shoot fresh and dry biomass were recorded. For anatomical studies, a 2 cm piece from the base of third leaf, a 2 cm piece from the thickest seminal root, and a 2 cm piece from the centre of third internode was selected. The material was immediated put into a fixative formalin aceto-alcohol (FAA) solution for 48 hr and then transferred to a preservative acetic alcohol for long-term preservation. Free hand sectioning was used to prepare permanent slides by a standard double-stain technique using fast green and safranin dyes.

Results

Experiment I

Fresh and dry weights of both shoots and roots and leaf area of two different populations of *P. antidotale* decreased markedly under saline conditions applied for 46 days (Figure 14.1). Comparison of the populations showed that under salt stress the population collected from saline soil (hereafter the population will be referred to as salt tolerant) had significantly higher fresh and dry masses of both shoots and roots and leaf area than the populations from non-saline soil.

Chlorophylls *a* and *b* decreased significantly in both populations at stress treatment (Figure 14.2). However, populations did not differ significantly for chlorophyll *a*. Cholorophyll *b* was higher in the normal population at control than that in the other population. Chlorophyll *a/b* ratios in the normal were higher at stress conditions than those at non-stressed conditions, whereas those in the salt tolerant population had no consistent pattern of increase or decrease.

Net CO_2 assimilation (P_n) of the two populations had a decreasing trend under stress conditions (Figure 14.6). The salt tolerant population had significantly higher net CO_2 assimilation rate than the normal non-saline population. Stress conditions of the growth medium caused a significant decrease in transpiration in both populations. The normal population had lower transpiration than that of the salt tolerant population under saline conditions (Figure 14.2). Stomatal conductance (Figure 14.2) of both populations also decreased significantly under stress conditions. The salt tolerant population had higher stomatal conductance than the normal non-saline population under saline conditions.

Relative intercellular CO_2 concentration (C_i/C_a=intercellular CO_2 concentration/ambient CO_2 concentration) generally decreased under stress condition in both populations (Figure 14.2). The salt tolerant population had significantly higher relative intercellular CO_2 concentration than the non-saline population under saline conditions.

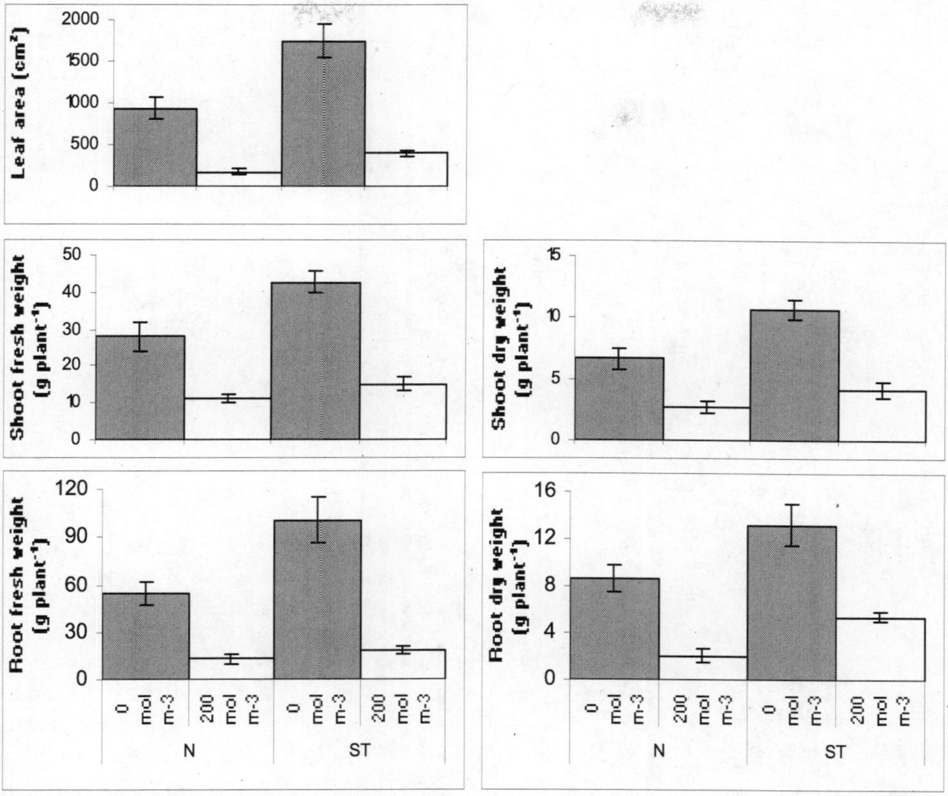

Figure 14.1: Growth Parameters of Two Populations of *P. antidotale* when 30-day Old Plants were Subjected for 46 Days to Salt Stress Conditions in Solution Culture (N: Normal non-saline, ST: Salt tolerant)

Water use efficiency (P_n/E) decreased significantly in both populations at both stress levels (Figure 14.2). Although intrinsic water use efficiency (P_n/g_s) decreased in both populations under stress conditions, the normal population was the highest in this attribute under stress condition (Figure 14.2).

Experiment II

All the growth parameter of all three populations, *i.e.*, normal non-salt tolerant, moderately salt tolerant and highly salt tolerant decreased under all salt levels applied for 45 days except root length, which increased slightly in the highly salt tolerant population (Figure 14.3) and considerably increased in the normal population. Both shoot fresh and dry weights of normal and moderately salt tolerant populations were the most affected parameters at both 100 and 200 mM NaCl of the growth medium, however, in the highly salt tolerant population, these were least affected. At moderate salt level, root dry weight of highly salt tolerant population was slightly increased, while the shoot dry weight remained almost unaffected.

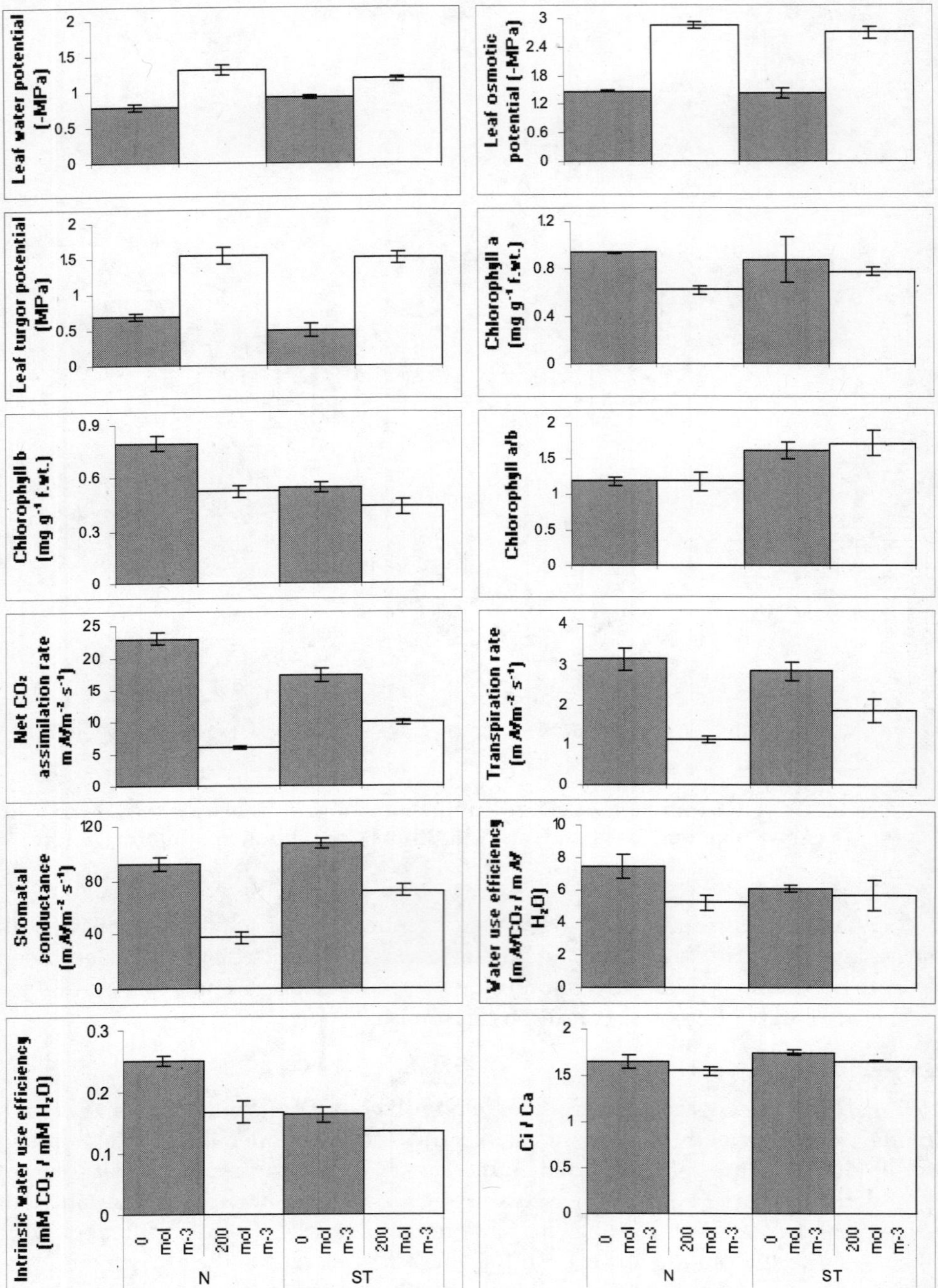

Figure 14.2: Growth Parameters of Two Populations of *P. antidotale* when 30-day Old Plants were Subjected for 46 Days to Salt Stress Conditions in Solution Culture (N: Normal non-saline, ST: Salt tolerant)

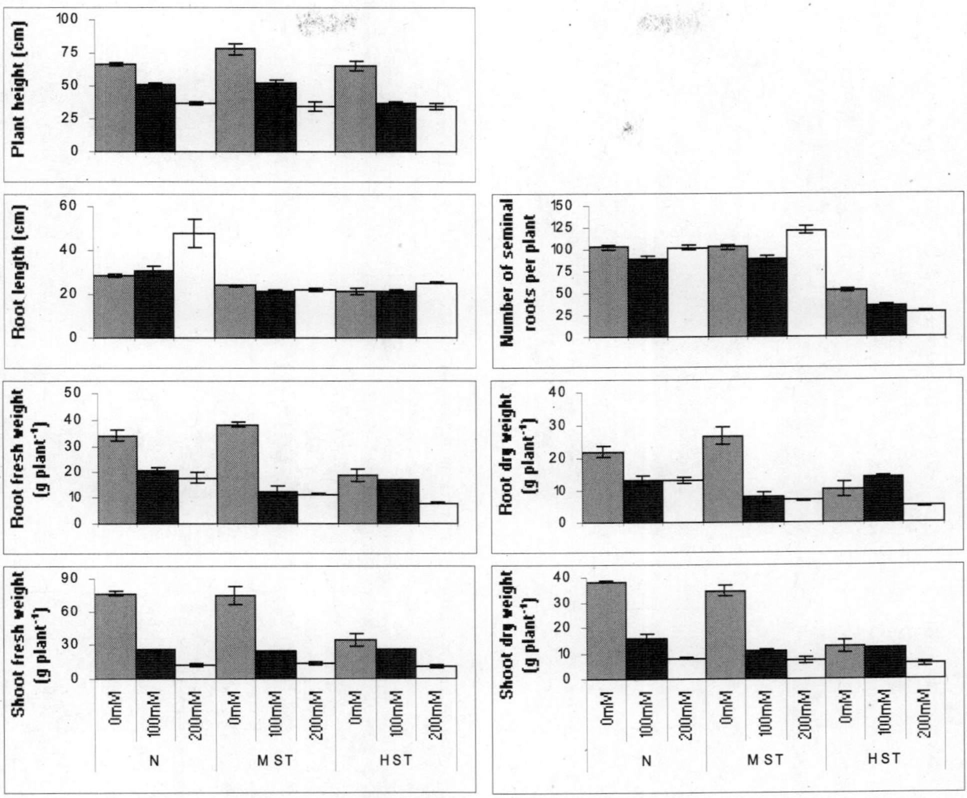

Figure 14.3: Growth Parameters of Three Populations of *P. antidotale* when 30-day Old Plants were Subjected for 46 Days to Salt Stress Conditions in Solution Culture (N: Normal non-saline, MST: Moderately salt tolerant, HST: Highly salt tolerant)

Root area (Figure 14.4 and Plate 14.1) was increased in normal population and decreased in moderately salt tolerant, but it was slightly affected in highly salt tolerant population at moderate salt level and thereafter it remained stable at the highest salt level (200 mM NaCl). Metaxylem area generally decreased with increase in salt level, however in the highly salt tolerant population it was considerably increased at the highest salt level. Similarly metaxylem area of the moderately salt tolerant population increased at the highest salt level, but this was decreased at 100 mM NaCl level. Phloem area, however, was the most responsive parameter showing considerable increase in highly tolerant population with increase in salt level.

Stem area generally decreased with increase in salt level in both normal and moderately salt tolerant populations, however, it was increased in the highly salt tolerant population (Figure 14.5 and Plate 14.2). Epidermis thickness was relatively unaffected by increased salt concentration in the rooting medium, though moderately salt tolerant population showed an increase in epidermis thickness. Sclerenchyma thickness was increased in both salt tolerant populations, but the normal population

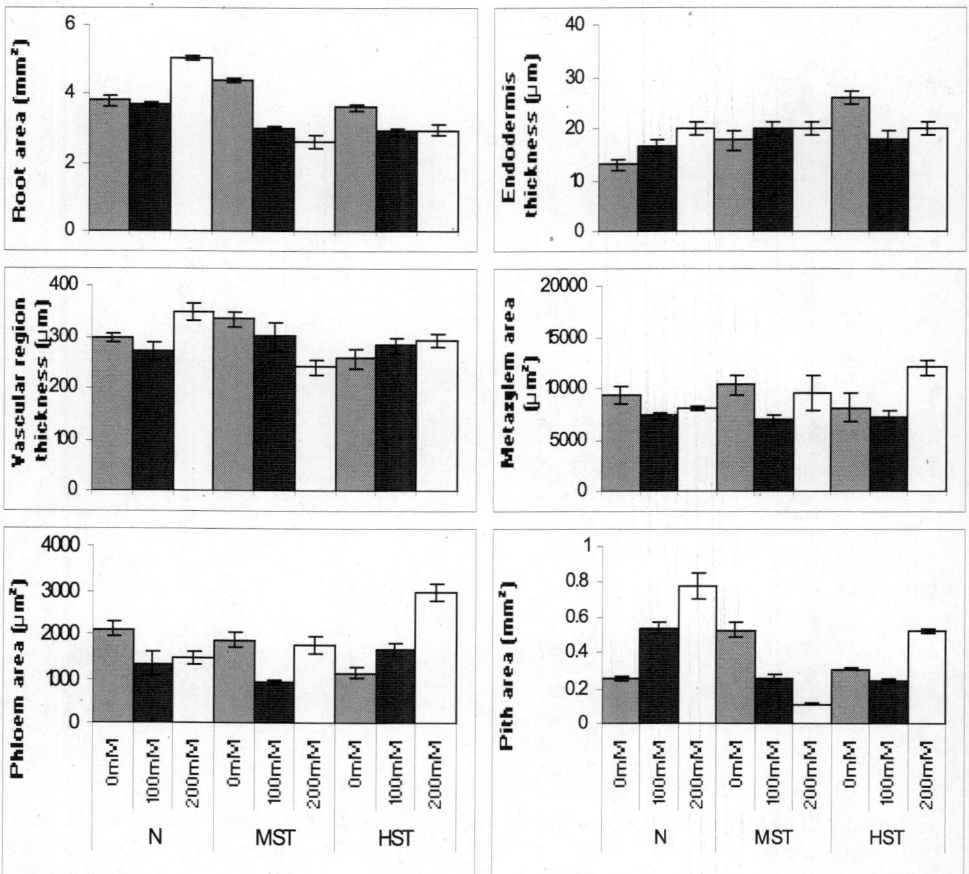

Figure 14.4: Root Anatomical Parameters of Three Populations of *P. antidotale* when 30-day Old Plants were Subjected for 46 Days to Salt Stress Conditions in Solution Culture (N: Normal non-saline, MST: Moderately salt tolerant, HST: Highly salt tolerant)

showed decreased sclerification at the highest salt level. The highly salt tolerant population, however, showed a considerable increase in its sclerenchyma thickness at both salt levels. Vascular bundle area, metaxylem area and phloem area, all increased with increase in salt level in the growth medium with a few exceptions. However, the highly salt tolerant surpassed all the other populations in relation to these characteristics.

Epidermis thickness at both leaf surfaces gradually increased with increase in salt level in both normal and moderately salt tolerant population, but the highly salt tolerant population showed a slight decrease under all salt treatments (Figure 14.6 and Plates 14.3 and 14.4). Number of hairs/trichomes per unit area increased only in highly salt tolerant population with an increase in salt level, however, the other two

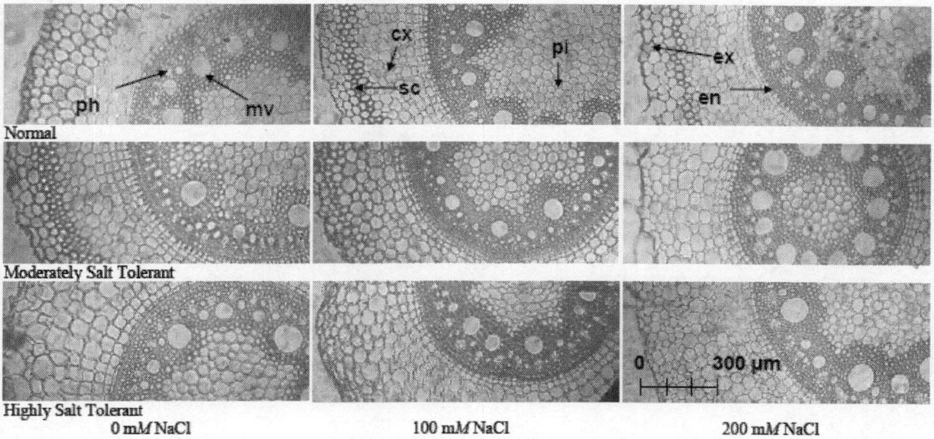

Plate 14.1: Root Transverse Section of Three Population of *Panicum antidotale* when 30-day Old Plants were Subjected for 46 Days to Salt Stress in Solution Culture (cx: Cortex, en: Endodermis; mv: Metaxylem vessel; ph: Phloem; pi: Pitch; sc: Sclerenchyma)

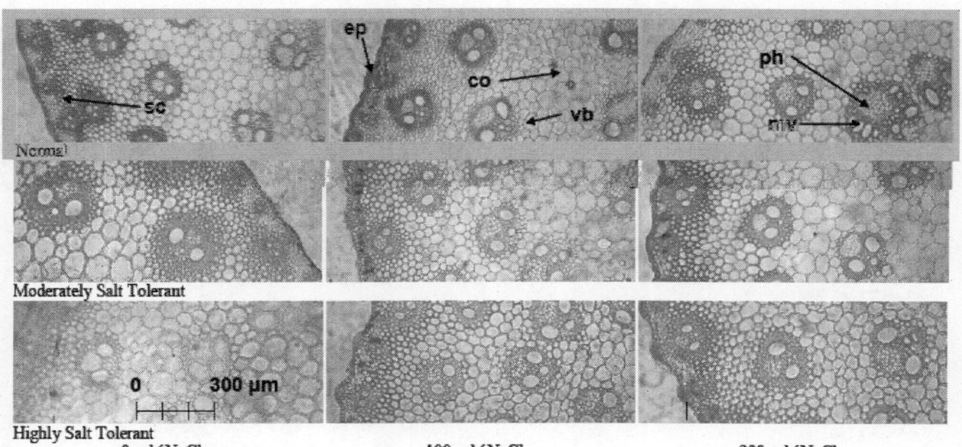

Plate 14.2: Stem Transverse Section of Three Population of *Panicum antidotale* when 30-day Old Plants were Subjected for 46 Days to Salt Stress in Solution Culture (cx: Cortex, en: Endodermis; mv: Metaxylem vessel; ph: Phloem; pi: Pitch; sc: Sclerenchyma)

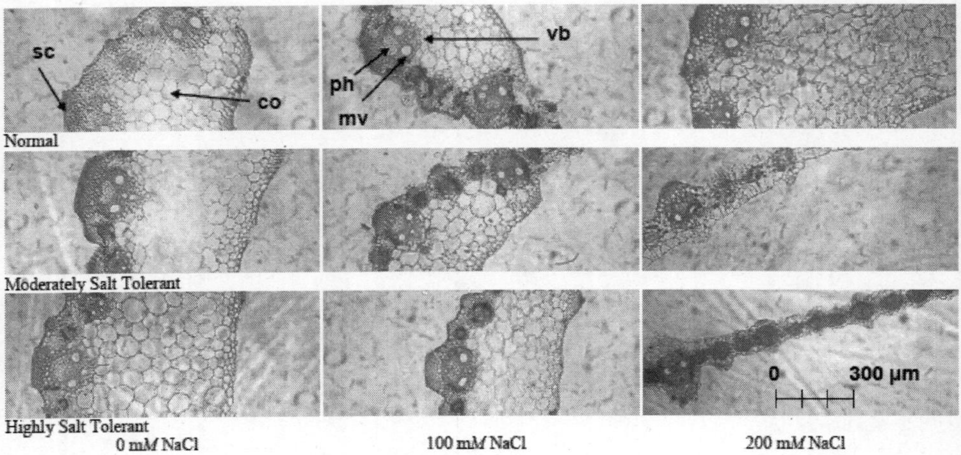

Plate 14.3: Leaf Transverse Section (Midrib) of Three Population of *Panicum antidotale* when 30-day Old Plants were Subjected for 46 Days to Salt Stress in Solution Culture (cx: Cortex, en: Endodermis; mv: Metaxylem vessel; ph: Phloem; pi: Pitch; sc: Sclerenchyma)

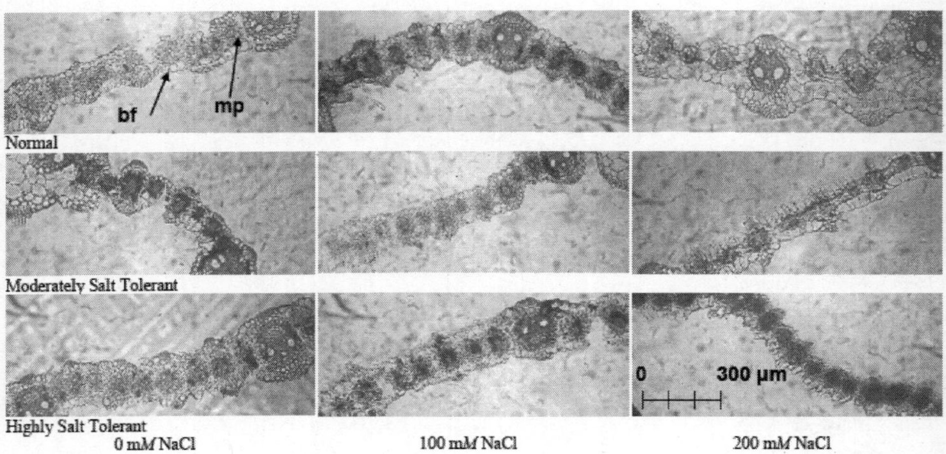

Plate 14.4: Leaf Transverse Section (Laimna) of Three Population of *Panicum antidotale* when 30-day Old Plants were Subjected for 46 Days to Salt Stress in Solution Culture (cx: Cortex, en: Endodermis; mv: Metaxylem vessel; ph: Phloem; pi: Pitch; sc: Sclerenchyma)

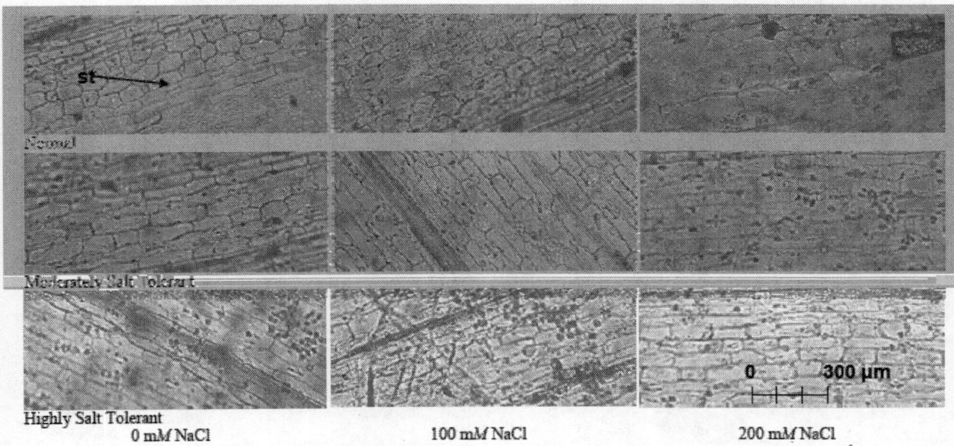

Plate 14.5: Leaf Surface View (Abaxial) of Three Population of *Panicum antidotale* when 30-day Old Plants were Subjected for 46 Days to Salt Stress in Solution Culture (st: Stomata)

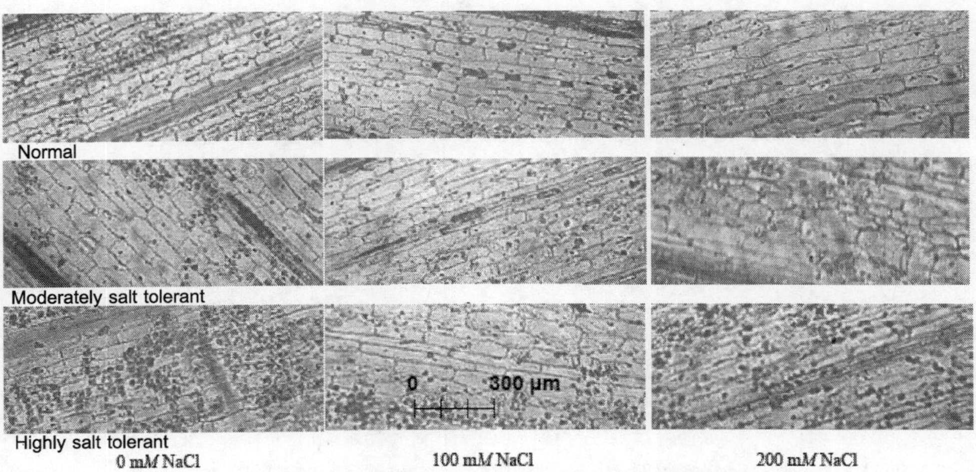

Plate 14.6: Leaf Surface View (Adaxial) of Three Population of *Panicum antidotale* when 30-day Old Plants were Subjected for 46 Days to Salt Stress in Solution Culture

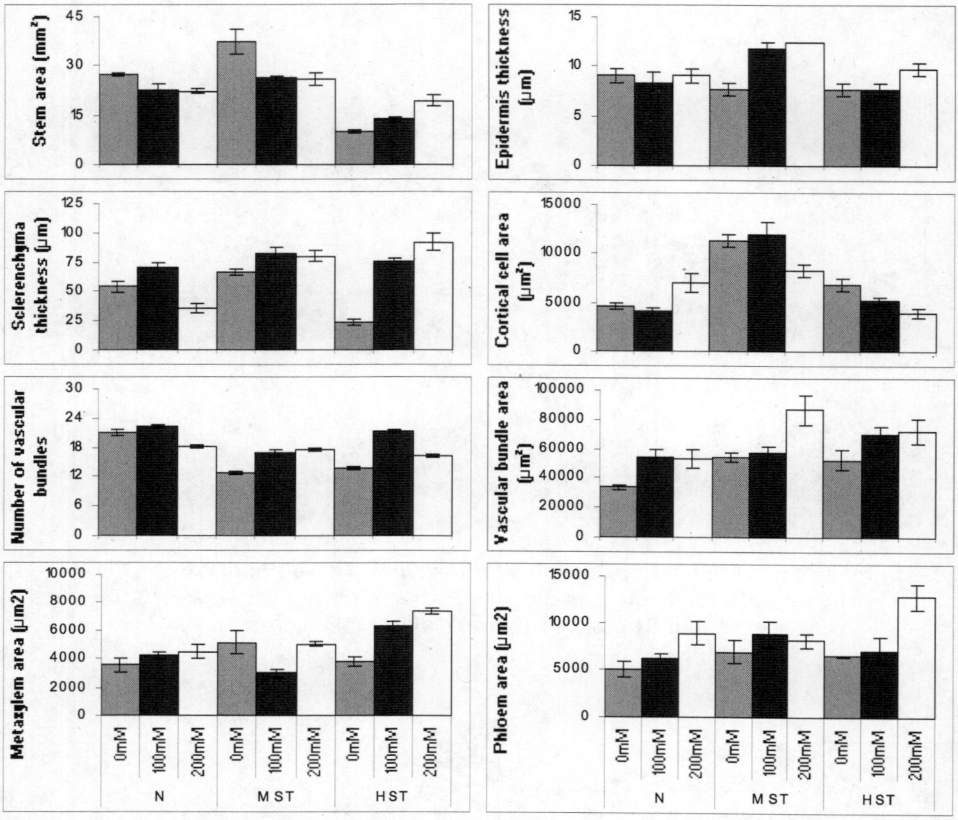

Figure 14.5: Stem Anatomical Parameters of Three Populations of *P. antidotale* when 30-day Old Plants were Subjected for 46 Days to Salt Stress Conditions in Solution Culture (N: Normal non-saline, MST: Moderately salt tolerant, HST: Highly salt tolerant)

populations showed a decrease in this parameter. Bulliform cells were increased with increasing salt level in the rooting medium in all the populations, although the highly salt tolerant population was more responsive to salt stress showing a considerable increase in its bulliform cells that have almost fully covered lower epidermis at the highest salt level (200 mM NaCl). Leaf thickness (midrib), sclerenchyma thickness and mesophyll cell area decreased in most of the cases, whereas leaf thickness in normal population showed a reverse pattern by showing an increase with increase in salt concentration. Stomatal density (Plate 5 and 6) gradually decreased with increase in salt level, however, its area increased in highly salt tolerant population at abaxial leaf surface and in normal population at the adaxial surface.

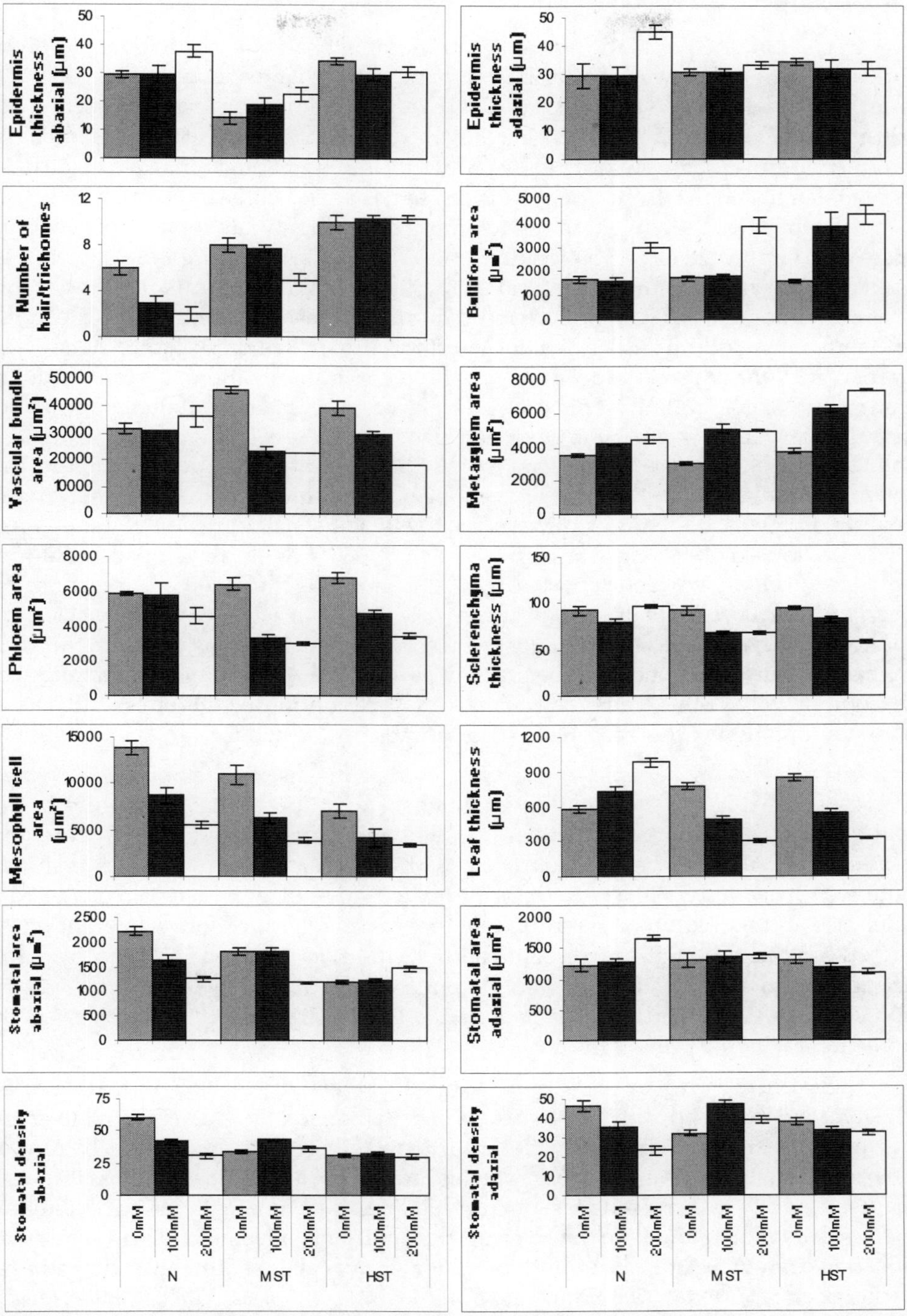

Figure 14.6: Leaf Anatomical Parameters of Three Populations of *P. antidotale* when 30-day Old Plants were Subjected for 46 Days to Salt Stress Conditions in Solution Culture (N: normal non-saline, MST: Moderately salt tolerant, HST: Highly salt tolerant)

Discussion

In general, the population collected from highly salt affected habitat at Forest Research Institute, Gatwala, Faisalabad surpassed the other two populations collected from moderately salt soil affected at the University Campus, Khadeja Hall and normal soil at the University Campus, Lalazar Colony agricultural fields in terms of biomass production and other growth attributes. Thus, the former population was categorized as salt tolerant and that from along roadside near Khadeja Hall as moderately tolerant. Growth attributes showing high salt tolerance of the highly salt tolerant population were fully supported by various anatomical adaptations. For example, leaf thickness (succulence) was very much reduced in highly salt tolerant population under the saline environment. That might have been due to the reason that thin leaves are more efficient in in-rolling and this characteristic can prove to be quite beneficial particularly under limited moisture conditions by minimizing water loss through transpiration. Leaf rolling seems to be an efficient mechanism in tolerant population as bulliform cells were found to be spread all over the lower epidermis at the highest salt level in both moderately tolerant and highly tolerant populations. In highly salt tolerant population, bulliform cells were even more developed. There are several reports on increased leaf thickness in relatively salt sensitive plants (Raafat *et al.*, 1991; Kozlowski, 1997; El-Araby and Hegazi, 1999). Highly developed bulliform epidermal cells have been reported by Gielwanowska *et al.* (2005) in response of *Deschampsia antarctica* plants to salt stress. Adaptation to saline environment is also found to be associated with stability in mesophyll cell area so as to ensure normal exchange of gases and photosynthetic activities (Jafri and Ahmad, 1995). Least affected mesophyll cells in the highly salt tolerant population under salt stress indicate a sound adaptation of this ecotype to unfavourable saline habitat.

Leaves with thicker epidermis may provide a means of checking undue water loss, especially under water limiting conditions. The unaffected epidermis on adaxial leaf surface and enlarged thickness on abaxial surface in highly salt tolerant population makes it a better selection to withstand harsh saline climates than the others. For example, Bahaji *et al.* (2002) related large size epidermal cells to salt tolerance and regarded it a selection tool for *in vitro* identification of tolerant lines. Increased thickness with increasing salinity was also reported in different plant species, *e.g.*, tomato (Ponomar'ova and Kabuzenko, 1978), *Ziziphus* cultivars (Awasthi and Pathak, 1999), and *Ziziphus mauritiana* (Awasthi *et al.*, 1999), which are in agreement with the present findings.

Bulliform cells are supposed to be distinctively related to salt tolerance (Bahaji *et al.*, 2002) and increased bulliform thickness in both moderately and highly salt tolerant populations may have been very helpful in leaf rolling, particularly under physiological droughts caused due to high salinities so as to avoid water loss through leaves. In addition, increased metaxylem, phloem and sieve area under high salinities in the salt tolerant population may be crucial for efficient water and nutrient transport, which is even more critical under limited water availability. Similar findings have earlier been reported in rice (Datta and Som, 1973), *Arabidopsis thaliana* (Baloch *et al.*, 1998), and *Zizyphus mauritiana* (Awasthi *et al.*, 1999).

Reduction in stomatal density and size might help check undue water loss via transpiration during limited moisture availability under high salinities. However, the highly salt tolerant population showed decreased stomatal area and density under saline conditions. Because of this, it can be regarded as the best adapted population against highly saline environments. A number of earlier reports (Omer and Schlesinger, 1980; Gill and Dutt, 1982; Kayani and Mujeeb-ur-Rahman, 1988; Walsh, 1990; Botti *et al.*, 1998; González *et al.*, 2000; Bray and Reid, 2002) very much support this argument.

Generaly, saline conditions can reduce plant stem area as was earlier observed, *e.g.*, in rice (Datta and Som, 1973) and *Prosopis strombulifera* (Reinoso *et al.*, 2004). However, in the highly salt tolerant population, stem area was markedly increased under saline regime. This increased succulence in stem may aid to store additional water and hence better survival under harsh environments. Furthermore, increased sclerenchyma thickness was observed in the stem of this population. This characteristic may offer some resistance to water loss through stem and again may play a crucial role in adaptation to unfavourable conditions. Earlier reports on *Spartina alterniflora* (Walsh, 1990) and *Puccinellia tenuiflora* (YuJing *et al.*, 2000) support the present findings, but those in rice (Chandramony and George, 1975) and *Deschampsia antarctica* (Gielwanowska *et al.*, 2005) are in disagreement.

Regarding physiological adaptations, salt stress caused a marked decline in net CO_2 assimilation rate in both normal and salt tolerant populations, but salt treatment proved to have more inhibitory effect on photosynthetic capacity of normal population than that of the salt tolerant population. There are many earlier reports, which show that plant photosynthetic capacity is inhibited by salinity (Robinson *et al.*, 1983; Ball and Farquhar, 1984; Seemann and Critchley, 1985; Ashraf, 1999). Relationship of the data of growth variables with those of leaf gas exchange characteristics at the vegetative stage clearly shows that vigorous growth of salt tolerant population under salt stress can be ascribed to its higher net CO_2 assimilation rate under these conditions. Such a positive association between growth and leaf photosynthetic rate has earlier been reported in a numberof crops, *e.g.*, *Zea mays* (Crosbie and Pearce, 1982), *Spinacia oleracea* (Robinson *et al.*, 1983), *Phaseolus vulgaris* (Seemann and Critchley, 1985), *Vigna mungo* (Babu *et al.*, 1985), *Gossypium barbadense* (Cornish *et al.*, 1991), *Gossypium hirsutum* (Pettigrew and Meredith, 1994), *Asparagus officinalis* (Curtis and Läuchli, 1986), and six *Brassica* diploid and amphiploid species (Ashraf, 2001).

The salt tolerant population had higher WUE under saline conditions as compared to that of normal population. The high water use efficiency in the salt tolerant population is expected in view of its natural adaptation to saline conditions where osmotic effect of salts shares the effect of water deficit. It is known that drought or salt stress triggers a number of physiological responses in plants, which may ultimately affect water use efficiency, and high water use efficiency is one of the major adaptive characteristics of plants growing under osmotically affected environment (Hsiao, 1973; Morgan, 1984).

Chlorophylls *a* and *b* were reduced in both populations under salinity, but they had no positive association with net CO_2 assimilation or the growth of different populations under varying stress conditions. Absence lack of association between the two attributes is not in conformity with some earlier reports in which a positive relationship between chlorophyll content and net CO_2 assimilation was found (Keller and Koch, 1964; Kozlowski, 1982; Ashraf and Rehman, 1999).

Relationships between data for leaf water potential and that of turgor pressure of both salt tolerant and normal populations under varying stress conditions were not possible to be drawn, as well as between these two attributes and leaf stomatal conductance. For example, leaf water potential of salt tolerant population was higher under saline condition, but in contrast, leaf turgor pressure was low under saline condition. However, stomatal conductance at stress treatment was higher in the salt tolerant population as compared to that of the other population. Such an inconsistent relationship among the three water relation attributes can be related to some earlier findings. In cowpea no positive association between leaf stomatal conductance and leaf water potential was found (Bates and Hall, 1981 and Osonubi, 1985). Similarly, Blackman and Davies (1985) did not find any consistent relationship between leaf stomatal conductance and water potential or turgor potential in maize.

In view of a number of anatomical and physiological adaptive components of salt tolerance observed in the salt tolerant population, such types of populations could be potential colonizers of salt affected soils. Since *P. antidotale* inherently produces a large quantity of biomass, it could be used as a potential fodder and forage crop for livestock and small ruminants.

References

Açýkgöz, N. and M.N. Gevrek. 1994. Investigations on salt tolerance of rice mutants. Turkish Journal of Agriculture and Forestry18: 179-186.

Ashraf, M. 2004. Some important physiological selection criteria for salt tolerance in plants. Flora 199: 361–376.

Ashraf, M. 2003. Relationships between leaf gas exchange characteristics and growth of differently adapted populations of Blue panicgrass (*Panicum antidotale* Retz.) under salinity or waterlogging. Plant Science 165: 69-75.

Ashraf, M. 1999. Interactive effect of salt (NaCl) and nitrogen form on growth, water relations and photosynthetic capacity of sunflower (*Helianthus annuus*). Annals of Applied Biology 153: 509-513.

Ashraf, M. 2001. Relationships between growth and gas exchange characteristics in some salt-tolerant amphidiploid *Brassica* species in relation to their diploid parents. Environmental and Experimental Botany 45: 155-163.

Ashraf, M. and H. Rehman. 1999. Interactive effects of nitrate and longterm waterlogging on growth, water relations, and gaseous exchange properties of maize (*Zea mays* L.). Plant Science 144: 35-43.

Awasthi, O.P. and R.K. Pathak. 1999. Effect of salinity levels on survival and anatomy of four scion cultivars budded on Indian jujube. Horticultural Journal 12: 53-59.

Awasthi, O.P., R.K. Pathak and S.D. Pandey. 1999. Anatomical variation in leaf lamina of ber seedling and budded plants grown at different sodicity levels. Indian Journal of Horticulture 56: 29-33.

Babu, R.C., P.S. Srivinisan, N. Natarajaratnam and S.R.S. Rangasamy. 1985. Relationship between leaf photosynthesis and yield in blackgram (*Vigna mungo* (L.) Hepper) genotypes. Photosynthetica 19: 159-163.

Bahaji, A., I. Mateu, A. Sanz and M.J. Cornejo. 2002. Common and distinctive responses of rice seedlings to saline- and osmotically-generated stress. Plant Growth Regulation 38: 83-94.

Baker, N.R. 1996. Photosynthesis and the Environment, Kluwer, Dordrecht. The Netherlands.

Ball, M.C. and G.D. Farquhar. 1984. Photosynthetic and stomatal responses of two mangroves, *Aegiceras corniculatum* and *Avicennia marina*, to long-term salinity and humidity conditions. Plant Physiology 74: 1-6.

Baloch, A.H., P.J. Gates and G.M. Baloch. 1998. Anatomical changes brought about by salinity in stem, leaf and root of *Arabidopsis thaliana* (L.) Heynh (thale cress). Sarhad Journal of Agriculture 14: 131-142.

Bates, L.M. and A.E. Hall. 1981. Stomatal closure with soil water depletion not associated with changes in bulk leaf water status. Oecologia 50: 62-65.

Begum, F. and J.L. Karmoker. 1999. Effect of salinity stress on the accumulation and distribution of proline in wheat. Rachis 18: 22-25.

Blackman, P.G. and W.G. Davies. 1985. Root to shoot communication in maize plants of the effects of soil drying. Journal of Experimental Botany 36: 39-48.

Bokhari, U.G., F. Alyacesh and M. Nori. 1987. Adaptive strategies of desert grasses in Saudi Arabia. Journal of Range Management 40: 19-22.

Botti, C., D. Palzkill, D. Muñoz and L. Prat. 1998. Morphological and anatomical characterization of six jojoba clones at saline and non-saline sites. Industrial Crops and Products 9: 53-62.

Bozcuk, S. 1991. Determination of the effects of salinity on germination and limits of salt tolerance in some crops. Doga Turkish Journal of Biology15: 144-151.

Bray, S. and D.M. Reid. 2002. The effect of salinity and CO_2 enrichment on the growth and anatomy of the second trifoliate leaf of *Phaseolus vulgaris*. Canadian Journal of Botany 80: 349-359.

Chandramony, D. and M.K. George. 1975. Nutritional effects of calcium, magnesium, silica and sodium chloride on certain anatomical characters of rice plant related to lodging. Agricultural Research Journal of Kerala 13: 39-42.

Chaudhary, S.A. 1989. Grasses of Saudi Arabia, National Herbarium, National Agriculture and Water Research Center, Ministry of Agriculture and Water, Riadh, Kingdom of Saudi Arabia.

Cope, T.A. 1982. Poaceae. Vol. 143. In Flora of Pakistan; Nasir, E. and S.I. Ali (Eds.); University of Karachi: Karachi, Pakistan.

Cornish, K., J.W. Radin, E.L. Turcotte, Z. Lu and E. Zeiger. 1991. Enhanced photosynthesis and stomatal conductance of Pima cotton (*Gossypium barbadense* L.) bred for increased yield. Plant Physiology 97: 484-489.

Crosbie, T.M. and R.B. Pearce. 1982. Effects of recurrent phenotypic selection for high and low photosynthesis on agronomic traits in two maize populations. Crop Science 22: 809-813.

Curtis, P.S. and A. Läuchli. 1986. The role of leaf area development and photosynthetic capacity in determining growth of kenaf under moderate salt stress. Australian Journal of Plant Physiology 13: 353-365.

Datta, S.K. and J. Som. 1973. Note on the effect of salinity on the structural changes in the stem of rice varieties. Indian Journal of Agricultural Sciences 43: 614-617.

El-Araby, M.M. and A.Z. Hegazi. 1999. Growth responses and hormonal changes of kidney beans at different levels of soil salinization. J. Union Arab Biol., Cairo, 7(B) Physiology and Algae 327-343.

Farquhar, G.D. and T.D. Sharkey. 1982. Stomatal conductance and photosynthesis, Annual Review of Plant Physiology 33: 317-345.

Gielwanowska, I., E. Szczuka, J. Bednara and R. Górecki. 2005. Anatomical features and ultrastructure of *Deschampsia antarctica* (Poaceae) leaves from different growing habitats. Annals of Botany 96:1109-1119.

Gill, K.S. and S.K. Dutt. 1982. Effect of salinity on stomatal number, size and opening in barley genotypes. Biological Plantarum 24: 266-269.

Gohl, B. 1975. Tropical feeds. Animal Production and Health Division paper 98. FAO, Rome.

González, L.M., R.C. López, I. Fonseca and R. Ramírez. 2000. Growth, stomatal frequency, DM yield and accumulation of ions in nine species of grassland legumes grown under saline conditions. Pastos y Forrajes 23: 299-308.

Gorham, J. and R.G. Wyn Jones. 1990. A physiologist's approach to improve the salt tolerance of wheat. Rachis 9: 20-24.

Gorham, J., R.G. Wyn Jones and E. McDonnell. 1985. Some mechanisms of salt tolerance in crop plants. Plant Soil 89: 15-40.

Gould, F.W. and R.B. Shaw. 1983. Grass systematics. 2nd ed. Texas A&M Univ. Press, College Station, TX.

Hsiao, T.C. 1973. Plant response to water stress, Annual Review of Plant Physiology 24: 519-570.

Jafri, A.Z. and R. Ahmad. 1995. Effect of soil salinity on leaf development, stomatal size and its distribution in cotton (*Gossypium hirsutum* L.). Pakistan Journal of Botany 27: 297-303.

Kayani, S.A. and Mujeeb-ur-Rahman. 1988. Effects of NaCl salinity on shoot growth, stomatal size and its distribution in *Zea mays* L. Pakistan Journal of Botany 20: 75-81.

Keller, T. and W. Koch. 1964. The effect of iron chelate fertilization of poplar upon CO_2 uptake, leaf size, and content of leaf pigments and iron. Plant Soil 20: 116-126.

Kozlowski, T.T. 1997. Responses of woody plants to flooding and salinity. Tree Physiology Monograph No. 1, Heron Publishing, Victoria, Canada. 1-29 pp.

Kozlowski, T.T. 1982. Water supply and tree growth. II. Flooding/ Forest Abstract 43: 145-161.

Leopold, A.C. and R.P. Willing. 1984. Evidence for toxicity effects of salt on membranes. In: R.C. Staples and G.H. Toenniessen (eds.) Salinity Tolerance in Plants. Strategies for Crop Improvement. A Wiley-Interscience Publication, Toronto, Singapore, 67-76 pp.

Lipschitz, N. and Y. Waisel. 1974. Existence of salt glands in various genera of the Gramineae. New Phytology 73: 507-507.

Magness, J.R., G.M. Markle and C.C. Compton. 1971. Food and Feed Crops of the United States.

Marcum, K.B. and C.L. Murdoch. 1990. Salt glands in the Zoysieae. Annals of Botany 66: 1–7.

Marcum, K.B., S.J. Anderson and M.C. Engelke. 1998. Salt gland ion secretion: a salinity tolerance mechanism among five Zoysiagrass species. Crop Science 38: 806–810.

Marschner, H. 1995. Mineral nutrition of higher plants. 2nd ed. Academic Press, San Diego, CA.

Morgan, J.M. 1984. Osmoregulation and water stress in higher plants. Annual Review of Plant Physiology 35: 299-319.

Niknam, S.R. and J. McComb. 2000. Salt tolerance screening of selected Australian woody species-a review. Forest Journal of Ecology and Management 139: 1-19.

Omer, L.St. and W.H. Schlesinger. 1980. Field and greenhouse investigations of the effect of increasing salt stress on the anatomy of *Jaumea carnosa* (Asteraceae), a Salt Marsh Species. American Journal of Botany 67: 1455-1465.

Oross, J.W. and W.W. Thomson. 1982. The ultrastructure of the salt glands of *Cynodon* and *Distichlis* (Poaceae). American Journal of Botany 69: 939–949.

Oross, J.W., R.T. Leonard and W.W. Thomson. 1985. Flux rate and a secretion model for salt glands of grasses. Israel Journal of Botany 34: 69–77.

Osonubi, O. 1985. Responses of cowpeas (*Vigna unguiculata* (L.) Walp.) to progressive soil drought. Oecologia 66: 554-557.

Pettigrew, W.T. and W.R. Meredith. 1994. Leaf gas exchange parameters vary among cotton genotypes. Crop Science 34: 700-705.

Pitman, M.G. 1984. Transport across the root and shoot/root interactions. In: R.C. Staples and G.H. Toenniessen (eds.) Salinity Tolerance in Plants. Strategies for Crop Improvement. A Wiley-Interscience Publication, Toronto, Singapore, 93-123 pp.

Poljakoff-Mayber, A. 1975. Morphological and anatomical changes in plants as a response to salinity stress. In: Poljakoff-Mayber, A. and J. Gale, (eds.), Plants in Saline Environment. Ecological Studies 15. Springer, Berlin, 97–117 pp.

Poljakoff-Mayber, A. 1988. Ecological-physiological studies on the responses of higher plants to salinity and drought. Scientific Review of Arid Zone Research 6: 163–183.

Ponomar'ova, S.O. and S.M. Kabuzenko. 1978. Anatomical structure of tomato leaves in conditions of chloride salinity. Ukraine Botanichnii Zhurnal 35: 237-240.

Prasad, M.V.R. and R.P. Singh. 1973. Select your crop for Western Rajasthan. Indian Farming 22: 13-15.

Qiu, L.-Z., Y.-J. Huang, J.-Q. Huang, G.-H. Xia and N. Gong. 2006. Comparative study on vegetal and physiological characteristics of different salt-tolerant plants under salt stress. Journal Zhejiang University of Agriculture Life Science 32: 420-427.

Raafat, A., S.A. Habib, I.Z. El-Shami and H.M. El-Antably. 1991. The effect of salinity on the anatomical features of tomato plant. Annals of Agricultural Science (Cairo) 36: 307-321.

Rana, R.S. 1986. Evaluation and utilisation of traditionally grown cereal cultivars on salt affected areas in India. Indian Journal of Genetics and Plant Breeding 46: 121-135.

Reinoso, H., L. Sosa, L. Ramírez and V. Luna. 2004. Salt-induced changes in the vegetative anatomy of *Prosopis strombulifera* (Leguminosae). Canadian Journal of Botany/Review of Canadian Journal of Botany 82: 618-628.

Robinson, S.P., W.J.S. Downton and J.A. Millhouse. 1983. Photosynthesis and ion content of leaves and isolated chloroplasts of salt-stressed spinach. Plant Physiology 73: 238-242.

Ruyle, G.B. and D.J. Young. 1997. Arizona Range Grasses. Cooperative Extension Publication AZ97105. University of Arizona, College of Agriculture, Tucson.

Ryan, J., S. Miyamoto and J.L. Stroehlein. 1975. Salt and specific ion effects on germination of four grasses. Journal of Range Management 28: 61-64.

Sarwar, M., Mahr-un-Nisa, M.A. Khan and M. Mushtaque. 2006. Chemical composition, herbage yield and nutritive value of *Panicum antidotale* and *Pennisetum orientale* for Nili buffaloes at different clipping intervals. Asian-Australasian Journal of Animal Science 19: 176-180.

Seemann, J.R. and C. Critchley. 1985. Effects of salt stress on the growth, ion content, stomatal behaviour and photosynthetic capacity of a salt-sensitive species *Phaseolus vulgaris* L. Planta 164: 151-162.

Sen, K.C. and S.N. Ray. 1964. Nutrient values of Indian cattle feeds and the feeding of animals. New Delhi Indian Council. Agricultural Research Bulletin No. 25.

Sinha, A., S.R. Gupta and R.S. Rana. 1986. Effect of soil salinity and soil water availability on growth and chemical composition of *Sorghum halepense* L. Plant Soil 95: 411–418.

Spychalla, J.P. and S.L. Desborough. 1990. Superoxide dismutase, catalase, and alpha-tocopherol content of stored potato tubers. Plant Physiology 94: 1214-1218.

Suhayda, G.G., R.E. Redmann, B.L. Harvey and A.L. Cipywnyk. 1992. Comparative response of cultivated and wild barley species to salinity stress and calcium supply. Crop Science 32: 154-163.

Volkmar, K.M., Y. Hu and H. Steppuhn. 1998. Physiological responses of plants to salinity: A review. Canadian Journal of Plant Science 78: 19-27.

Walsh, G.E. 1990. Anatomy of the seed and seedling of Spartina alterniflora Lois. (Poaceae). Aquatic Botany 38: 177-193.

Whyte, R.O., T.R.G. Moir and J.P. Cooper. 1959. Grasses in Agriculture. Plant Production and Protection Division, Food and Agriculture Organization of the United Nations, Rome,

Wilson, C. and J.J. Read. 2006. Effect of mixed-salt salinity on growth and ion relations of a barnyardgrass species. Journal of Plant Nutrition 29: 1741-1753.

Wise, R.R. and A.V. Naylor. 1987. Chilling enhanced photooxidation: Evidence for the role of singlet oxygen and endogenous antioxidants. Plant Physiology 83: 278-282.

Wright, L.N. and A.K. Dobrenz. 1970. Water use in relation to management of blue panicgrass (*Panicum antidotale* Retz.). Journal of Range Management 23: 193-196.

YuJing, Z., Z. Yong, H. ZiZhi and Y. ShunGuo. 2000. Studies on microscopic structure of *Puccinellia tenuiflora* stem under salinity stress. Grassland of China 5: 6-9.

Zhong, G.Y., J. Dvorak and G.Y. Zhong. 1995. Evidence for common genetic mechanisms controlling the tolerance of sudden salt stress in the tribe Triticeae. Plant Breeding 114: 297-302.

15

Salt Sensitivity of Wheat at Germination Stage

G.H. Ranjbar, S.A.M. Cheraghi and M.H. Banakar
National Salinity Research Centre, Azadegan Blvd, Yazd, Iran

ABSTRACT

Salinity stress is a limiting factor to crop production in many parts of the world. Seed germination like other plant growth phases is affected by salt stress. To evaluate the salt sensitivity of wheat at seed germination, two experiments were conducted with natural (NSW) and artificial saline water (ASW) in a controlled germinator. Salinity levels for both water sources were 0.5, 5, 10, 15, 20 and 30 dSm^{-1} and local wheat cultivars were Kavir, Roshan and Marvdasht. Results showed that, the effect of salinity on germination was different for all cultivars. Seed germination of Kavir was affected at all level of salinity. Increasing salinity levels had less effect on seed germination of Roshan and Marvdasht cultivars. Salinity tolerance threshold of Kavir, Roshan and Marvdasht at germination stage were 7.34, 12.49 and 15.74 in NSW and 11.05, 14.58 and 18.90 in ASW, respectively. The variation observed among the cultivars suggests that although wheat is tolerant at germination stage, wide variation exists among the cultivars. In both NSW and ASW treatments, Roshan and Kavir had the highest and the lowest percentage of germination, respectively. Marvdasht had an intermediate response compared with Kavir and Roshan.

Introduction

Salinity stress is a limiting factor for crop production in many parts of the world. It is especially prevalent in marginal lands associated with poor drainage or high water tables. There are many approaches to overcome salinity stress, *i.e.* soil reclamation, leaching, and use of salt tolerant crops. It seems, however, that use of tolerant crops is the simplest and most economical means for using saline lands and waters. In addition, relatively small increase in salt tolerance may greatly reduce the leaching requirement in areas that need to be reclaimed by leaching. Salt sensitivity of crops is often different from one growth stage to another. Most crops are tolerant during seed germination but most of them become sensitive during seedling emergence and initial establishment (Maas, 1990).

Wheat is considered moderately tolerant to salinity (Maas and Hoffman, 1977). Considerable variation in salt tolerance appears among and within wheat population (Norlyn and Epstein, 1984). Results have shown that wheat is more tolerant during seed germination (Francois *et al.*, 1986). Generally, wheat seed can germinate in saline media, but at a delayed rate (Aceves *et al.*, 1975). On other hand, salt tolerance at germination seems not to be related to tolerance during post germination under salt condition (Francois *et al.*, 1986; Maas, 1990). This study was conducted to determine salt sensitivity of wheat cultivars at seed germination and to determine salt tolerance threshold during seed germination.

Materials and Methods

Wheat seeds (cvs. Kavir, Roshan and Marvdasht) were sterilized with 5 per cent (wt/vol) calcium hypochlorite. Thirty surface sterilized seeds were placed on Whatman filter paper No. 1 in petri dishes (90 mm). Three replicates were used for each treatment. Two experiments were conducted using natural saline water (NSW) and artificial saline water (ASW) with 2:1 molar ratio of $NaCl:CaCl_2$.

In the first experiment filter papers were soaked in ten ml of NSW with salinity levels of 0.5, 5, 10, 15, 20 and 30 dSm^{-1}. Different salinity levels lower than 17 $dS\ m^{-1}$ were made by adding adequate tap water to base NSW (Table 15.1). Other salinity levels higher than 17 dSm^{-1} were obtained by natural evaporation. In the second experiment different salinity levels (0.5, 10, 20 and 30) were made by addition of NaCl and $CaCl_2$ in ratio of 2:1 on equivalent basis. Seeds were allowed to germinate in dark conditions at 25±2℃ daily temperature in a germinator. Seeds were considered germinated when radicle and plumnule could be clearly distinguished (Begum *et al.*, 1992). After seven days, the germinated seeds were recorded and, 50 per cent reduction in germination was measured.

Table 15.1: Natural Saline Water Composition

CO_3^{2-} (Meq/liter)	CO_3H^- (Meq/liter)	Cl^- (Meq/liter)	SO_4^{2-} (Meq/liter)	Ca^{2+} (Meq/liter)	Mg^{2+} (Meq/liter)	Na^+ (Meq/liter)	ECe (dSm^{-1})
0.7	3.2	153.75	27.6	18.4	39.6	154.0	17.61

Results and Discussion

Effect of salinity levels on germination were not significantly the same for all cultivars (Table 15.2).

Table 15.2: Effects of NSW Treatments on Seed Germination of Wheat Cultivars

Cultivar	*Salinity (dSm^{-1})*					
	0.5 (Control)	*5*	*10*	*15*	*20*	*30*
Kavir	77.4aB†	60.01bC	60.01bC	55.94bC	44.04cC	14.96dB
Roshan	90.00aA	84.74aA	82.52aA	82.52aA	82.52aA	59.32bA
Marvdasht	75.04aB	73.30aB	71.81aB	70.07aB	55.77bB	15.2cB

† Values followed by different capital letter in column and small letters in rows represent differences for among cultivars and salinity levels within cultivars, respectively.

Germination of Kavir was affected by increasing salinity levels of NSW (Table 15.2). In NSW treatment with 30 dSm^{-1} salinity, the percentage of seed germination for Kavir was reduced by 79 per cent. Seed germination of Roshan was not affected by NSW up to 20 dS m^{-1} salinity. At 30 dSm^{-1} its germination was significantly reduced (Table 15.2). Seed germination of Marvdasht was not influenced by NSW up to 15 dSm^{-1} salinity. At 20 and 30 dSm^{-1} salinity the percentage of seed germinated was significantly reduced by 26 and 86 per cent respectively compared with control.

Seed germination of Kavir was affected by ASW similar to NSW (Table 15.3). With each increase in salinity levels of ASW, the germination percentage of Kavir decreased significantly compared with the control (Table 15.3). The seed germination percentages of Roshan and Marvdasht in ASW treatment were not significantly affected up to salinity level of 10 and 20 dSm^{-1}, respectively (Table 15.3). These results are consistent with other germination tests under saline conditions (Qavami *et al.*, 2004 and Niazi *et al.*, 1992). Reduction in germination due to salinity is known for many species (Francois *et al.*, 1986; Francois *et al.*, 1988; Norlyn and Epstein, 1984).

Table 15.3: Effect of Artificial Saline Water on Seed Germination of Wheat Cultivars

Cultivar	*Salinity (dSm^{-1})*			
	0.5	*10*	*20*	*30*
Kavir	75.04aB†	68.74bB	42.02cC	5.26dC
Roshan	84.74aA	84.74aA	77.26bA	63.51cA
Marvdasht	73.30aB	73.30aB	70.07aB	41.27bB

† Values followed by different capital letter in column and small letters in rows represent differences for among cultivars and salinity levels within cultivars, respectively.

Salt tolerance threshold at germination stage in NSW treatment, were 12.49, 15.47 and 7.34 dSm^{-1} for Roshan, Marvdasht and Kavir, respectively (Table 15.4). Each unit EC increase in salinity above the threshold reduced seed germination by 1.84, 6.02 and 3.32 per cent for Roshan, Marvdasht and Kavir, respectively (Table 15.4).

Table 15.4: Salinity Tolerance Threshold Value (CT), Slop Reduction Percentage and 50 per cent Reduction (C50) of Kavir, Roshan and Marvdasht at Germination Stage

Cultivar	*NSW*		*ASW*	
	Slop per cent	C_T *(dSm⁻¹)*	*Slop*	C_T
Kavir	3.32	7.34	4.91	11.05
Roshan	1.84	12.49	1.59	14.58
Marvdasht	6.02	15.74	3.93	18.90

The tolerance threshold for seed germination in ASW treatment were 14.58, 18.90 and 11.05 for Roshan, Marvdasht and Kavir, respectively (Table 15.4). On other hand, in each salinity level, seed germination rate for cultivars were different significantly in both NSW and ASW treatments. Roshan and Kavir had the most and least germination, respectively, while Marvdasht represented an intermediate response (Tables 15.2 and 15.3). Froncois *et al.* (1986) also observed that germination of probred (a wheat cultivar) in trays containing fine, washed sand at 8.8 $dS.m^{-1}$ and above was significantly delayed and the final germination percentage was markedly reduced.

These data indicated that both Roshan and Marvdasht cultivars are more tolerant to salinity than Kavir at germination stage. These variations also supported this idea that wide variation exists among wheat cultivars at germination stage under salt stress.

As it has been shown in Table 15.4, the levels of salinity tolerance in ASW for all cultivars were higher than for NSW experiment. The differences between tolerance threshold in NSW and ASW experiment may be due to the lower ratio of sodium to calcium in ASW. Calcium is known to have a significant effect on improving the rate and percentage of seed germination (Grattan *et al.*, 1999). Norlyn and Epstein (1984) noted that calcium had differential effect on salinity responses among lines within a species. They concluded that addition of Ca^{2+} can reduce the hazardous effect of Na^+ and can inhibit absorption of Na^+, therefore improving the water imbibitions of seeds. When external calcium concentration is high (ASW treatment), they may mitigate the effect of salinity. High ratio of sodium to calcium in the medium (NSW treatment) tends to be deleterious. Inadequate concentrations of Ca^{2+} may adversely affect membrane function and growth (Lauchli and Epstein, 1990).

This experiment dealt only with salt tolerance at the critical first stage of life cycle, which may or may not be associated with the ability to cope at later stages.

References

Aceves, N.E., L. H. Stolzy and G.R. Mehuys. 1975. Combined effects of low oxygen and salinity on germination of semi-Dwarf of Mexican wheat. Agron. J. 67: 530-532.

Begum, F., J. L. Karmoker, Q. A. Fattah and A. F. M. Maniruzzman. 1992. The effect of salinity on germination and its correlation with K^+, Na^+, Cl^- accumulation in germinating seeds of *Triticum aestivm* L. CV. Akbar. Plant Cell Physiol. 33: 1009-1014.

Francois, L. E., E. V. Maas, T. J. Donovan and V. L. Youngs, 1986. Effect of salinity on grain yield and quality, vegetative growth, and germination of semi-dwarf and durum wheat. Agron. J. 78: 1053-1058.

Francois, L. E., T. J. Donovan, E.V. Maas and G. L. Rubenthaler. 1988. Effect of salinity on grain yield and quality, vegetative growth, and germination of triticale. Agron. J. 80: 642-647.

Grattan. S. R. and C. M. Grieve. 1999. Mineral nutrient acquisition and response by plant grown in saline environments. pp. 203-229. In: Pessarakli, M. Hand book of plant and crop stress. Marcel Dekker, Inc. New York. 1254 P.

Lauchili, A., and E. Epstein. 1990. Plant responses to saline and sodic condition. pp. 113-137. In : K.K. Tanji. Agricultural Salinity Assessment and Management. ASCE. Publication. 619 P.

Maas, E. V. 1990. Crop salt tolerance. pp. 262-303. In: K.K. Tanji. Agricultural Salinity Assessment and Management. ASCE. Publication. 619. P.

Maas, E.V. and G.J. Hoffman, 1977. Crop salt tolerance–current assessment. J. Irrig. Drainage Div. ASCE: 103: 115-134.

Niazi, M. L. K.; K. Mahmood,; S.M. Mujtaba and K. A. Malik. 1992. Salinity tolerance in different cultivars of barley (Hordeum vulgare L.). Biologia plantarum. 34: 465-469.

Norlyn, J. D. and E. Epstein. 1984. Variability in salt tolerance of four triticale lines at germination and emergence. Crop. Sci. 24: 1090-1092.

Qavami, F., Melboei, M. A., Qanadha, M. R., Yazdi Samadi, B., Mozafari, J., and Aghaei, M. J. 2004. Responses of Iranian salt tolerant wheat cultivars at germination and seedling stages. Iran Agric. Sci.35: 464-453.(In Persian).

16

Agriculture Sector of the Benenese Economy

A.D.T.E. Christophe

Journalist/Reporter (Freelance), 07 LP: 784 Cotonou, Benin

E-mail: assochrist2002@yahoo.fr, egbadiran@uva.org, don.aristote.le.sage@caramail.com

ABSTRACT

Benin is an underdeveloped country whose economy depends on agriculture, the agricultural production is however, generally low. Adoption of modern methods of crop production is needed to improve the situation. The new leaders of the country have promised to do so. There is also a need for expoloiting the potential use of brackish water and salt affected lands for agricultural production in the semi-arid areas of the country.

Introduction

According to the census of 2006, The Republic of Benin, Western Africa has a population of 7,513,946 and covers an area of 114,763 km² while extending on 670 km, of the Niger river (in north) and at the Atlantic coast (in the south). The country has Togo in the west, Nigeria in the east and Niger and Burkina Faso in the North-West. Benin gained independence in 1960, under the denomination of Republic of Dahomey, before taking its current name in 1975. The official capital is Porto-Novo; Cotonou being the economic capital. The country (called at a time the "Latin Quarter of Africa") has French as official language and its currency is CFA franc. Benin has presidential form of government; the present president is Yayi Boni who succeeded Mathieu Kérékou after the elections of March 19, 2006. Since the end of the Marxist-Leninist regime in 1989, Benin has a very strong image of a democratic country in sub-Saharan Africa; and is indeed one of the pioneers of the African multiparty system.

Climate

The climate is tropical, hot and wet with seasonal and geographical variations. The mean annual rainfall varies from 900 to 1300 millimetres; the wet areas are in the south-eastern angle, of Cotonou in Oporto Novo, Atakora between Natitingou and Djougou, the areas of Dassa and Ndali in the north of Parakou. The maxima of precipitations are in the south (equatorial climate), during mid-March to mid-July, and slightly less in November-December. Moisture laden air blows from April to November in the south-west due to the effect of monsoon and vicinity to ocean. The continental and dry air, coming from north- the Sahel, blows in the opposite direction of monsoon from November to May, bringing an orange colour dust. Water the content, always significant, ranges between 65 and 95 per cent. The average temperature ranges between 22 and 34 °C. Summer lasts for about six months but April and May are the hottest months before monsoon brings some relief.

Geography

Geographically, Benin is spread between the Niger River in north and the coastal plain in the south. Sagboroa is the highest mountain of 658 meters. The north of the country mainly consists of savannas and semi-arid mountains while low coastal plain strewn with marshes, lakes and lagoons for example the lake Nokoué or the lagoon of Porto Novo prevail in the south. The major part of the population lives in the southernmost coastal plains, where the largest cities of Benin are situated *e.g.* Porto Novo and Cotonou. The country being located in an inter-tropical zone, its climate is hot and wet, but with relatively low rainfall during April to July and September to November. The North of the country (area bordering with Niger) is constantly under the threat of advancing desertification from the Sahel.

Short Description of Beninese Agriculture

Agriculture has always been the most significant sector of the Benenese economy, with approximately 40 per cent of the PIB. Benin has two principal agricultural areas. The south and southern central areas, which cover 20 per cent of the country, are suitable from the agricultural point of view because of favourable climate and availability of labour. Here, the climate allows two crops per annum. Corn, oil palm, manioc, yam, groundnut and vegetables are the principal crops. In the northern Benin (it is advisable to distinguish here the North-East, department of Borgou and the North-West, department of Atacora) only 10 per cent of the lands are cultivated. The principal crops are cotton, corn, yam and groundnut. Cotton, the oil palm and to a lesser extent, groundnut, coffee and tobacco, constitute the crops of traditional export of Benin. The production of cotton accounts for 80 per cent of the receipts of the exploitations and more than 24 per cent of the receipts of state. The oil palm tree was for a long time the principal crop of export of Benin but since ten years, there is a considerable fall of production of the palm oil and this situation brought the State to a rehabilitation of the plantations which involves experiments on 20.000 ha of selected palm trees. The production area of rice covers a approximately 320.000 ha, including 200.000 ha of hollows distributed on the extent of the own territory with a strong concentration in the centre and the North of the country. It is possible to obtain a number of products from manioc such as the flour, tapioca, starch, waste and the food of cattle. The corn is one of the agricultural produce which constitutes the base of the food. The pineapple is on the way to become a crop of export. The production of banana has been approximately 13.000 tons per annum for a few years. The cashew nut becomes a significant source of income, behind cotton. It is cultivated in the North and the centre of Benin on a total area of more than 5000 ha. In 1997, exports rose to 20.000 tons against 15.000 in 1995.

The Market–Gardening

Benin has a coastline of more than 120 km length. In the coastal cities like Ouidah, Porto-Novo and especially Cotonou, water is saline. In its cities, because of the lack of ground for the food crops, large populations rely on market-gardening activities on dry empty spaces. They use brackish water salted for watering the plants. The market-gardening production is not significant to meet the needs for the national market. In the interior of the country in plains, hills and mountains, the market gardeners use the water of the wells or the rain which is less saline than sea water. However, it has been noted that the produce of these coastal localities are prone to diseases and attack of insects. There is a need to educate these people in proper methods of cultivation and ways to avoid or reduce losses due to diseases.

17

Biosaline Agriculture in Pakistan

M. Ajmal Khan
Department of Botany, University of Karachi – 75270, Pakistan
E-mail: ajmalk@super.net.pk, khanm74@yahoo.com

ABSTRACT

Fresh water resources are under severe pressure along with the agricultural irrigation system which suffers from a steady increase in salinity. In order to maintain good supply of water for human consumption and for agriculture, efforts should be made to find alternate source of water and utilization of saline land for economic benefits. A number of halophytes could be used as cash crop (forage, fodder, fuel, medicine, chemicals, ornamentals etc). Pakistan spans a distance of 1,600 kilometers from the Arabian Sea to plains and prairies, the playas of temperate northern mountains across deserts, covering an area of 800,000 square kilometers. The varied climatic conditions have resulted in a rich diversity of flora including halophytes. Compared to the total 2500 species reported worldwide, Pakistan alone has about 410 halophytes and 178 of them have not been reported before. The efforts made to combat salinity and utilization of halophytes as cash-crop in Pakistan would be discussed.

Introduction

The rapidly increasing human population in the arid and sub-arid regions of the world has tremendously increased the pressure on the availability of good quality water for human consumption, industry and for agriculture. In addition, defective irrigation practices have increased the level of under ground water and large areas have become water logged which eventually results into higher salinities of the soil

(Yensen, 2006). It is becoming increasingly difficult to grow any conventional crop in hyper saline soils and therefore, it is imperative to look for other solutions. In this paper I will discuss in reference to the Pakistan about the extent of degradation due to salinity, various approaches adopted to address salinity problems and efforts to introduce cash crop halophytes to sustained management of saline areas.

Pakistan covers an area of about 796,100 million hectares, stretching about 1600 kilometers from north to south and about 900 kilometer from east to west (Agricultural Statistics of Pakistan, 2004). The country can be divided into seven major landscape units:

1. Coast
2 Balochistan Plains
3. Indus Plains
4. Potwar Plateau
5. Deserts including Thal, Thar and Cholistan
6. Arid and Semi-arid Mountains including the northern Balochistan, Sulemania range, Waziristan, Kurram agency, Gilgit, and Chitral (Hindukush)
7. Moist Mountains including the Western Himalayas, Swat, Kashmir, Murree and Kaghan (Khan, 2003)

The Indus is a major river, which passes through Pakistan with an approximate annual water flow of 115 billion cubic meters (Ahmed and Chaudhary, 1988). It originates from the Tibetan plateau at an altitude of about 5500 meters and flows south to the Arabian Sea. In addition other major tributaries, the Chenab, the Jehlum, the Sutlej, the Beas, and the Ravi join the Indus at the upper Indus plain.

Pakistan is primarily arid and semiarid, except for a narrow belt in the north, with low and variable rainfall. Annual precipitation ranges from 1500 mm on the southern step of the Himalayas to less than 100 millimeters on the south western Balochistan. About 69 per cent of the country receives rainfall of less than 250 millimeters per year. Most of the rains take place during the monsoons (June–September). However, the southwestern Balochistan receives winter rain with a Mediterranean trend, and some northwestern areas have both winter and summer rains. The areas which receive more than 500 millimeters of rains are about 9 per cent of the country's area. Pan evaporation ranges from 2800 mm in Sindh to 1300 mm in Punjab which is lowest in winter and highest in summer. The rainfall in most part of the Pakistan is not sufficient and often erratic; therefore, about 75 per cent of the agricultural land is irrigated through canals.

Status of Salinization in Pakistan

Surface and ground water in agricultural areas in Pakistan is rapidly becoming saline. Furthermore, salt deserts (caused by a lack of fresh water) and saline inland basins (caused by the rising level of saline ground water as a result of leakage of drainage water) are being created. About 6.2 out of 23 million hectares of cultivable lands are affected by salinization where 2.8 million hectares have some kind of cultivation while 3.4 million hectares could not sustain any conventional crops (Agriculture Statistics of Pakistan 2004). Salinization is caused by either natural cause (Primary salinity) or due to human activities (Secondary salinity). Primary salinity is a minor player in creating salinity problems in Pakistan. Modern irrigation system is a major cause of the secondary salinities in Pakistan through:

1. Seepage from irrigation canals
2. High salt concentrations in irrigation water
3. Insufficient leaching of salt, and
4. Type of salts in the irrigation water.

The soil salinity increase thus results into the following type of soils. Slightly saline and sodic (0.2 MH), moderately saline (2.3 MH), highly saline (1.5 MH), and very highly saline (2.0 MH).

Approaches to Manage Saline, Sodic and Water-logged Soil

There are several strategies involved in the management of the saline soils like engineering approach, reclamation approach and Bio-saline agriculture approach (Qureshi and Barrette-Lennard, 1998). In Pakistan all these approaches were practiced, with varied results.

Engineering Approach

Where logic behind was that the salinization of the soil could be reversed by various kinds of drainages:

1. Proximity of irrigation canals results in seepage of about 0.21 million cubic meters of waters per day. This increase in water table ultimately causes soil salinization. Three major techniques were used in Pakistan:
   (*a*) Surface drainage
   (*b*) Lining of canals, and
   (*c*) Pumping of ground water known as SCARP (Salinity Control and Reclamation Projects).

   They helped in pumping out ground water work however, the disposal of this water of usually bad quality, posed serious questions on the sustainability of the program.
2. With the help of USDA a project was initiated in 1957 to install tube wells for pumping underground water.
3. Left bank outfall drain.

A canal was dug in the lower Indus valley in order to dump saline waters collected from tube wells or by shallower surface drains to 290 kilometers away in the sea. This project had a cost 26 billions Pakistani rupees and expected to dump 25-30 million tones of salt into the sea each year. This project has questionable success record. It was a complete failure in sodic soils which allows little penetration, caused destruction of good quality soil, advantage gained in some areas were short lived and expensive.

Reclamation Approach

This includes small interventions at a local scale to leach out the surface salts. Tactics used were: (*a*) leaching with higher level of irrigation: It is least cost effective, inefficient, and may convert saline soil into sodic soil. (*b*) Use of gypsum or acids: Generally used in sodic soil is cost effective, efficient, with rapid response. The gypsum is getting more expensive and it requires extra irrigation water for washing and soils rapidly goes back to the saline state in few years time. Acid are effective but expensive and could leach important nutrients.

Bio-saline Agriculture

This approach was to utilize saline soils by growing genetically manipulated salt tolerant crops. A number of conventional method using radiation, breeding etc. were used widely to introduce salt tolerant cultivars of wheat, barley, rice, cotton, mustard and other crops. The salt tolerant varieties were able to produce better crop in the soil with very low salinity, therefore providing relative better production in comparison to their salt sensitive cousins. However, the soil salinity of about 3.4 million hectare of agricultural land is in the range that is beyond the capacity of any conventional salt tolerant crop plants (Agricultural Statistics of Pakistan, 2004).

Utilization of Halophyte as Non-conventional Cash-crops

Given the submissions made above we are left with little choices but to seek alternate methods to ease the pressure on the requirement of good quality water and the utilization of the land degraded by high soil salinity. There are growing indications that cultivation of crops with a high salt tolerance can be realized as an economically feasible option for utilizing saline soils and conserving fresh water (Boyko, 1964; Glenn *et al.*, 1998; Aronson, 1991; Lieth, 1994). Potential halophytic crops could be broadly grouped into three categories:

1. Plants with a high salt tolerance: they grow in water with salt contents equal to or even higher than that of seawater.
2. Crops with an average salt tolerance: they grow in brackish water and
3. Crops with a moderate salt tolerance: They grow in slightly brackish water that is not suitable for conventional agriculture.

Khan and Qaiser (2006) reported that 410 halophytic species are distributed in Pakistan. Considering that about 2500 halophytes are reported for the world (Aronson, 1985, 1989; Menzel and Lieth, 1999) 410 in Pakistan indicate a higher diversity as well as great opportunity to find cash crop halophytes (Khan, 2003).

Utilization of halophytes in Pakistan is not a new concept. Groups in various parts of Pakistan have been trying to find cash-crop halophytes with limited success. Research groups from NWFP Agricultural University, Forest Research Institute from Peshawar, Soil Salinity Research Institute, Pindi Bhattian, International Waterlogging and Salinity Research Institute, Lahore; Nuclear Institute for Agriculture and Biology, University of Agriculture, Faisalabad, Nuclear Institute for Agriculture, Tandojam and Department of Botany, University of Karachi have been active in exploring the potential of various plant species and conducting applied ecological/

ecophysiological studies in this context. The research work done in these laboratories varies from the utilization of salt tolerant conventional crops, non-conventional forage and fodder plants including grasses, shrubs and trees.

Research on salt tolerant trees was conducted in different parts of Pakistan with the help of Council of Scientific and Industrial Research Organization (CSIRO), Australia. Many species and provenances of Australian salt tolerant *Acacia, Casuarina* and *Eucalyptus* species were cultivated in saline areas with few local control species. It was found that *Acacia ampliceps, Casuarina equisetifolia* and *Eucalyptus occidentalis* were more salt tolerant and relatively better adapted to various ecological systems in Pakistan (Marcar *et al.*, 2003). *Prosopis juliflora* was also grown successfully in Bhawani, near coast of Balochistan and it was argued it could be utilized as source of fuel wood (Khan *et al.*, 1986). In a parallel project CSIRO also provided funding to grow several Australian *Atriplex spp.* at various locations in Pakistan. *Atriplex amnicola, A. lentiformis* and *A. nummularia* were relatively more salinity tolerant and produced higher biomass in comparison to other *Atriplex spp.* studied (Davidson and Gallaway, 1993).

Kallar grass (*Deplachne fusca*) is big success story for waterlogged saline soils of central Punjab. Malik *et al.* (1976) have done a series of experiments and successfully demonstrated that the species could be grown in highly saline waterlogged soil and arguably used as fodder.

Halophyte Biology Laboratory in the University of Karachi is also involved in developing the technology to provide complete feed for the live stock production using brackish water irrigation (Khan and Ungar, 1995; Khan and Weber, 2006; Khan *et al.*, 2006; Ozturk *et al.*, 2006). We have successfully demonstrated that high quality fodder grass could be raised by controlled brackish water irrigation. In addition we are using leaf succulent salt sucking halophytes to remove the salt from the system. Succulent halophytes are harvested periodically and burned to produce soda which is utilized in soap industry. Through managing the use of brackish water using drip irrigation and succulent species to achieve salt balance in the soil we believe that we have created a sustainable system for green fodder production. We have learned that ration for the animals require dry fodder more than the green fodder. There are few grass candidates available in the area that could produce high biomass with brackish water irrigation. Experiments are under progress to produce a large amount of grass to use as dry fodder. We know that their nutrition level is quite high. We also need to provide some high sugar and protein source to make a balance meal. We are investigating the possibility to grow fruit trees with brackish water irrigation and use them as sugar source. At the same time we are working with an industry and their involvement forces us to think about the economic viability of the project. The process of taking laboratory research to mass scale profitable industrial production is long, arduous and painful. However, one has to go through this patiently. We believe that most of the failure lies in crossing this bridge.

References

Ahmed, N. and G.R. Chaudhary. 1988. Irrigated Agriculture of Pakistan. Shahzad Nazir Publications, Lahore. 810 pp.

Aronson, J. 1989. HALOPH; Salt Tolerant Plants for the World A Computerized Global Data Base of Halophytes with Emphasis on their Economic Uses. University of Arizona Press. Tucson, USA.

Aronson, J. 1991. The present and future roles of salt tolerant trees and their associated microsymbionts in arid lands. In: Choukr Allah, R. (ed.), Plant Salinity Research New Challenges. pp. 363 376. Proc. Intnl. Conf. Agril. Mgt. Salt Affected Areas, Agadir, Morocco.

Aronson, J.A. 1985. Economic halophytes - a global review. In: Wickens, G.E., Goodin, J.R. and Field, D.V. (eds.), Plants for arid lands, pp. 177-188. George Allen and Unwin, London.

Boyko, H. 1964. Principles and experiments regarding irrigation with highly saline and seawater without desalinization. In: Transactions of the New York Academy of Sciences Series II /26. pp. 1087-1102. New York Academy of Sciences.

Davidson, N. and R. Gallaway. 1993. Productive use of saline land. Proceeding of a Workshop held at Perth, Western Australia, 10–14 May 1991. ACIAR Proceedings No. 42, 124.

Glenn, E.P., R. Tanner, M. Miyamoto, K. Fitzsimmons and J. Boyer. 1998. Water use productivity and forage quality of the halophyte Atriplex nummularia grown on saline waste water in a desert environment. Journal of Arid Environment 38: 45-62.

Khan, D., R. Ahmad and S. Ismail. 1986. Case history of Prosopis juliflora plantation at Makran coast raised through saline water irrigation. In: R. Ahmad and A. San Pietro (eds.), Prospect for Biosaline Research, Proceeding of US-Pak Bio-saline Research Workshop. Department of Botany, University of Karachi.

Khan, M. A. 2003. Halophytes of Pakistan: Distribution and Ecology. In H. Lieth and M. Moschenko. Cash crop halophytes: Recent Studies: 10 years after the Al-Ain meeting (Tasks for Vegetation Science, 38). 167-188. Kluwer Academic Press, Netherlands.

Khan, M. A. and M. Qaiser. 2006. Halophytes of Pakistan: Distribution, Ecology, and Economic Importance In: M. A. Khan, H. Barth. G. C. Kust and B. Boer (eds.), Sabkha Ecosystems: Volume II: The South and Central Asian Countries. pp. 129-154. Kluwer Academic Press, Netherlands.

Khan, M.A. and Ungar, I.A. 1995. Biology of Salt Tolerant Plants. Department of Botany, University of Karachi, Pakistan. 424 pp.

Khan, M. A. and D. J. Weber. 2006. Ecophysiology of High Salinity Tolerant Plants. Springer, Netherlands.

Khan, M. A., Barth, H., Kust, G. C. and Boer, B. 2006. Sabkha Ecosystems: Volume II: The South and Central Asian Countries. Springer, Netherlands.

Malik, K.A., Aslam, Z. and Naqvi, M. 1986. Kallar grass: a plant for saline land. Nuclear Institute for Agriculture and Biology, Faisalabad, Pakistan, 93 pp.

Marcar, N.E., Ansari, R., Khanzada, A.N., Khan, M.A. and Crawford, D.F. 2003. Performance of several tree species on a saline site in southern Pakistan Journal of Tropical Forestry Science 15: 457-468.

Menzel, U. and Lieth, H. 1999. Halophyte Database Vers. 2.0. Halophyte Uses in different climates I: Ecological and Ecophysiological Studies. Progress in Biometeriology, Vol. 13. pp. 77-88. In: H. Lieth, M. Moschenko, M. Lohman, H.–W. Koyro and A. Hamdy (eds.), Backhuys Publishers, The Netherlands.

Ozturk, M., Waisel, Y., Khan, M.A. and Gork, G. 2006. Biosaline Agriculture and Salinity Tolerance in Plants. Birkhuser, Germany.

Qureshi, R.H. and Barrett-Lennard, *E.G.* 1998. Saline agriculture for irrigated land in Pakistan: A handbook. Australian Centre for International Agricultural Research Canberra, Australia.

Yensen, N.P. 2006. Halophyte uses for the twenty-first century. In: M. A. Khan, and D. J. Weber (eds.), Eco-physiology of High Salinity Tolerant Plants. pp. 367-396. Springer, Netherlands.

18

Crop and Forage Production Using Saline Waters in Dry Areas

L.Y. Win

Biotechnological Department, Government Technical College, Kyaukse, Department of Technical & Vocational Education, Yangon, Myanmar

E-mail: blakitan@ristek.go.id, blakitan@yahoo.com.sg

ABSTRACT

According to land stack of Myanmar, total cultivable land is 28.06 million acres. Out of which problem soil occupies only about 7.8 per cent (2.371 million acres) of the total cultivable land. Again out of that problem soil area 69.17 per cent (1640 million acres) is occupied by the saline and alkaline soils. Saline and alkaline soils do exists in the agriculturally important agroecological zones, namely, R_3S_1, R_3S_5 and R_5S_5. Most of the saline and alkaline soils are currently under crop production. Therefore, the soil needs prior attention.

The main soil types of saline soils are Meadow Gley (Gleyic Cambisol) and Meadow (Eutric Cambisol). On this soil rice is grown intensively. Alkaline soils are Dark Compact (Vertisol) and Brown Compact (Chromic Vertisol) which is agriculturally important for rice, cotton wheat, pulses and oil seed crops production.

To define more accurately the extent of saline and alkaline soils 1) the delineation of saline and alkaline soil area 2) the amount of crop land affected 3) the potential for future development of saline and alkaline soil 4) the identification of land that could potentially become saline or alkaline- are taking into account.

Saline and alkaline soils have actual and potential uses suited to existing soil condition other than crop production. It is necessary to develop the alternative land use options for particular environmental units. To combat saline and alkaline problem, various possible solutions are available. Drainage–leaching combination approach is expensive and technically difficult for most of the developing country. Next to irrigation–leaching combination approach, another possible way is biological approach. Biological approach improves soil structure, its permeability, its biological activities, fertility and adds biomass and lower soil pH. In addition to simplicity and inexpensive in implementation of the approach, the affected individuals can take initiative and can get return in a relatively short period of time.

In order to effective implementation of the biological approach: (*i*) Introduce the salt tolerant plants through international organization (IAEA, FAO, UNDP, NIAB-Pakistan, etc. (*ii*) Invite technical experts on salinity and alkalinity. (*iii*) Promote technical exchange through TCDC program-are necessary. Currently saline water is not used for irrigation of crops in central arid zone of Myanmar. Only river water lifted from a major river, Ayeyawady and its tributaries, or saline groundwater are used on salt tolerant crop species. In addition, our government is implementing drilling of artisan wells for local use in the arid zone.

Many native plants exist in the central arid zone, such as aloe, physic plant, toddy palm and neem. Aloe vera is not cultivated but is of medicinal value. Physic nut plant is now widely and commercially planted for extraction of oil for use as biodiesel. Toddy palm trees are numerous but wild and are exploited commercially for sugar and jaggery extraction from its juice and for roofing using its leaves. Neem tree are also wild populations and it has many use such as medicine and pesticide preparations. Tissue culture and genomic studies projects are planned. As of 2006, we have collaboration with international organizations such as FAO, UNDP, UNESCO and USDA for greening of the central arid and education of its local populace.

Introduction

Myanmar is one of the developing countries in the South-East Asia region richly endowed with natural resources. The overall development of national economy is highly reliant in Agriculture. Agriculture plays primary role for self-sufficiency in national food requirements, for adequate production of raw material for local agro-based industries and for the generation of substantial surplus for export. Since the economy of the country shifted from centrally planned reforms, several achievements have been accomplished in agriculture, however, there are potential in agricultural development of the country.

Location

Myanmar is geographically located between 9" 50° and 28 32′ North latitude and 92"10° and 101"11° East longitude. It is bounded on the north by People's Republic of China, on the east by Thailand and Laos, on the west by Bangladesh and India and on the south by Bay of Bengal. The total area of the country is about 261228 square miles.

Topography

Myanmar is a forest-clad mountainous country. General slope of the country is from north to south, and all rivers flow in the same direction. Its altitude varies from 3000 to 7000 feet.

Myanmar can be divided into different natural regions based on the different topography, land forms, climate, natural vegetation and agriculture. They are:

1. Coastal strip
2. Western mountain region
3. Northern hilly region
4. Central Myanmar dry zone region
5. Ayeyarwady delta
6. Bago Yoma
7. Sittang river valley
8. Shan plateau
9. Tanintharyi coastal strip

Though country is classified into nine natural regions as mentioned above, it is also divided into seven states and seven divisions based on the administrative convenience.

Climate

There are three well defined seasons. They are rainy season (from May to October) winter season (from November to February) and summer season (from March to April). In the Central Myanmar plains seasonal temperature variation in the magnitude of 40.6-43.3 °C in hot season and 10.0–15.6 °C in cold season are common. In Chin hills and Shan Plateau, on account of higher altitude, maximum temperature will not exceed 29-4 °C and the minimum temperature will be as low as 7.2 °C. The mean maximum temperature ranges from 30.8 °C to 41.8 °C and mean minimum temperature from 0.3 to 13.4 °C. (Yegong *et al.*, 1978).

Rainfall

Myanmar receives its annual rains mainly from the south west monsoon from mid May to mid October. The rainfall, however, varies depending on the locality, elevation and months of the year. July and August are generally regarded as rainy months of the year. The heaviest annual rainfall varying from 2500-5000 mm are recorded in the Rakhine, parts of Kachine, Kayin states and Tanintharì coastal regions including part of Mon State. In the Central dry zone the annual rainfall is below 1000 mm.

Agro-ecological Zone Classification

Rainfall pattern of Myanmar can be classified into three groups based on the system employed by Dr. C.R. Panabokke. R5: It has a moderate to high degree of variability between years and as well as within the main cropping season but rainfall usually exceeds total evaporation during the wet season and is followed by a well expressed dry season at least three consecutive dry months. The annual rainfall pattern is bimodal. This type of rainfall can be found in the Central Myanmar dry zone regions of Mandalay, Magwe and southern parts of Sagaing divisions.

Using the above mentioned soil and rainfall groups, Nyi Nyi classified Myanmar into 10 agroecological zones, these along with location, description, agricultural importance, major crops and constraints are expressed in table. Only four out of ten agroecological zones, namely, $R_3 S_1$, $R_4 S_1$, $R_5 S_4$ and $R_5 S_5$ are classified as very important agriculturally. The area covered by these four zones is only 24 per cent of the total area of Myanmar. Out of ten agroecological zones, salinity and alkalinity problem soils are located in 3 agroecological zones, two in agriculturally important zones, R_3 S_1, and R_5 S_5 and one in R_3 S_5 zone.R_3 S_1, agroeological zone is situated along the coastal regions and Ayeyarwady deltas. R_5 S_5 and R_3 S_5 zones represent the Central Myanmar dry zone and Taninthari coastal strips respectively.

Dry Zone Project

UNDP assisted project under the name of Agricultural Development and Environmental Rehabilitation in the Dry Zone is also being implemented. The project recommended preventive measures to soil losses, water losses and fertility losses, by protecting the land against erosion, reducing run-off and increasing infiltration and stopping nutrient losses due to erosion. Specific targets are set to demonstrate soil and water conservation techniques such as soil bunds, gully plugs, check dam, weirs, ponds, tube wells, pumping wind breaks, woodlots etc.

Problem Soils of Myanmar

Problem soils are characterized by soil, agroclimatic constraints to sustainable agricultural production which limit the range of crops that can be grown successfully.

Table 18.1

Sl.No.	*Type of Problem Soils*	*Area Acre*	*Per cent of the Total*
1	Salinity	1510	63.69
2	Alkalinity	130	5.48
3	Acid sulfate	20	0.84
4	Hyper acid	20	0.84
5	Degraded	100	4.22
6	Peat	1	0.40
7	Swampy	600	25.31
	Total	2371	

Source: Land resource in Myanmar for agriculture purpose (Hla Aye, 1990).

About 69.17 per cent (1.610 million acres) is occupied by the saline and alkaline soils, most of which are currently cultivable. Therefore, these soils need prior attention.

Table 18.2: Location of Salt-affected Agroecological Zones and its Agricultural Importance

Problem	*Affected Agroecological Zone*	*Location*	*Agricultural Importance of Zones*
Saline	R_3S_1	Rakhine coastal strip	Very important
		Taninthari coastal strip Ayeyarwady delta	
	R_3S_5	Taninthari coastal strip	Important
	R_5S_5	Central dry zone *	Very important
Alkaline	R_5S_5	Central dry zone *	Very important

Causes: A soil becomes saline when influx of salt is greater than efflux. In coastal areas, salinity is due to direct inundation of salt water or upward or lateral movement of saline groundwater.

Saline soils very widely in their chemical, physical and hydrological properties. Important variables include (Table 18.3).

Table 18.3

1.	Land	(land form, physiography and land use)
2.	Salt	(source, content and distribution–lateral and vertical)
3.	Soil properties	(pH, OM content, fertility status and nature of clay)
4.	Hydrology	(amount and distribution of rainfall, rates of seepage and percolation, depth of water table and drainage density)
5.	Management	(land preparation, crop establishment, and care) practices
6.	Season	(dry season and wet season)

Saline soil area may be subjected to influence by:

1. Rivers and creeks
2. The upward movement of saline water

In the Central Myanmar dry zone area, crop lands are irrigated with saline water or affected by the evapotranspiration of salty groundwater which is rather high in some area.

Alkaline soils are mostly restricted to the arid and semi-arid area which is situated in Sagaing, Magwe and Mandalay divisions

Saline and Alkaline Soils: Present Use

Arid and Semi-arid Region (R_5S_5)

Main soil types of the saline area of this region are Brown Compact (Chromic Verticals) and Dark Compact (Pellic Verticals). These soils occupy the flat and level areas of the rolling and undulating topography of the regions bounded the Red Brown (Chromic and Ferric Luvisols) soils on the gentle slopes.

These soils are the most agriculturally important soils of the region next to the Luvisols. They can be sown with varieties of crops such as rice, cotton, sesame, chilli and pulses, oil crops etc. and can be irrigated. Irrigation, however, is one of the main causes of salinity and alkalinity of the soils.

The major cropping patterns in this area were as follow (Table 18.4).

Table 18.4

Agroecological Zone	*States/Divisions*	*Cropping Patterns*	*Cropping Intensity*
R_5S_5	Sagaing	Rice-Rice	140
	Mandalay	Sesame-Rice	
		Rice-Vegetable	
		Groundnut- Sesame	
		Cotton-Pulses	
		Cotton-Sesame	
		Maize-Sesame	
		Sesame-Pulses	
		Sesame-Wheat	

Central Dry Zone

Districts Greening Program

A special program of the Government is being implemented for greening 9 districts in central dry zone of the country. This area contributes 12 per cent by area and 20 per cent by population to the whole country. The area was rather heading to its severe climatic conditions. It is a multi-sectoral integrated project involving several sectors.

Table 18.5

Sl.No.	*District*	*Land Area*
1.	Sagaing	340
2.	Monywa	1004
3.	Myingyan	698
4.	Meiktila	579
5.	Yamethin	1088
6.	Pakokku	1388
7.	Magway	963
8.	Minbu	932
9.	Thayet	6992
	Total	13989

The situation is now improved with the provision of irrigation water; proper farming techniques, soil conservation and planting quick growing trees and also with the special project for the use of alternate fuel for wood. Being the major oilseed crops growing area, a special project for increased production of oilseed crops is also being implemented in this region with the provision of high lift water pumps along the rivers, Ayeyarwady, Chindwin, Myitnge, etc., in the region. This project would also result into the greening of the area as well.

Saline and Alkaline Soils: Alternative Use

Saline and alkaline soil has actual and potential uses suited to existing soil conditions other than crop production. Land uses such as mangrove production, nipa palm leaf for roofing, brackish water fish culture, salt making and local soap making–all formed an important part of local economics and social preferences.

Reclamation of Saline and Alkaline Soil

There are two approaches in reclamation of saline and alkaline soils. They are:

Drainage and Leaching Combination Approach

The solution to the problem of salinity and alkalinity has started with the provision of groundwater drainage. However, this method is not easy. There are a number of engineering and other difficulties that make drainage difficult. Therefore, this approach is expensive and difficult to implement. The groundwater is to be drained into the sea or nearby disposal site which in turn raises secondary salination problem. Also, there may be hardly any other shallow area where the drained saline water could be stored.

This may be a bit easier in area where the lands to the drained are not too far from the disposal site and gradient is available for gently flow.

If drainage canals are not concrete lined, some of the drainage water seep back into the surrounding aquifer. So, the construction of the drainage system and disposal of the drained water are formidable problems. After the draining of groundwater the surface salt are leached down through application of water. This application must be properly managed. In case of alkaline soil, application of water for leaching is ineffective. The soil should be amended with gypsum and organic materials to restore structure and permeability that the dissolved salt be leached down. Soil amendments are very expensive now.

The drainage is to be provided over a large area and can be done only by the government or by large cooperatives. If affected population comprise poor farmers, collective action of that magnitude is almost impossible. The affected individuals and communities, therefore, have to suffer until some thing is done by the government.

Drainage–leaching combination is not only prohibitively expensive and technically difficult for most of the developing country. It is therefore, necessary that cheaper and quicker alternatives must be looked into even if provision of drainage may be the ultimate answer.

Biological Approach

A possible approach could be to tailor plants to suit the environment. This approach is known as biological approach.

Biological approach is essentially based on the rationale that the soil is not just a mass of dead chemicals but is a living system undergoing numerous chemical and biological reactions and is in constant interaction with numerous environmental factors. The alkaline soil has excess of sodium and has little or no organic matter. It is impermeable and is biologically almost dead. Introduction of salt tolerant crop will provide a green cover and will improve the environment of fertility. The penetrating root will provide crevices for downward movement of water and thus help leaching of salt from the surface. Further, the biomass produced could also be used as green manure which lowers pH. The soil structure, its permeability, its biological activities and fertility could be restored. With extra irrigation, the surface salt could be leached down.

In Myanmar, in order to effectively implement the biological approach, work has been carried out on the following aspects:

1. Introducing of salt tolerant plants through International organization (IAEA UNDP, FAO, NIAB, Pakistan etc.)
2. Screening of these plants in the prevailing conditions in different location of salt-affected areas.
3. Recommendation of cost-effective location specific technology
4. Training on biological approach for staff of Myanmar Agriculture Service.

References

Aye, H. 1987. Land resource in Burma for agriculture purpose.

Oo, T.T. 1992. Problem soils: Country report, union of Myanmar.

Malik, K.A. Crop production in salt-affected soils: A biological approach.

RAPA REPORT. (1988/1). The Management of Saline/Sodic soils.

SECID: Annex 1, Soils and agro-climatic condition (1988).

Singh, V.P., H.U. Neue and M. Akbar. Coastal saline soils for rice cultivation.

Win, K. and N. Nyi. 1979. Rainfed farming system and practices in relation to agro-ecological zones (Burma).

19

Crop Production in Salinity Affected Areas in Sri Lanka

A.A.Y. Amarasinghe

Faculty of Agricultural Sciences, Sabaragamuwa University of Sri Lanka, Belihuloya – 70140, Sri Lanka

E-mail: amre3@sab.ac.lk, amare3@hotmail.com, amare3@163.com

ABSTRACT

Although high degree of salinity is not a serious issue in Sri Lanka, prevalent and growing low to moderate salinity directly and indirectly affects country's agriculture. Rice is the commonly affected crop by salinity. In dry areas including arid zones, high evaporation causes the rising up and deposition of salts in the topsoil layers. Massive irrigation projects like Mahaweli river irrigation system develop salinity in most of the North Central areas in Dry zone due to improper management of irrigation water and drainage systems. Coastal salinity is experiencing in the areas near to the sea. Coastal areas in Hambantota and Puttlum districts show high salinity problems. Salinity in coastal areas in Galle, Matara, Hambantota, Ampara and Mulathivu districts has been increased severely after the tsunami disaster. Ampara district, which records higher production of rice, was hardly hit by massive waves and high salinity due to tsunami.

To mitigate salinity in Mahaweli areas, improvement of drainage and prevention of excessive irrigation are suggested as prompt measures. In addition, introducing agronomic practices such as incorporation of organic matter, growing salt tolerant cover crops, mulching, ploughing lands to break hard pans and proper management of soil fertility can reduce the development of salinity.Micro irrigation systems like drip irrigation is recommended to dry areas to avoid heavy removal of groundwater. Coastal salinity is not easy to mitigate, but maintaining it below a threshold level is possible by monitoring crops, soil and water. Use of salt tolerant varieties is the best way to tackle coastal salinity. Tsunami affected cropping areas are rehabilitated by flooding and flushing the lands with freshwater using soil amendments like dolomite. Further development of salt tolerant rice and other crop varieties are recognized as important tasks to confront the growing problem of salinity in Sri Lanka.

Introduction

Increasing crop production to meet the food requirements of the world's growing population will exert great pressure on global water resources. There should be plenty of fresh water for cultivation of crops and forage for increasing demand in the future. However, unfortunately, available fresh water supply may be exploited fully in the near future, leaving behind only polluted water for the use of crop production. In most of the dry areas in the world, salinization is one of the major problems in agriculture where water stress drops the production and productivity of crop plants. Hence, it is right time to introduce methods and technologies to grow crops using saline waters. Introduction of salt tolerant plants is of utmost important in this context.

Sri Lanka is an agricultural country in Asia with population of 18.5 million. The agriculture sector earns 22 per cent of foreign exchange to the country, especially from plantation crops, which cover an area of 889,000 hectares. Rice being the staple food, is grown on 0.89 million hectares. Sixty-six per cent of rice cultivation is carried out during the major season or wet season (Maha), while 34 per cent, during the minor season or dry season (Yala). Although, high degree of salinity in Sri Lanka is not frequently reported, some degree of inland and costal salinity have vigorously affected the cropping systems in Dry zone and low land coastal areas in Wet zone where most of the paddy fields are located in Sri Lanka. Salinity in Sri Lanka mainly affects rice production. Use of salt tolerant rice varieties is one of the vital solutions for such areas in Sri Lanka so that they could be grown even using saline waters without experiencing any yield drop.

Climatic and Soil Parameters in Sri Lanka

Climate: Climate in Sri Lanka is basically determined by the geographical location of the country in the equatorial belt and its position in the inter-tropical convergence zone. Wide range of climatic and soil conditions are available in Sri Lanka. Rainfall and temperature are the major determinants of climate in Sri Lanka. The annual rainfall ranges from 600 to 6,000 mm. Elevation ranges from mean sea level (MSL) to 2,575 m above MSL and the average temperature ranges from 27.5 ℃ at the MSL to 15 ℃ at the upper most elevations. Diurnal variation of temperature is obvious and the range increases with altitude and distance from the sea. Only a little variation in mean monthly temperatures can be seen throughout the year in most parts of the country. There are 3 major climatic zones, Wet, Dry and Intermediate, which have been divided according to annual rainfall in Sri Lanka. In addition to these 3 climatic zones, 2 areas in North west and South east of the island, where the annual rainfall is too low (< 600 mm) and having a prolonged and intense drought, have been identified as Arid zones in Sri Lanka (Table 19.1). The Dry zone experience a prolonged dry period from May to September with drought conditions and dry winds, prevailing from June to August. The relative humidity in the country varies from 70–95 per cent.

Table 19.1: Important Parameters Considered Dividing the Island to Zones

Zones/Country	*Parameters*		
	Rainfall (mm)	*Elevation (m from MSL*)*	*Soil Type*
Dry Zone (DZ)	< 1,500	-	Reddish brown earth Humic gley alluvials Red yellow latosols Non-calcic brown soil Sandy regosols Alkaline and saline soils Grumosols
Intermediate Zone (IZ)	1,500 - 2,500	-	Reddish brown earth Red yellow podzolic soils Non-calcic brown loam
Wet Zone (WZ)	> 2,500	-	Red yellow podzolic soils Bog, half bog soils Sandy regosols
Low Country (LC)	-	0 - 300 m	-
Mid Country (MC)	-	300 - 800	-
Up Country (UC)	-	> 800	-

*: Mean Sea Level.

Three types of rainfall, monsoonal, convectional and depressional, are available in different time periods of the year. The two monsoonal periods, the South west (May–September) and the North east (December - February) are the major rainy seasons in the country. Rainfall distribution is determined by the topography in Sri Lanka. The South west monsoon provides rain mostly to South western parts and the central highlands of the country. The North east monsoon along with inter-monsoonal depressional activity in October - November is stronger and precipitation distributes throughout the island. Less rainfall is available during the other inter-monsoonal period in March - April.

Soils: Sri Lanka possesses nine out of the ten major soil orders of USDA soil taxonomic systems. These are distributed throughout the country in a mosaic pattern. In the Dry zone, the predominant soil group is the well-drained reddish brown lands in association with poorly drained humic gley alluvials and red yellow latosols. In the Dry zone coastal areas, non-calcic brown soil with sandy regosols, alkaline and saline soils, and grumosols are distributed in patches. In the Wet zone, red yellow podzolic soils form the major soil group with bog, half bog soils, and sandy regosols along the South west coast. The Intermediate zone displays a transition from reddish brown earths to red yellow podzolic soils, with non-calcic brown loam in patches (Panabokke, 1975).

By combining the two parameters, rainfall and elevation, seven major agro-ecological zones (AEZ) have been identified (LCDZ, LCIZ, LCWZ, MCIZ, MCWZ, UCIZ and UCWZ). These AEZ were further subdivided into 24 agro-ecological regions, considering the rainfall distribution, soil type and the landform (Panabokke, 1996).

Characteristics of Saline Soils

Saline soils contain mostly soluble neutral salts like chlorides and sulphate of Ca, Mg and Na in concentration what interfere with the growth of most crops. The electrical conductivity (EC) of soil saturation extract is normally 4 dSm^{-1} or more while pH is 8.2 or less and exchangeable Na is less than 15 per cent. In dry soils, a white layer of salts is usually obvious. These soils are usually flocculated and the soils have good permeability due to the presence of excess salts. Table 19.2 illustrates the way of classifying saline soils based on the threshold level up to which plant growth may not be severely affected.

Table 19.2: Salinity in Saturated Soil Extract Based on Scale of ECe

Relative Salt Level	*EC (dSm^{-1})*	*Plant Condition*
Low	0 – 2.5	Salinity effects mostly negative
Medium	2.5 – 5.0	Very sensitive plants affected
High	5.0 – 7.5	Many plants affected
Excessive	> 7.5	Only salt tolerant plants grow

Source: USDA Information Bulletin, 194, 1994.

Because crop plants differ quite markedly in their level of salt tolerance, the effect of salinity on yield is a function of the threshold salinity above which yield declines, and the percentage of yield decrease per unit of salinity increases above the threshold (Thiruchelvam and Pathmarajah, 2000). Adverse effects on plant growth could be seen when the excess salts are present in soil. Salts make the nutrients less available to the crop plants because of osmotic pressure. Furthermore, excess salts become toxic to plants. The long-term presence of excess salts can damage the soil irreversibly (Thiruchelvam and Pathmarajah, 2000).

Salinization of Soils and Waters in Sri Lanka

Inland and coastal salinity in Sri Lanka is reported to have developed mainly due to the following reasons.

1. Dryness in the area and non-monsoonal periods
2. Excessive irrigation and poor drainage systems
3. Excess use of fertilizers
4. Natural disasters like tsunami

Dryness in the area and non-monsoonal periods: The areas, which receive the annual rainfall less than 1,500 mm is known as Dry zone in Sri Lanka. In addition, 2 arid areas receive the annual rainfall less than 600 mm. Most of these dry areas show mild or moderate salinity due to rising up and deposition of salts in upper layers of the soil, probably the areas of the root zones of most shallow rooted plants, with the evaporation of soil water. Arid and semi-arid areas in southern Sri Lanka namely Matara and Hambantota Districts, and some dry areas in Northern and North-eastern Sri Lanka experience salinity because of dryness. Long dry spells are more common in February to April and October to November during non-monsoonal periods. In which, development of salinization has been reported even in Wet and Intermediate zones in Sri Lanka.

Excessive irrigation and poor drainage systems: Salinization is the major problem in crop and livestock production, approximately in 40 per cent of the world's irrigated areas due to rising up the water table. In semi-arid areas associated with large river irrigation systems in Southeast Asia, the salinity problem both in soil and water is serious. Both wet and dry season crops are affected with the development of salt level due to excess use of irrigation water creating dramatic effects of salinity on Agriculture, leading to increase the extents of abandoned saline lands. The total land extent of irrigated lands in Dry zone of Sri Lanka is about 0.5 million hectares. In Sri Lanka, although irrigation projects like Mahaweli river irrigation project have initially been launched with the aim of improving Agricultural production, most of these projects are now facing the developing salinity problems. Unless authorities will follow proper ways of managing irrigation systems, these areas are in danger of becoming highly saline due to inadequate irrigation management.

Thiruchelvam and Pathmarajah (2000) have carried out studies to measure the impact of soil salinization on rice production and environment in Mahaweli River System H Irrigation scheme, the major irrigation system in Sri Lanka associated with Mahaweli Development Project. The Mahaweli System H is located in Dry zone of Sri Lanka and contributes 25 per cent, 30 per cent and 20 per cent of the national production of rice, chilies and big onion, respectively. Nearly 28,000 families are dependent on agriculture in the area. They have conducted their studies in the paddy lands of Nochchiyagama and Madatugama. Nochchiyagama area is located far end of the irrigation scheme, while Madatugama is an area closest to the main reservoir called Kalawewa tank. Investigators in 2000 reported that crops and land in Madatugama and Nochchiyagama are damaged significantly by 10 per cent and 25 per cent respectively due to salinity. They have also found that large numbers of farmers in the affected areas cannot cultivate their land fully, while many are suffering low productivity as the salinization increases with irrigation water.

The centrally placed mountain mass encircled by coastal plains provides for radial pattern of surface drainage to all rivers except the river Mahaweli. Drainage is usually associated with irrigation. When managing large irrigation schemes, better drainage systems must be adopted. Otherwise, salts accumulated in the top layers will not be washed out. Most of the farming areas in Dry zone of Sri Lanka have not launched proper drainage systems even though farmers used to irrigate excessive amount of water to their fields. Further, continuous application of fertilizers increases the salinity due to poor drainage facilities in these areas.

Excess use of fertilizers: Since Ca, Mg, Na, K, Cl, S, C, N and B are found in saline soils or waters, they are considered as the major elements causing salinity. Therefore, excessive and continuous use of most of the inorganic fertilizers such as Urea, Ammonium Sulphate, Muriate of Potash, Triple Supper Phosphate and Rock Phosphate, without considering proper drainage facilities, especially in rice growing areas in Dry zone develops the salinity. The danger again is the development of salinity over time, changing the degree of salinity from moderate to high in such areas.

Natural Disasters Like Tsunami: Salinity problems in Sri Lanka are basically associated with coastal areas. Coastal salinity and salinity near coastal areas are caused mainly by seawater, which contaminates the fresh water resources. Higher tides or flooding of seawater back to the inland through the rivers and tsunami disasters develop this salinity. Seawater contains considerable quantities of sodium salts so that its intrusion would create soil salinity.

The tsunami disaster, took place on 26th December 2004, affected the most of coastal lands (Figure 19.1) by seawater intrusion, which caused to develop salinity. The disaster also damaged the present crop and made these lands unsuitable for cultivation due to the accumulation of soluble salts in soils in these areas. Due to the tsunami, extensive damages to croplands have been occurred in North-eastern area, especially in Ampara District in Sri Lanka. Not only rice, but also tomato, chillies, cowpea and brinjals grown in highland areas of Vinaygapurum in Thrukkovil have also been damaged with the intrusion of seawater. There are few areas identified as highly saline areas where much cannot be done to neutralize the effect of soil salinity.

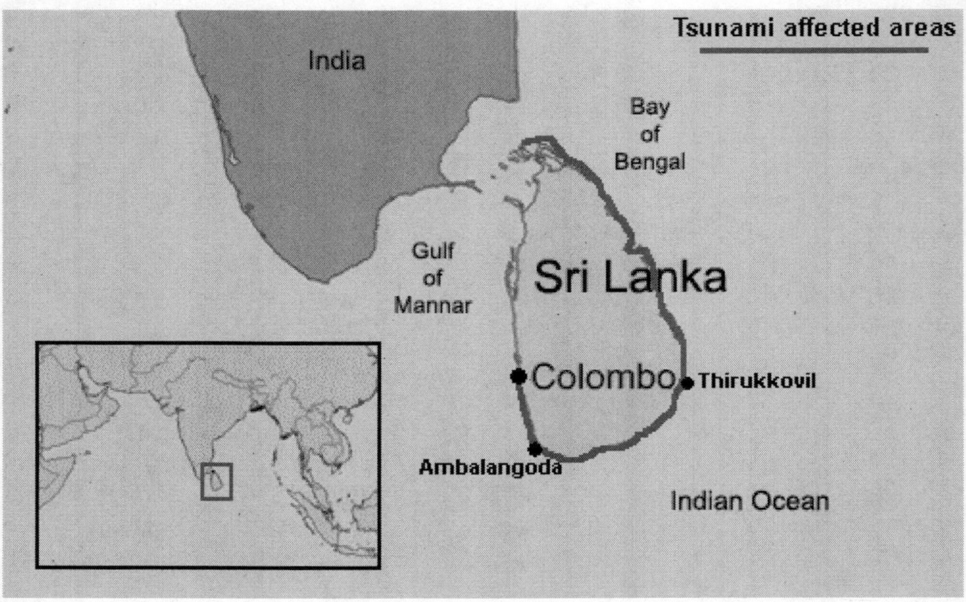

Figure 19.1: Tsunami Affected Areas in Sri Lanka

Effects of Salinization on Crop Production in Sri Lanka

Crop production in Sri Lanka depends on many environmental and soil factors. Some of the factors directly influence the final yield of crops. There are 2 obvious ways of increasing the production such as expanding the cultivated area and increasing the productivity of unit land area. However, both of these face number of constraints. One of the major constraints and the commonest soil problem in the world is salinity (IRRI, 1975; Ponnamperuma, 1977; Chandler, 1979; Akbar and Ponnamperuma, 1982). Therefore, saline soils are a most critical factor in Agriculture. It is estimated that, more than 150 million hectares of rice lands in tropics and sub-tropics are affected by salinity. As in other countries, salinization is a major environmental problem, which directly and indirectly affects Agriculture, causing the reduction of crop yield, especially in rice production in Sri Lanka. Rice lands are distributed in almost all the agro-ecological environments except elevations above 1,200 m MSL in Sri Lanka. Hence, compared to many other rice growing countries, Sri Lanka grows rice under a wide range of environmental conditions.

Inland and coastal salinity is one of the major constraints to the delivery of the full potential of rice yields of new varieties in certain areas of Sri Lanka. Gangodawilla (1990, 1994) reported that about 0.7 million hectares of land area in Sri Lanka are with saline/ alkaline limitations. Of which 18,000–45,000 hectares of land is under inland salinity. The newly irrigated majority of land areas, which are under major irrigation programs like Mahaweli river system project, in Dry zone are identified as the salinity affected areas and some lands are also identified as those with potential for future saline limitations. The problem of salinity is also more commonly encountered in semi and arid areas and in some of the dry areas in Sri Lanka during non-monsoonal periods because the salt rises to the soil surface through capillary action during dry periods. Hence, in such areas salinity damage is less severe during monsoonal or wet seasons.

Followed by salinity measurements in Sri Lanka, Thiruchelvam and Pathmarajah (2000) showed that in affected areas, soil salinity was the principal factor determining rice productivity. In moderately saline areas, the yield loss ranged from 10 per cent to 15 per cent; in high and severe salinity areas, yield was reduced by one third. Net income from rice fell by about 22 per cent and 43 per cent in moderate and high saline areas respectively. In addition, the study found that salinity has affected the quality of drinking water, human health, and vegetation. Nutrient deficiencies and toxicities are frequently being reported with the crops grown in these areas as Phosphorous and Zinc deficiencies, and Iron and Aluminium toxicities are the major nutrient problems associated with salinity.

Furthermore, some of the other field crops grown in some areas in Dry zone are affected by the increasing problem of salinity due to poor drainage facilities and excessive use of irrigation water. Of which, Chilli, Onion, Maize and Sorgum are more susceptible to the problem and the complications can be identified in their growth even at the early stages of vegetative phases before flowering, leading to the severe reduction of the yield.

Effect of Salinity on Rice

Rice is considered as a salt-tolerant crop. Maas and Hoffman (1977) showed that threshold EC of rice is 3 dS/m and increasing of salinity by 1dS/m reduces rice yield by 12 per cent. Moorman and Breeman (1978) reported that 6-10 dSm^{-1} decreases the yield by 50 per cent. Pearsons and Ayres (1960) have shown that tolerance varies with growth stages of rice. The plant is tolerant during germination while young seedlings are sensitive until the age of four weeks. An increase in salt tolerance then occurs up to tillering. However, the plant becomes sensitive again during flowering. Tolerance increases again during the maturation period. IRRI (1978) found that during the reproductive stage, salts adversely affect the number of spikelets per panicle. Yield under salinity is further reduced due to adverse effects on panicle formation and grain setting.

Effects of Tsunami on Rice and Other Highland Crops

Mapa *et al.* (2005) reported that about 575 acres of rice growing fields in Nindavur area where the soil type is Alluvial (Entisols) have been affected by saline water after the tsunami disaster in 2004. The rice variety, BG 94-1, cultivated in most of the fields in this area has been totally damaged by salinity and the paddy grains became empty since the plants are near maturity when the disaster took place. Although some of the unaffected soils in this area show mild acidity, the damages cannot be recognized by visual or external morphology. Hence, the limit where the crop was visually affected can be determined as the limit of seawater intrusion by tsunami disaster.

The salinity level in unaffected areas is 0.04 dSm^{-1} of EC in saturated soil paste while it is reported as an increased value by 20–200 times in affected areas, depending on the degree of intrusion and the distance from the sea. The highest EC recorded is 6.2 dSm^{-1} in the rice fields nearest to the sea. The pH values have been measured as 6.4 and 7.1 in non-affected and affected fields respectively. These 2 factors clearly showed that the affected lands in this area are now under high salinity becoming the area unsuitable to grow rice except salt tolerant plants because the crop will be damaged permanently if the EC is higher than 4 dSm^{-1}. Even though, some traditional rice varieties that tolerant to salinity can be grown up to EC of 8 dSm^{-1}, most of the improved modern varieties except tolerant varieties like At 354 and At 401 cannot be cultivated beyond 3 dSm^{-1} without loosing their yield (Appendix A). Mapa *et al.* (2005) recorded that even 2 weeks after the tsunami disaster followed by heavy rains in this area, stagnated water in the fields showed higher EC of 5.8 to 11.5 dSm^{-1} leaving the high salinity due to the poor seepage of water from rice growing fields where the hard pan below the plough depth has been developed, and blocked drainage channels, which do not allow salts to be washed away from the paddy fields.

By the investigation conducted in Ampara District, Mapa *et al.* (2005) reported that about 460 acres of highland annual crops like tomato, brinjals, chillies, maize and cowpea and even perennials like banana and Palmyra palms in Vinayagapurm village, where the soils belong to sandy Regosols, have been damaged by sea waves and salinity after tsunami disaster in 2004. Due to the sandy soil, which provides good drainage, accumulated salts have washed away with heavy rains after the tsunami incident in this area. With these situations, EC in the well-drained highland areas has been reported at very low levels ranging from 0.13 to 0.60 dSm^{-1}. However, in some eroded areas due to massive sea waves, EC of stagnated water was 3 dSm^{-1}, which is high enough to damage most of the crops. With the evaporation of water in some areas, whitish patches of salt crust on the surface of the soil could be seen. Water in the wells in this area is polluted with salts and cannot be used for domestic or agricultural activities. The ground water level in this area has been raised up making a shallow ground water table. The danger of this is the increasing of salinity over time due to the higher evaporation rates in the area

Growing Crops in the Areas Experiencing Salinity

Areas affected with salinity problems can be used to cultivate crops mainly in 2 ways. First, crops could be cultivated in these areas after mitigating the salinity using several treatments, management and agronomic practices. Second, there are some situations, where it is either impossible or uneconomical to attain desirably low soil salinity levels. Halophytes or salt tolerant plants would be introduced to such areas and could be grown with saline waters.

Mitigating Salinity: Crop production in medium to high saline soils or using saline water is not an easy task in Sri Lanka without introducing methods to mitigate the problem. In high saline areas, much cannot be adopted to counteract the effect of soil salinity on crop production. Therefore, adaptive research is recommended to reclaim affected soil. However, the possibilities of reclaiming the soil are much less once the soils become excessively saline. While mitigating the problem, Thiruchelvam and Pathmarajah (2000) strictly recommended that the lands experiencing moderate levels of salinity must be monitored to avoid becoming it worse. This is the main challenge in the areas developing salinity due to improper irrigation and drainage systems associated with large river projects in Dry areas. It takes long time to implement proper and necessary drainage system by the authorities. However, as a short-term measure, investigators suggested encouraging farmers to improve drainage in their own fields and to manage excessive irrigation to prevent the rising up of water table, which will increase surface salinity due to evaporation. Even though, short-term measures are introduced, long term monitoring in this region is required by developing proper drainage channels. Some of the Mahaweli areas are now required immediate actions or reclamation for the growing salinity problem, which would be more critical in the near future. Proper use of corrective fertilizers, efficient drainage and irrigation management, cultivation of tree crops such as neem, eucalyptus, and other shrub species and use of proper mulching to reduce intensities of evaporation would be useful to mitigate the salinity in these areas. Mahaweli authority in Sri Lanka is now redesigning drainage channels associated with the irrigation systems with ongoing rehabilitation programs in Mahaweli areas. Furthermore, micro-irrigation systems like drip irrigation methods are recommended introducing for cultivation of banana in highlands in Jaffna peninsula, Northern part of the island, to prevent the development of salinity in the area as heavy removal of groundwater and excessive use of irrigation water are more responsible for growing salinity in this dry area.

It was reported that the total estimated damage to crops is about 4,217 ha in Sri Lanka after tsunami disaster (Table 19.3). Even though some of the areas in Wet and

Intermediate zones were also affected by tsunami, the areas may not experience much problems longer because natural flushing and other remedial measures could be done with the rainy season so that about 90 per cent of the affected land would be able to produce at least 80 per cent of normal production capacity. However, the coastal areas of Dry zone may experience prolonged effects of tsunami on salinity. Tsunami affected areas with high salinity must be rehabilitated using 3 strategies as short term, medium term and long term strategies proposed by Mapa *et al*: (2005). One of the best short term strategies identified was to improve drainage immediately, especially by cleaning drainage channels, removing blocked sands and other debris from the fields and drainage systems prior to concentration of salts in the field due to evaporation. Leaching the soil with clean water and use of soil amendments like Gypsum, which would make soil reclamation faster, were some other short-term strategies suggested. In addition, cultivation of salt tolerant varieties such as At 353, At 354 and At 401 in these fields were also recommended.

Table 19.3: Extent of Crop Damages by Tsunami in December 2004

District	*Number of Farm Families*	*Extent of Damage (ha.)*					
		Paddy	*Vegetables*	*OFC**	*Fruits*	*Others*	*Total (ha.)*
Kaluthara	88	0.1	0.2	0	4.4	0.1	4.8
Galle	1041	160.0	32.6	0.2	35.4	39.1	267.1
Matara	337	113.6	7.7	32.6	1.2	0.0	122.5
Hambantota	696	261.2	23.9	7.7	22.6	4.9	314.7
Ampara	2614	1737.2	283.0	23.9	0	0	2020.2
Batticaloa	2355	62.7	116.5	283.0	59.7	89.1	369.1
Trincomalee	803	84.0	8.7	116.5	12.8	0	222.4
Mulaitivu	556	706.1	10.0	8.7	0	0	755.1
Jaffna	558	92.3	0	47.8	0.6	0.4	141.2
All areas	9048	3541	483	247	137	134	4217

* Other Field Crops

Source: FAO tsunami assessment report, 2005.

In highlands, water table must be lowered by improving drainage through constructing drains. Furthermore, investigators have stressed to empty and clean all the affected wells in highlands so that salts will not enter again to the subsoil. Then these wells should be used for domestic and irrigation purposes, as they will provide freshwater again. Out of the medium term strategies suggested, training farmers to acquire knowledge and skills about special cultural and agronomic practices, which may be useful in mitigating salinity and the ways of monitoring the level of salinity in their fields, was of utmost important. As the salt tolerant rice varieties are not available in large quantities, more salt tolerant varieties must be produced for these salt affected fields for few seasons. Long term strategies included advisory services to clean and maintain irrigation and drainage channels regularly in the long run, to wash out salts with monsoon rains and to flood the fields followed by ploughing them for the improvement of drainage so that sodium ions would be displaced. Eroded highland areas must also be developed by adding organic matter, which would provide better drainage to the area as well.

Introduction of Salt Tolerant Crops

Today, the production of more and more salt tolerant plants has become a challenge for the developing salinity problem in the major fields of Agriculture. Selection, breeding or hybridization and biotechnology are more important in producing such varieties, for instance, some of the selected salt tolerant shrubs and grass species have already been taken into consideration to use in pasture improvement programs in salt affected areas throughout the world. Selection of species showing high biomass and high level of protein production, which have been combined with another important feature of having the ability to survive in a wide range of environmental conditions including salinity, has now become a vital process in further development of halophytes.

In Sri Lanka, rice is the major crop threatened by salinity as the most of rice fields are developing salinity. Rice varieties vary in their tolerance of salt levels in soil. It is an advantage of producing high yielding varieties, which show greater tolerance to salinity in saline soils and in potential lands of salinity. Emphasis is now being given to develop high-yielding, non-lodging varieties, which tolerate in saline soil, through hybridization and other innovative breeding approaches. It is important to introduce precise and accurate identification procedures to select better saline tolerant varieties in developing high yielding rice crops with high tolerance for salinity. Rice Research Institutes in Sri Lanka, especially at Ambalanthota conducts research to develop varieties with saline tolerant. Different screening methods are commonly used to differentiate tolerant rice varieties from non-tolerant varieties at different growth stages under field and greenhouse conditions. All of these screening procedures used for rice in the field and greenhouses are time consuming. Dissanayake and Wijeratne (2006) have however developed a laboratory screening method, which has not been taken much time for screening rice at germination and early growth stages.

Today, production of saline tolerant rice through breeding or hybridization and biotechnological methods stands out although the methods are time consuming. After the tsunami disaster, Rice Research Institutes (RRI) at Bombuwela and Ambalanthota with the collaboration of Department of Agriculture in Sri Lanka are taken much effort to develop varieties suitable for salt affected areas. Sri Lanka has some traditional rice varieties like Pokkali and Podiwi A8, which are more resistant to salinity. In addition, there are some pure lines developed for salinity by RRI in different locations in Sri Lanka (Table 19.4). It has been recommended that, salt tolerant varieties must be introduced to rice fields only after at least some extent of soil reclamation because salt tolerant varieties are only a short-term solution unless other amelioration methods are used to control the increasing salinity. Otherwise, this may increase the salinity over the longer period of time due to improper management practices etc.

Table 19.4: Salt Tolerant Rice Varieties in Sri Lanka

Variety Name	Growth Duration (Days)	Parents	Varietal Type	Recommended Location	Other
At 353	90	BG 94-1 (R), BG 400-1	Pureline indica	Saline areas in Nilwala scheme	Resistant to bacterial leaf blight, blast and tolerance to salinity
At 354	105	BG 94-1, Pokkali	Pureline indica	Southern province	White pericarp, tolerance to salinity
At 401	120	BG 94-1, Pokkali	Pureline indica	Southern province	Red pericarp, tolerance to salinity
At 95-26-4	120	-	Pureline indica	-	High yielding, red pericarp, tolerance to salinity
BW 266-7	105	BW 242-5-5, OB 677, BG 90-2	Pureline indica	-	Long grain, white pericarp, tolerance to salinity and acidity
BW 302	90	BW 259-3, BW 242-5-5	Pureline indica	Wet zone	Tolerance to salinity and acidity
BW 351	105	BG 90-2, BG 400-1	Pureline indica	Wet zone	Red pericarp, tolerance to salinity
BW 400	120	BW 259-2, BW 242-5-5	Pureline indica	Wet zone	Red pericarp, tolerance to salinity, blast resistant
BW 451	135	BG 400-1, BG 11-11	Pureline indica	Wet zone	White pericarp, tolerance to salinity and acidity
IR 2151	-	-	Pureline indica	-	Tolerance to salinity
PODIWI A-8	-	Traditional variety	Pureline indica	-	Tolerance to salinity
Pokkali	120	Traditional variety	Pureline indica	-	Low yielding, red pericarp, tolerance to salinity,

As another solution for environmental and soil problems such as salinity, toxicity etc., Department of Agriculture, Universities and other private organizations in Sri Lanka have introduced polythene tunnels or poly-tunnels to cultivate valuable crops with or without using soils where hydroponic systems are used with fertilizer solutions. These are now being popularized in almost all the 3 climatic zones and can be used as a solution for salinity problem. Furthermore, these structures in some coastal and dry areas are being used to grow high value salt tolerant crops using saline waters in Sri Lanka.

Discussion

Salinization is one of the critical problems encountered in Agriculture. This restricts the cultivation of crops and reduces yield especially in arid and semi arid areas in Sri Lanka. Although, high degree of salinity is seldom reported in Sri Lanka's soils, moderate salinity problem in arid and semi arid areas and rising salinity during non-monsoonal periods and due to excess irrigation in Dry zone areas, and other unexpected situations in Sri Lanka dramatically affect Agriculture. However, there is a danger of soils becoming highly saline under large-scale irrigation projects like

Mahaweli river irrigation project in Dry zone in Sri Lanka due to inadequate water management practices. With the rising level of salts both in soil and water, wet and dry season crops are affected and the area becomes marginal as abandon lands with no further crops cultivated. This is a real threat in such areas if remedies or precautions are not recommended. Coastal salinity level is also considerable in Sri Lanka. In addition to these causes, tsunami affected areas are now experiencing high degree of salinity in both soil and ground water. Therefore, farmers in the affected areas cannot cultivate their lands fully or effectively and experience low productivity. In most of the areas, the problem is much serious with rice, chilies and other field crops. However, rice is the mostly affected crop by salinity in Sri Lanka because most of the rice fields in Dry zone are located in such areas experiencing salinity. Water is stagnated in these fields due to poor infiltration and seepage. This has been caused by the formation of hard pans and poor water management system with irrigated water. Soil salinity is considered as one of the principal factors determining rice productivity in Dry zone. Sri Lanka was devastated by the tsunami along 1200 km (68 per cent) of its 1770 km costal belt. Twelve of the fourteen coastal districts were affected and only the damage was minimal in the North Western region. Due to the tsunami disaster, croplands in the coastal areas of Sri Lanka were inundated by seawater leading to development of high salinity. In most of the areas, rice fields and crops in some of the highland areas were flooded by seawater and increased the level of salinity, which was high enough to damage the crops completely in the area.

Considering the causes for the development of salinity in Sri Lanka, the most important strategy is to implement measures to mitigate the problem rather than using salt tolerant varieties. Especially in Mahaweli irrigation areas, the drainage must be improved although it is a long-term strategy. However, individuals can practice the same in their cultivating fields by themselves as a short-term measure. At the same time, farmers must manage soils and irrigation water properly in the area. Excessive irrigation must be prevented so that rising of the water table could be avoided. Use of agronomic practices is also important to mitigate the salinity problem. Drainage and soil texture and structure are improved by applying organic matters to the fields. During off-season, salt and drought tolerant crops like sunflower and minor millets are recommended to grow because plant cover reduces the evaporation, thereby reduces the development of salinity. In addition, improved crops especially rice varieties withstanding under the stress conditions like salinity is used in these areas. Varieties tolerant to salinity are available, but their use does not substitute for proper water and irrigation management. Breeders will unlikely be able to produce varieties with ever-increasing tolerance of salinity. A variety adapted to present levels of salinity may no longer survive if salinity continuously increases. Selection and breeding techniques and non-conventional methods like biotechnology are more vital of producing highly improved rice varieties adapted to salt stresses. However, frequent release of salt tolerant crops and their cultivation using saline waters are still new for Sri Lankans and most of these techniques are still under research level. Nevertheless, tsunami affected lands need to be rehabilitated urgently implementing possible short-term, medium-term and long-term strategies for the agricultural community to restart their farming activities. Eventually, global and national policymaking, regarding the mitigation of salinity, and their implementation with close supervision by experts are timely important in affected areas. However, after implementing these methods to tackle salinity, monitoring of the same is important for their sustainability.

Conclusions

High degree of salinity is not common in Sri Lankan Agriculture. Low to moderate salinity developed in dry areas including the areas of Mahaweli irrigation project, coastal areas and arid areas of Sri Lanka should be mitigated by recommended strategies. Management of irrigation and drainage systems for sustainable crop production and lowering of evaporation rates using agronomic practices etc in these areas are more vital of mitigating salinity. Short-term strategies must be implemented immediately to prevent further development of salinity in the future. Development and introduction of halophytes such as salt tolerant rice varieties are also a possible solution for salinity in such areas. Prompt actions must be taken to rehabilitate tsunami-affected areas in Sri Lanka before the area becoming unsuitable for cultivation of crops.

Acknowledgements

The author acknowledges the NAM S&T Centre in India, NASTEC in Sri Lanka and University of Birjand in Iran for their support. Further, the author is grateful to Prof. R.P. Mahaliyanaarachchi, the Vice Chancellor in Sabaragamuwa University of Sri Lanka for his kind assistant and Dr. (Mrs.) S.T.C. Amarasinghe, Ms S.A.P.I Piyasena, Ms. W.A.L. Arunodini, and Ms. H.K.S.G. Gunadasa, academic staff members, and Mrs. M.L.K.M. Dissanayake for providing useful information and support in preparing this paper.

References

Akbar, M. and F.N. Ponnamperuma. 1982. Saline soils of South and South East Asia as potential rice lands, Rice research strategies for the future. IRRI, Manila, Philippines. 265-281 pp.

Ayers, R.S. and D.W. Westcot. 1976. Water Quality for Agriculture, Irrigation and Drainage. FAO, Rome. 29 pp.

Chandler, R.F.Jr. 1979. Rice in the Tropics; A guide to the development of national programes. West view press, Colarado.

Dissanayake, P.K. and A.W. Wijeratne. 2006. Development of a varietal screening procedure for salt tolerance of rice (*Oryza sativa* L.) varieties at germination stage. The Journal of Agricultural Sciences, Faculty of Agricultural Sciences of Sabaragamuwa University of Sri Lanka, 2: 63-72.

FAO, 2005. Tsunami assessment report. Rome, FAO.

FAO/UNESCO. 1992. Irrigation, Drainage and Salinity. Rome, FAO.

Gangodawilla, C.D. 1990. Country report–Land resources with problem soils in Sri Lanka. Problem soils in Sri Lanka, Problem soils of Asia and Pacific Regional office for Asia and Pacific (RAPA) FAO.

Gangodawilla, C.D. 1994. The collection and analyzing land degradation data in Sri Lanka. The collection and analysis of land degradation data, RAPA publication, Regional office of Asia and pacific, Food and Agricultural Organization of the United Nation. 215-225 pp.

IRRI, 1975. Annual report 1974. Los Banos, Philippines. 127-130 pp.

IRRI, 1978. Proceedings of the workshop on the genetic conservation of rice, IRRI-IBPGR, Los Baños, Philippines. 54 pp.

Maas, E.V. and G.J. Hoffman. 1977. Crop Salt Tolerance: Current Assessment. Proceedings of the International Conference on Management of Saline. Water for Irrigation, Lubbock, Texas, Texas Tech. University.

Mapa, R.B., W.M. Wickramasinghe, A.D.B., D.N. Sirisena and K.M.A. Kendaragama. 2005. Salt in the soil. A Report, Soil Science Society of Sri Lanka.

Moormann, F.R. and N. Van Breedman. 1978. Soils and Land Properties that Affect the Growth of Rice. Los Banos, Philippines, IRRI. 121-121 pp.

Panabokke, C.R. 1996. Soils and Agro-ecological environments of Sri Lanka. Natural Resources series - 2, Natural Resources and Energy Authority of Sri Lanka. 140-163 pp.

Panabokke, C.R. and R. Kannangara. 1975. The Identification and Demarcation of the Agro-ecological Regions of Sri Lanka. Proc. Sec. B. Ann. Sessions, Assoc. Advmt. Sci., 31: 49.

Pearson, G.A. and A.D. Ayres. 1960. Rice as a Crop of Salt Affected Soil in Process and Reclamation. USDA Prod., Research Report.

Ponnamperuma, F.N. 1977. Screening rice for tolerance to mineral stress. IRRI, Research paper series 6, Scardaci S.C., M.C. Shannon, S.R. Grattan, A.U.Eke.

Thiruchelvam, S. and S. Pathmarajah. 2000. An Economic Analysis of Salinity Problems in the Mahaweli River System H Irrigation Scheme in Sri Lanka. Research Report, International Development Research Centre, Ottawa, Canada.

http://idrinfo.idrc.ca/Archive/Corpdocs/114522/Economic.htm

USDA, 1994. Information Bulletin, 194 pp.

APPENDIX

Crop Yield Decreases as Salinity Increases

Table 1: Salt Tolerance of Field

	Relative Yield Decrease–%			
	0	*10*	*25*	*50*
Barley	8.0	10.0	13.0	18.0
Cotton	7.7	9.6	13.0	17.0
Sugarbeet	7.0	8.7	11.0	15.0
Wheat	6.0	7.4	9.5	13.0
Safflower	5.3	6.2	7.6	9.9
Sorghum	4.0	5.1	7.2	11.0
Soybean	5.0	5.5	6.2	7.5
Rice(Paddy)	3.0	3.8	5.1	7.2
Broadbean	1.6	2.6	4.2	6.8
Corn	1.7	2.5	3.8	5.9
Flax	1.7	2.5	3.8	5.9
Peanut	3.2	3.5	4.1	4.9
Cowpea	1.3	2.0	3.1	4.9
Fieldbean	1.0	1.5	2.3	3.6

Table 2: Salt Tolerance of Forage

	Relative Yield Decrease–%			
	0	*10*	*25*	*50*
Tall wheatgrass	7.5	9.9	13.3	19.4
Wheatgrass	7.5	9.0	11.0	15.0
Crested wheatgrass	3.5	6.0	9.8	16.0
Barley hay	6.0	7.4	9.5	13.0
Perennial ryegrass	5.6	6.9	8.9	12.2
Tall fescue	3.9	5.8	8.6	13.3
Beardless wildrye	2.7	4.4	6.9	11.0
Sweet clover	1.5	3.2	5.9	10.3
Orchard grass	1.5	3.1	5.5	9.6
Vetch	3.0	3.9	5.3	7.6
Alfalfa	2.0	3.4	5.4	8.8
Corn fodder	1.8	3.2	5.2	8.6
Love grass	2.0	3.2	5.0	8.0
Meadow foxtail	1.5	2.5	4.1	6.7
Clover- alsike, red Ladino, strawberry	1.5	2.3	3.6	5.7

Table 3: Salt Tolerance of Vegetable Crops

	Relative Yield Decrease–%			
	0	*10*	*25*	*50*
Beets	4.0	5.1	6.8	9.6
Broccoli	2.8	3.9	5.5	8.2
Tomato	2.5	3.5	5.0	7.6
Cucumber	2.5	3.3	4.4	6.3
Cantaloupe	2.2	3.6	5.7	9.1
Spinach	2.0	3.3	5.3	8.6
Cabbage	1.8	2.8	4.4	7.0
Potato	1.7	2.5	3.8	5.9
Sweet Corm	1.7	2.5	3.8	5.9
Pepper	1.5	2.2	3.3	5.1
Lettuce	1.3	2.1	3.2	5.2
Radish	1.2	2.0	3.1	5.0
Onion	1.2	1.8	2.8	4.3
Carrot	1.0	1.7	2.8	4.6
Beans	1.0	1.5	2.3	3.6

Table 4: Salt Tolerance of Fruit Crops

	Relative Yield Decrease–%			
	0	*10*	*25*	*50*
Date palm	4.0	6.8	10.9	17.9
Fig, Olive	2.7	3.8	5.5	8.4
Grape	1.5	2.5	4.1	6.7
Grapefruit	1.8	2.4	3.4	4.9
Orange	1.7	2.3	3.2	4.8
Lemon, Apple	1.7	2.3	3.3	4.8
Pear, Walnut	1.7	2.3	3.3	4.8
Plum	1.5	2.1	2.9	4.3
Peach	1.7	2.2	2.9	4.1
Almond	1.5	2.0	2.8	4.1
Apricot	1.6	2.0	2.6	3.7
Blackberry	1.5	2.0	2.6	3.8
Boysenberry	1.5	2.0	2.6	3.8
Raspberry	1.0	1.4	2.1	3.2
Strawberry	1.0	1.3	1.8	2.5

Source: Ayers and Westcot, 1976.

20

Current Status of Saline Agriculture in Iran

M. Kafi[1*], M. Jami Al Ahmadi[2] and G.R. Zamani[2]

[1]*Faculty of Agriculture, Ferdowsi University of Mashhad, Iran*

[2]*Faculty of Agriculture, Birjand University, Birjand, Iran*

**E-mail: mkafi2003@yahoo.com, m.kafi@ferdowsi.um.ac.ir*

ABSTRACT

Iran is one of the most arid countries of the world. Even if all the renewable water resources could be utilized, excluding the incoming international rivers, the total annual volume of water would not be more than 117 billion cubic meters (bcm). Considering that about 88.5 bcm are currently used each year, the country is left with about 28.5 bcm of additional water capacity for future use. Agricultural land availability is not a major constraint in the development of Iranian agriculture. The major constraint is the availability of water for the development of these lands. In more than 70 per cent of the Iran plateau, people are facing shortage of rainfall, receding the groundwater level, salinization of soil and water resources, growing demand of water from other sections such as industry, desertification of their lands (rangelands, and arable lands) and day to day increasing pressure in the natural resources. On the other side population of the country is growing rapidly, living standards of society is improving, with advancement of better utilization of resources. Thousand hectares of land are left unused because of accumulation of excess salt on them, or due to dryness of the water resource. The plant coverage of rangelands and forests of Iran particularly in desert areas is reducing regularly. Many villages have been abandoned due to inefficient agricultural activities. This is not limited to Iran; the situation is the same in many other countries in the region. What we are able to do for our nations as agricultural specialists?

I think we should looking for the key of this lock in nature in the region, how, the camel adapted to the desert conditions and how did human explore that camel could be used as a mode of transportation in the harsh desert conditions? How did human find date palm trees, in the hot and saline desert? There are thousands of plants and animal species that have became acclimatized to such conditions and these can be more effectively utilized as crops, forages or domestic animals. As far as the subject of this workshop is concerned, working on halophytes could be one of the best ways to produce vegetation for the deserts, fuel for the villagers, forage for domestic animals, vegetable for humans and much more. The production of halophytes using saline water and soil, and feeding livestock with them, is one of the most sustainable ways of conservation desert ecosystems and food production for people living in these areas.

Introduction

The Islamic Republic of Iran covers a total area of approximately 1.65 million km^2 and is bordered by Azerbaijan, the Caspian Sea and Turkmenistan to the north, Afghanistan and Pakistan to the east, the Gulf of Oman and the Persian Gulf to the south, and Iraq and Turkey to the west (Keshavarz *et al.*, 2005; Statistic Centre of Iran, 2001). About 52 per cent of the country consists of mountains and deserts, and nearly 16 per cent of the country has an elevation of more than 2,000 m above sea level. The largest mountain is that of the Zagros, which runs from northwestern Iran first southwards to the shores of the Persian Gulf and then continues eastwards till the most southeastern province. Other mountain ranges extend from the northwest to the east along the southern edge of the Caspian Sea. The Central or Interior Plateau is located in between these mountain chains and covers over 50 per cent of the country. It is partly covered by two remarkable salt swamps ('kavir') and partly by areas of loose sand or stones, with stretches of better land near the foothills of the surrounding mountains (Badripour, 2004; Ghazi, 2002; Siadat, 2004).

The climate of Iran is one of great extremes due to its geographic location and varied topography. The summer is extremely hot with temperatures in inland reaching more than 55 °C. In winter, however, because of the high altitude of major part of the country and its continental location the temperatures far lower than minus 30 °C can be recorded in the northwest and minus 20 °C in many regions (Alizadeh and Keshavarz, 2005; Committee to rehabilitate rangelands and mitigate drought effects, 2004).

Annual rainfall ranges from less than 50 mm in the deserts to more than 1,600 mm on the Caspian Plain. Table 20.1 shows the distribution of rainfall in the country. The average annual rainfall is 250 mm and approximately 90 per cent of the country is arid or semiarid. Overall, about two-thirds of the country receives less than 250 mm of rainfall per year (Alizadeh *et al.*, 2001; Keshavarz, 2005, Siadat *et al.*, 1997).

The total population is about 70 million (2005), of which 33 per cent is rural. The average population density is 42 inhabitants/km (2002), but it ranges from less than 10 in the eastern part of the country up to more than 150 in the Gilan province. The annual demographic growth rate was estimated at 3.4 per cent over the period 1980-1990 and at 2.6 per cent over the period 1990-1994 and 1.2 per cent in the period between 2000-2005 (FAO).

Table 20.1: Distribution of Precipitation in Iran

Annual Precipitation (mm)	*Area (Km^2)*	*Percent*
<50	100,000	6
50-100	285,000	17
100-200	465,000	28
200-300	370,000	23
300-500	280,000	17
500-1,000	130,000	8
>1,000	18,000	1
Total	**1,648,000**	**100**

Agriculture plays a pivotal role in a country's national economy and employment; therefore, Iran would gain tremendously from further development of agricultural section. In 2003, agriculture employed around 25 per cent of the total labour force and accounted for 11.3 per cent of GDP. (Ghazi, 2002). Abiotic stresses, mainly drought, and subsequent high temperature and salinity, are the major challenges faced by Iranian agriculture. Unpredictable drought coinciding with critical growth stages of crops, especially in dry land areas, threatens crop production. Even in irrigated lands because of shortage of water, drought stress is common at different stages of plant growth. (Alizadeh and Keshavarz, 2005; Committee to rehabilitate rangelands and mitigate drought effects, 2004).

Soil Resources

The cultivable area is estimated at about 51 million ha, which is 31 per cent of the total area. Almost 18.5 million ha, or 36 per cent of the cultivable area, is considered usable for agriculture, while 14.4 million ha are actually cultivated (Table 20.2). Of this area, 12.8 million ha consists of annual crops and 1.6 million ha of permanent crops. About 70 per cent of the land holders possess less than 5.5 ha (of which on an average 2.13 ha are irrigated and 3.25 ha are rain fed) (Official Web site of Ministry of Jihad-e-Agriculture of Iran, FAO).

Approximately 11 per cent of the total land area of the country is used for crop production. Of this area, annually, about 12.7 Mha is cultivated (irrigated plus rain fed) and nearly 5.8 Mha is left as fallow. According to Table 20.3, an estimated area of 32 Mha of unused land of Iran is potentially suitable for crop production, but shortage of water resources on these lands limits their role in agriculture. More than half of the country is covered by rangelands (mostly of low to medium productivity), wastelands, and mountains. Sand dunes, salt flats, and other non productive land, mostly in the central plateau and the southern coastal area, comprise 11.6 Mha. Major soil types of

Iran are Xerosols, Arenosols, Regosols, Solonchalks, and Lithosols (Badripour, 2004; Bureau of Information and Statistics, 1994; Bybordi, 1989).

Table 20.2: Main Agricultural Criteria of Iran at Index Year 2000 and Short and Long Horizons (Keshavarz *et al.*, 2005)

Year Criterion	*2000*	*2005*	*2020*
Population (million)	63.0	70.0	100
Water allocated to the agriculture sector (bcm)	82.5	90.0	100
Total quantity of products of irrigated lands (tone)	56.0	85	189
Water productivity (kg/m^3)	0.7	1.0	1.9
Increase in allocated water (per cent)	–	9.7	22.0
Expected increase in products of irrigated lands (per cent)	–	150.0	337.0
Total water allocated to the agriculture sector (per cent)	93.0	93.0	93.0

Table 20.3: Agricultural and General Land Use in Iran

Sl.No.	*Land Use*	*Area/Million ha*
1.	Irrigated Land	
	Annual crops	5.2
	Orchards	1.2
	Fallow	2.2
	Subtotal	8.5
2.	Rainfed	
	Annual crops	6.4
	Fallow	3.6
	Subtotal	10.0
3.	Unused but potentially productive land in arid environments	32.5
4.	Forest and scattered woodland	12.4
5.	Rangelands, wasteland and mountains	90.0
6.	Sand dunes, salt flats, others	11.6
	Total Area	165.0

The soils of salt-affected regions are heavy (clay) without proper drainage, having high-level aquifer and high salinity, particularly in the surface and among soil different layers. In general, these lands have bare soils and are improper for cultivation. However, in some parts, due to having lower soil salinity concentrations, salt-tolerant crops such as cotton, sugarcane, sugar beet, wheat and barely could be cultivated. Therefore, perhaps there is no limitation in land resources for crop and forage production, particularly in dry areas (Badripour, 2004; Keshavarz *et al.*, 2005; Siadat, 2004).

Water Resources for Irrigation

Both systems of irrigated and rain fed farming (dry farming), are practiced in different parts of the country, while the area devoted to each system varies considerably depending on agroclimatical conditions. Rain fed agriculture and dry farming are most successful in the western and northwestern of Iran, as well as the sloping lands in the Caspian coast. In other parts of the country, dry farming is also practiced in hilly areas, but the yields obtained are limited. In the Central Plateau, as well as the southern plains and the southern coastal areas of Iran, crop production is mostly possible only under irrigation. This is because of low rainfall and high evaporation rate. In some low-lying plains of the central plateau, the annual rainfall is about 50 mm, while the annual evaporation may exceed 4000 mm (Agricultural statistics of Iran, 1996 97; Alizadeh and Keshavarz, 2005; Bybordi, 1989, Ghazi, 2002, Siadat *et al.*, 1997).

Annual precipitation in Iran is estimated at 413 billion cubic meters (bcm). Out of this volume, nearly 283 bcm directly evaporate from the soil surface or are evapotranspired from forest canopies, rain fed crops, etc. The annual surface streams with internal origin are about 105 bcm, including 13 bcm of surface water, which enter the country from the neighboring states. Direct recharge of the alluvial aquifers amounts to 25 bcm/year and 29 bcm of utilized water returns to the surface streams and groundwater annually. This means that the total potential volume of water for annual use is 159 bcm (130 bcm of renewable water resources and 29 bcm of return flows). Presently, the annual volume of water used in different sectors is as follows: agricultural sector 82.1 bcm, urban and industrial sectors 7.4 bcm (Alizadeh *et al.*, 2001; Ghazi, 2002, Keshavarz *et al.*, 2005).

Table 20.4: Water Availability and Use in Iran

Component	*Volume (bcm)*	*Percent of Total*
Precipitation	413.0	100.0
Evaporation	283.0	70.0
Renewable water	130.0	30.0
Surface water	105.0	–
Groundwater	25.0	–
Total water use	88.5	100.0
Agriculture	82.1	92.77
Domestic	5.6	6.33
Industry (etc.)	0.8	0.9

There are several large rivers, but the only navigable one of which is Karun, the others being too steep and irregular. The Karun River, with a total length of 890 km, flows in the southwest of the country to the Shatt El-Arab, which is formed by the Euphrates and the Tigris after their confluence. The few streams that empty into the Central Plateau dissipate into the saline marshes. All streams are seasonal and variable. Water, however, is stored naturally underground, finding its outlet in subterranean water canals (qanats) and springs. It can also be tapped by wells (Badripour, 2004; Siadat, 2004).

Internal renewable water resources are estimated at 130 km^3/year. Surface runoff represents a total of 105 bcm/year, of which 5.4 bcm/year comes from drainage of the aquifers, and groundwater recharge is estimated at about 49.3 bcm/year, of which 12.7 bcm/year is obtained from infiltration in the rivers bed (Keshavarz *et al.*, 2005; Statistic Centre of Iran, 2001).

Salinity of Soil and Water

Nearly 10 per cent of the earth's land is salt-affected and an estimated 10 million ha of agricultural land are lost annually due to salinization and water logging. Salinity of soil and water resources is a serious threat in many parts of Iran. The estimated land area affected by salinity varies between 16 to 23 Mha (Siadat *et al.*, 1997). These figures include both cultivated and barren lands. No exact data is available on intensity of this problem in the arable lands. The main difficulty in this regard is the temporal variations of salinity during the growing season, due to the effects of irrigation water, which add or leach the salts. In general, 50 per cent of all growing areas of Iran are encountered by high-salt concentration conditions and drought (Siadat, 2004).

As to the quality of the surface water resources is concern; the major rivers of the country have a low salt load (an electrical conductivity of about 0.3 0.7 dSm^{-1}). However, variations along the stream occur as these rivers receive drainage water from the neighbouring farmlands. Besides, with the decrease in flow rate during the summer months, salt concentration may increase to limiting levels (Ghazi, 2002; Siadat *et al.*, 1997).

Salinity of groundwater is more serious than that of surface waters. This has been increasing in recent years due to the overdraft and intrusion of the surrounding saline bodies of water. Considering the fact that nearly half of the water used in Iran's agriculture comes from the groundwater, the threat of the effects caused by salinity on the sustainability of crop production in the country becomes evident (Alizadeh and Keshavarz, 2005; Bybordi, 1989; INSRC, 2005).

Causes of Agricultural Salinity in Iran

Considering the diversity of geological, climatic, and hydrological conditions of Iran, it is no surprise that many factors are involved in the development of salinity problems in Iranian agriculture. According to the data extracted from 1:1,000,000 soil map of Iran (in digital format), slightly to moderately salt-affected soils cover about 25.5 million ha and soils having severe salinity occupy approximately 8.5 million ha. The slightly to moderately salt-affected soils are mostly formed on the piedmonts (normally, on the distal parts of the alluvial fans) at the foot of the Zagros and Alborz mountains. Lands having severe to extreme salinity are mostly located in the Central Plateau, the Khuzestan and Southern Coastal Plains, and the Caspian Coastal Plain.

There are two main group based on soil Salinization and water resources *i.e.* natural and human induced factors (Siadat *et al.*, 1997, Badripour, 2004).

Naturally Occurring Salinity

1. *Geological composition of the parent material of the soils*: Iran is rich in evaporate deposits, which were deposited episodically throughout the Phanerozoic. Stream salinity, salinization of the surface water resources, mainly due to natural conditions, is one of the main causes of salt accumulation in the soils of the Central Plateau.
2. *Wind-borne salinity*: Strong winds blows most of the time during the year in the Central Plateau, contribute to the expansion of soil salinity in the region by distributing the accumulated salts on the soil surface to a wider area.
3. *Seawater intrusion*: This phenomenon occurs mostly in the coastal areas where saline seawater enters the inland channels or inundates coastal lowlands by tidal waves. Seawater intrusion into the shallow groundwater in the coastal areas has also led to salinization of soils in the Caspian Coastal Plain, and in the Khuzestan and Southern Coastal Plains in southwestern Iran (Ghassemi *et al.*, 1995; UNCCD, 2003).

Human-induced Salinity

Land degradation and desertification in Iran have accelerated during recent decades due to the following factors:

1. *Mismanagement of soil and water resources*: Poor soil and water management, over-exploitation of groundwater and grazing, are amongst the main causes of human-induced soil salinity. In recent decades, use of groundwater for irrigation purposes has increased considerably. Continuous use of saline water in these areas has undoubtedly increased soil salinity, but there is no data available on the severity of the problem.
2. Population has doubled during the last 20 years. The need for more agricultural and pastoral products has forced people to use land extensively or to convert forests and rangelands to cultivated land, without any consideration for the real potential and capacity of such land.
3. Over-grazing and misuse of rangelands.
4. Over use of wood and plants as fuel for household cooking and heating, as well as irregular and uncoordinated exploitation of water resources.
5. Refugees have degraded 1.2 million hectares of natural resources to meet their needs for fuel, farming and breeding.

The policies and programs of the Government to rehabilitate and develop renewable natural resources, with special consideration for desertification control, are as follows:

1. Use of the mass media and other means to increase public awareness about the importance of renewable natural resources and the dangers of desertification;
2. Acceleration of socio-economic development in rural areas so as to prevent the migration of subsistence farmers and pastoralists to major cities and urban centres.

3. Protection of the environment and restoration of ecological stability.
4. Rehabilitation and reclamation of degraded land.
5. Industrial development using environmental sound technology.
6. Active demographic policies, which have decreased the population growth rate from 3.17 per cent to 1.2 per cent.
7. Sand dune stabilization to minimize negative effects on farm land, residential areas, strategic areas, roads and other valuable economic infrastructure.
8. Encouragement of popular participation in decision-making and implementation process.

Agricultural Production and Salinity

The agricultural sector is one of the main priorities in the national developmental plans. Cereal crops (*i.e.* wheat, barley, and rice) cover almost 70 per cent of the cultivated land, while wheat alone covers nearly 52 per cent. Among permanent crops, the most important ones are pistachio (354,000 ha), grapes (275,000 ha), dates (230,000 ha), citrus (230,000 ha) and apples (153,000 ha), respectively. All of these crops and trees are among relatively salt tolerant plants.

In Iran, it is a fact that the average yield of different crops is much higher under irrigated system than under dry farming. For example, the average yield of irrigated wheat is about 4,000 kg/ha while under dry farming this figure is below 1,000 kg/ha. This fact has led many policy makers and planners to lay special emphasis on the expansion of irrigated farming system. During the 1980's and early 1990s, irrigated area expanded at a rate of 3.8 per cent. For the fourth national development plan (2005–2010), an increase of 0.5 Mha in irrigated land, with emphasis on pressurized systems, has been set as an important target. According to published agricultural statistics, the area under irrigation in Iran is 7.80 Mha. Of this total irrigated land area, 3.0 Mha are devoted to cereal, 2.0 Mha to orchards, and 2.8 Mha to different field crops. For irrigating these areas, 82.1 bcm of water is used.

Overdraft of groundwater is the main problem associated with the expansion of irrigated farming. Presently, the total groundwater recharge of Iran is estimated to be 46 bcm while the annual extraction amounts to 55 bcm. A negative balance of nine bcm/year is alarming. This situation has resulted in both the lowering of groundwater tables and higher salinity of water (Siadat, 1994). Increase in irrigation water salinity would result in substantial increase in soil salinity.

Surface irrigation techniques are used on 98.75 per cent of the area equipped for irrigation and only the remaining area of 1.25 per cent benefits from a pressurized irrigation system. Almost, half the area is irrigated from groundwater resources, including spring water. According to the landholding and technologies which are used, the farming systems are grouped as: small farms (< 10 ha) 47 per cent, medium size farms (10-50 ha) 43 per cent, and large farms (> 50 ha) 10 per cent.

It is unfortunate that there is no established observation network for monitoring the changes in soil salinity over the years. Nevertheless, field observations of patchy crop stand, retarded growth, leaf burn (where sprinklers are used with poor water quality), as well as the increasing complaints by the farmers are evidence of the intensification of salinity problem. Another evidence, in this regard is the salinity trend in waters, particularly the well waters (Siadat, 1997).

Salinity in Iranian Literature

Iranians do not waste time, energy and seeds in a saline medium. Salinity of water and soil and its negative effects on crop production is one of the very obvious facts, which has also been quoted by many poets thousand of years ago. It indicates that not only were expert aware of this truth, but also the common people were familiar with the problem of soil salinity. Saadi Shirazi (1210 - 1290), one of the most famous poets of Iran, in one of his poems wrote:

Blossoms, a saline soil does not make/

Seeds and toil, they all go waste/

SaebTabrizi expressed the same meaning as below:

Merriment, a broken heart, O Saeb! Show?

Can a saline soil saffron grow?

Fakhroddin Asad Gorgani used salinity in his poems frequently, a few lines borrowed from some of his couples are cited below:

Seeds in this saline land thou sow/

Lose cradle and labour, nothing grow/

Advice that thou of me seek/

Seed in saline soil is not to reap/

There are many more poets such as moulavi Roomy, Saeb Tabrizi, Khaghani, Onsori and Hafiz Shirazi where verses about salinity be found.

There are some Iranian adages about salinity, which the natives use in their common languages and, of course, that too beautifully, and Iranian people use its consequences excellently. All these bear testimony to the fact that how important salinization was for people of this part of the world during the history.

Indigenous Knowledge

Salinity has a root in empowering and weakening of many dynasties in Iran and its neighbouring countries for thousands years. There is no sign from productive irrigated lands of the big civilization beside Tigris and Euphrates because of the salinization of their soils, and one of the main reasons for weakening of Sousanis dynasty might be the hyper-aridity of Khuzestan area where their troops accepted defeat in front of Arabs.

For many centuries people of Iran have been trying to overcome this major problem of salinity so that their soils remain productive for long, but because of the lack of proper technology, and nature of their climate their land became salty. They implemented many strategies for salinity management, which were comparable to those being used in the 21st century. Some of the strategies used are briefly mentioned below (Alizadeh *et al.*, 2001; Koocheki and Mahalati, 1992).

1. Fallow, the statistics in Table 20.3 indicate that Iranian farmers let their soil free of cultivation even year or a year out of three years. Apart from many other advantages of fallowing, one of the main objectives was to postpone the salinization of their soils and make corresponding exigency plans.
2. Use of salt tolerant crops, it is observed from the cropping pattern of Iranian agriculture, that the main part of the crops grown are salt tolerant. The main winter annual crops are wheat and barley where both of them are relatively salt tolerant. The main summer crops are cotton, sugar beet, turnip, millets, sorghum, etc. which are categorized as salt tolerant crops. Perennial trees pistachio and dates, which are the main fruits of Iran, are rated amongst the most salt tolerant trees.
3. Irrigation Method, Iranian farmers were aware that they should wash out excess salt from the root medium of their crops, therefore they implemented a flood irrigation method and with each irrigation they apply more water for drainage.
4. Invention of the subterranean ('Qanat'), since the history of agriculture, Iranian have recognized that the quality of water upstream is much better than that of downstream, hence, based on the natural slope of the surface they excavated an underground channel and diverted the underground fresh water to the surface, and irrigated their land with this water. After over drafting of the subsurface water many subterranean dried up and farmers had to pump the underground water from the site, but this water is not as fresh as the subterranean water and caused salinization of the soils (Alizadeh and Keshavarz, 2005; Keshavarz *et al.*, 2005).

Iranian Reserach Organisation from Saline Agriculture

There is no other administrative ministry larger than the Ministry of Jihad-e-Agriculture, which the largest research organization of the country placed in this ministry. Iranian Agricultural and Research organization have more than 344 research centres all over Iran that some of them doing research specifically on a particular crop or tree such as Pistachio Research Centre, Date Palm Research Centre, Rice Research Centre, and Citrus Research Centre, some are doing research on specialized areas such as dry land farming, salinity, biotechnology, while the rest are located in the centre of each province and based of the priorities of agriculture relevant to the respective regions. Conducting research on different aspects of crop production under salinity stress in pistachio, date palm is the main objective of national centres like Iran National Salinity Research Centre. For the other bodies as it shows in Table 20.5, total number of Iranian research centres are 246 that potentially or actually do research in management of stressful crop production particularly in salinity.

Table 20.5: Number of Research Centres Working Directly or Indirectly on Salinity

Type of Institution	*Number*
National Research Institutes	14
Faculties of Agriculture	50
Province Research Centres*	31
Iran National Salinity Research Centre	1
District Research Stations	150
Total	246

* These centres normally have 1-5 research stations.

Agricultural Biotechnology Research Institute of Iran (ABRII)

ABRII is one of the national agricultural research institutes working under the supervision of Iranian Agricultural Research Education and Extension Organization (AREEO). It has recently (year 2000) been upgraded to the level of an independent institute. ABRII offers research activities in the field of advanced plant molecular biology and biotechnology (Gharayazi, 2000).

Modern and advanced research facilities such as trial fields and lysimeters, several greenhouses and controlled rooms, 10 large phytotrons, cold rooms, tissue culture facilities, radioisotope facilities and an optical room have been established by ABRII and it is the most advanced agricultural biotechnology research institute in Iran. In addition, ABRII has several modern and well-equipped research facilities and laboratories such as genomics, plant physiology and proteomics, haploid breeding, microbiology, genetic engineering and analytical service laboratories, computer centre and access to Internet (ABRII, 2005).

The main objectives of ABRII are strengthening and expansion of research and promotion of national and international collaborations in the area of agricultural biotechnology and genetic engineering to solve the major agricultural problems such as biotic and abiotic (particularly salinity and drought) stresses and increase crop production (Gharayazi, 2000; ABRII, 2005).

Iranian National Salinity Research Centre (INSRC)

INSRC is one of the national agricultural research institutes working under the supervision of Iranian Agricultural Research, Education and Extension Organization (AREEO). INSRC was established in the year 2000 at the level of an independent centre in Yazd, Iran. INSRC offers research activities in the field of soil, water salinity and salinity management. As a national research institute, INSRC strives to promote applied and basic research in the main aspects of salt stress, and organizes workshops for advanced training of experts who are involved in R&D.

Modern and advanced research facilities, including field and laboratory equipments, have made INSRC the most technically advanced agricultural salinity research institute in Iran (INSRC, 2005).

The main objectives of INSRC are strengthening and expansion of research, and promotion of national and international collaborations in the area of agricultural salinity to increase crop production. Developing stress tolerant crops is a worldwide mission for alleviating poverty and providing world food security (Alizadeh and Keshavarz, 2005).

The main research projects of INSRC focused on the following four areas:

1. Database on salt affected soils and poor quality groundwater.
2. Management of soil and waterlogged soils.
3. Management of water resources and use of poor quality water.
4. Technology transfer, impact assessment and human resource development.

Academic Institutions

There are more than 105 state universities in Iran, out of these, around 50 are comprehensive universities and one faculty of agriculture is the main body of the University. There is one Agricultural University nationwide in Gorgan that due to importance of agriculture in Golestan province was established 15 years back. Major universities involved in research and education in salinity managements are:

Tehran University with two faculties in the name of Karaj and Aburaihan, Tarbiat Modarres University, Shiraz University, Razi University, Industrial University of Isfahan, Ferdowsi University of Mashhad, Tabriz University, Uromia University, Sharekord University, Birjand University, Mazandaran University, Shahid Chamran University, Bahonar University, Yazd University, Hormozgan University, Gorgan University, Zabol University, Buali Sina University, Islamic Azad University. Conducting research on salinity is not mandatory for staff members of Universities, but due to local importance, this branch of research is very popular amongst Iranian scientists.

Investments in the development of human resources and training of expertise in salinity management by the Iranian Universities have been substantial in the last decade. At present, agronomy students have to pass one three-unit course in the bachelor and one three-unit course in the Master on crop production and management under stressful environments. In addition, many postgraduate students opt for the topic of effects of salinity stress on different aspects of crop production as their research project.

Prospects

Iran is a promising place for development of biotechnologies for the following reasons:

1. Large number of trained personnel as biologists in several disciplines, in institutions throughout the country.
2. Richness in natural resources and biodiversity;
3. Young manpower in country, it gives the country a comparative advantage;

4. Although Iran has had a late start in biotechnology research and development, establishment of new organizations and research centres could bring country to an acceptable level in the world in short period of time.

Constrains

1. Saline agriculture in Iran is suffering from lack of a national long term plan. Neither researchers, nor farmers know which agricultural system should they implement, on which crop plant they should concentrate, and which method of salt amendment will be supported by the authority.
2. Most salinity research projects were limited and confined to conventional annual field crops. While there are many perennial and new plants that showed a good potential to grow in saline environments. There is limited number of laboratories in research centres having new biotechnologies facilities.
3. There is no adequate funding for basic research, all supporters are looking for projects where the results be available after a short time.
4. Most of institutions had no major role in formulating the national policy
5. Lack of interaction between institutions dealing with salinity and private industry
6. Research progress in salinity has not yet been adequately commercialized to make a significant contribution to the national economy.

Therefore the establishment of an Iranian national saline agricultural program is top priority for government. Decision should be taken to support long-term, basic research and advanced salinity projects by providing enough funds and engaging all skilled scientist. Government should help to identify these processes and products emerging out of research, which could be exploited commercially.

Acknowledgements

I would like to acknowledge the Centre for Science and Technology of Non-aligned and Other Developing Countries particularly its Director Professor Arun P. Kulshreshtha, and President of Birjand University, for making it possible to hold this workshop. I also would like to express my gratitude to Mr. Tajamul Showket for helping me in editing and typing this paper.

References

Agricultural statistics of Iran (1996–97). Ministry of Agriculture. Bureau of Statistics and Information. Publication No.77.01. March April, 1998. (In Farsi).

Alizadeh, A., G.H. Kamali and I. Malek-Mohammadi. 2001. Sustainable utilization of saline water in agriculture based on indigenous knowledge. International Symposium on Prospects of Saline Agriculture in the GCC Countries, Dubai, United Arab Emirates. Biswas, A.K. 1998. Deafness to global water crises. Causes and risks. Ambio 27: 93.

Alizadeh, A. and A. Keshavarz. 2005. Status of agricultural water use in Iran. Proceedings of an Iranian-American Workshop: Water Conservation, Reuse, and Recycling, Tehran, Iran, 94-105 pp.

Badripour, H. 2004. Rangeland management in Iran. Bureau of Rangeland–Forest, Rangeland and Watershed Management Organization (FRWO). Ministry of Jihad-e-Agriculture.

Bureau of Information and Statistics. 1994. Agricultural Statistics Yearbook. 1993. Deputy Ministry of Plan and Project, Ministry of Agriculture.

Bybordi, M. 1989. Problems in planning irrigation projects in Iran. In: Rydzewski, J.R., and Ward, C.F. (Eds.), Proceedings of the International Conference on: Irrigation: Theory and Practice. Uni. Southampton. London. Entsch Press.

Committee to rehabilitate rangelands and mitigate drought effects. 2004. Forest, Range and Watershed Management Organization, Tehran, I.R.Iran.

Dordipoura, I., H. Ghadirib, M. Bybordia, H. Siadatc, M.J. Malakoutia and J. Husseinb. 2004. The Use of Saline Water from the Caspian Sea for Irrigation and Barley Production in Northern Iran. ISCO 2004-13th International Soil Conservation Organization Conference–Brisbane, Australia.

Gharayazi, B. 2000. Agricultural biotechnology in Iran: present status, achievements Constrains and prospects. Proceeding of TWAS 12th general meets. Tehran.

Ghassemi, E., A.J. Jakeman and H.A. Nix. 1995. Salinization of land and Water resources. CABI.

Ghazi, I. 2002. Water resources management and planning in Iran. Report to the University of Isfahan.

Keshavarz, A., S. Ashraft, N. Hydari, M. Pouran and E. Farzaneh. 2005. Water Allocation and Pricing in Agriculture of Iran. Proceedings of an Iranian-American Workshop Water Conservation, Reuse, and Recycling, Tehran, Iran, 153-172 pp.

Koocheki, A. and M.N. Mahalati. 1992. Feed value of some halophytic range plants of arid regions of Iran. In Squires, V. R., and A. T., Ayoub (eds.), Halophytes as a resource for livestock and for rehabilitation of degraded lands. Kluwer Academic Publishers. Netherlands. 249-253 pp.

National Action Programme to Combat Desertification and Mitigation of Drought Impacts of I.R.Iran. 2003. UNCCD

Official Web site of Agricultural Biotechnology Research Institute of Iran (ABRII), Ministry of Agriculture, Iran, 2005. www.abrii.ac.ir/.

Official Web site of Iranian National Salinity Research Centre (INSRC), Ministry of Agriculture, Iran, 2005. www.abrii.ac.ir/.

Official Web site of Ministry of Jihad-e- Agriculture of Iran. www.maj.ir/.

Official Web site of Agricultural and Food Organization of United Nations (FAO). www.fao.org/.

Siadat, H., M. Bybordi and M.J. Malakouti. 1997. Salt affected soils of Iran: A country report. International symposium on "Sustainable Management of Salt Affected Soils in the Arid Ecosystem". Cairo. Egypt.

Siadat, H. 2004. Agriculture and Salinity in Iran. in WWW.unu.edu/env/land/iran-1/02-siadat per cent 20paper.doc/.

Statistic Centre of Iran. 2001. Year book 2000, Tehran, I.R.Iran

Szabolcs, I. 1992. Salt affected soils as the ecosystem for halophytes. In: Squires, V.R., and A.T., Ayoub (eds.), Halophytes as a resource for livestock and for rehabilitation of degraded lands. Kluwer Academic Publishers. Netherlands. 19-24 pp.

21

Ecology and Economic Potential of Halophytes: A Case Study from Turkey

M. Ozturk[1*], A. Guvensen[1] and S. Gücel[2]

[1]Ege University, Science Faculty, Botany Department, Bornova-Izmir, Turkey

[2]Near East University, Nicosia, North Cyprus

**E-mail: munirozturk@hotmail.com, munirozturk@gmail.com*

ABSTRACT

About 300 halophyte species belonging to 40 families and 150 genera are distributed in Turkey. The dominant families Poaceae, Chenopodiaceae, Asteraceae and Fabaceae include the maximum number of taxa. The genera with highest number of taxa are *Limonium* (17), *Juncus* (17), *Salsola* (16), *Plantago* (15) and *Trifolium* (12). The number of endemic taxa is about 41. Out of these 33 are in danger of extinction together with 11 non-endemic halophytes.

Introduction

The area of Turkey is around 7,835.77 km^2. It is geographically divided into seven divisions namely; North Anatolia, Marmara, West Anatolia, South Anatolia, Central Anatolia, Southeast Anatolia and East Anatolia. The climate is mediterranean in the west and south, oceanic in Marmara and north, and continental in the central and east Anatolian divisions. Southeast experiences a steppe climate with very hot summers and severe drought (DIE, 2000). A major part of nearly 4.3 million ha of the degraded and unproductive land in Turkey (Kaynak *et al.*, 2000; Ozturk and Guvensen, 2002) are mainly inhabited by halophytic taxa. Uptill now very few papers have been published on this group of plants notable among them being Birand (1960), Zeybek *et al*. (1976), Gehu and Uslu (1987), Guvensen *et al*. (1996), Ozturk *et al*. (1995, 2006a,b), Yurdakulol and Ercoskun (1990), Guvensen and Ozturk (2004) and Cetin *et al*. (2005). In this paper an attempt will be made to evaluate the ecology and economic potential of the halophytic plant cover in Turkey.

Halophyte Diversity

The general features, distribution areas and risk category of the important taxa with possible potential for an economic evalution are given below. The endemics have been included in the list (Table 21.1).

Table 21.1: List of Halophytic Taxa Together with Some Ecological Features

Families	*Taxa*	*Altitude (above msl)*	*Flowering Time (months)*	*Distribution*	*Threatened Categories*
Apiaceae	*Bupleurum semicompotisum*	200	3-6	WA	DD
"	*Bupleurum tenuissimum*	1-300	7-9	NWA, W.Med. and Black Sea	VU
"	**Bupleurum turcicum*	900-1100	6-7	W, CA	LR (nt)
Asteraceae	**Achillea sieheana*	950-1200	6-9	CA.	VU
"	*Artemisia campestris*	1-1500	6-9	NW, N, C, E A	
"	*Artemisia santonicum*	1-1300	7-10	NW, CA	
"	*Aster tripolium*	-	6-9	N, W, A	
"	*Bidens cernua* var. *radiata*	610-1060	8-9	N,C, E, S A	VU
"	*Carlina lanata*	10-800	6-8	E.Med.	
"	**Cousinia birandiana*	850-1200	6-8	CA	LR (nt)
"	**Cousinia humilis*	-	-	"	CR
"	*Cirsium alatum* subsp. *alatum*	900-1800	6-9	"	
"	*Inula aucherana*	550-1900	7-9	"	
"	*Inula crithmoides*	sea level	9-11	W, SA	
"	**Taraxacum farinosum*	800-1200	6-9	CA	lc
Boraginaceae	*Heliotropium europaeum*	0-1400	6-9	N, SA	
Brassicaceae	**Lepidium caespitosum*	800-1300	5-6	Turkey	
Caryophyllaceae	**Gypsophila oblanceolata*	950-1050	6-8	CA	VU
"	**Silene salsuginea*	900	7	"	EN
Chenopodiaceae	*Artrocnemum fruticosum*	1-1100	8	N, W, S, C.A.	
«	*Atriplex davisii*	900-1000	5-7	C, WA	
«	*Atriplex hastate*	1-50	5-8	W, NA	
"	*Atriplex laevis*	900-1200	5-7	C, EA	
"	*Atriplex lasiantha*	0-1800	5-7	W,E, N, Middle, SA	

Contd...

Table 21.1–Contd...

Families	Taxa	Altitude (above msl)	Flowering Time (months)	Distribution	Threatened Categories
Chenopodiaceae	*Camphorosma monspeliaca* subsp. *lessingii*	1-1400	6-8	C, E A	
"	*Chenopodium album* subsp. *album*	0-2000	5-8	E, N, Middle, E, SA	
"	*Chenopodium botrys*	0-1900	5-7	NW, Middle, SA	
"	*Chenopodium murale*	0-400	5-7	W, N, SE, SA	
"	*Chenopodium opulifolium*	sea level	5-8	WA	
"	*Halanthium kulpianum*	900-940	-	C, EA	
"	*Halimione portulacoides*	1-900	6-8	N, W, CA	
"	*Halimione verrucifera*	1-1000	6-8	N, CA	
"	*Halopeplis amplexicaulis*	1	6-8	SA	EN
"	*Halocnemum strobilaceum*	1-1100	7-9	N, W, S, CA	
"	**Kalidiopsis wagenitzii*	940	5-6	CA	EN
"	*Kochia prostrate*	1-1900	6-8	C, EA	
"	*Microcnemum coralloides* subsp. *anatolica*	940-1000	5-7	CA	VU
"	*Noaea minuta*	1300	6-8	Middle A	VU
"	*Noaea mucronata* subsp. *mucronata*	500-2000	5-7	Turkey	
"	*Panderia pilosa*	900-1200	5-8	N, C, EA	
"	*Petrosimonia brachiata*	0-900	6-9	C, SA	
"	*Salicornia europaea*	-	7-9	W, SA	
"	**Salsola anatolica*	1000	6-8	CA	LR (lc)
"	*Salsola crassa*	1000	5-7	«	
"	*Salsola inermis*	900-950	5-7	«	VU
"	*Salsola kali*	0-1010	5-7	SA	
"	*Salsola laricina*	950	6-8	CA	
"	*Salsola macera*	850-950	5-7	C, EA	
"	*Salsola soda*	1	5-7	N, W, SA	
"	*Salsola ruthenica*	1-1750	5-7	Turkey	
"	*Suaeda altissima*	900-1250	5-7	W, C, EA	
"	*Suaeda carnosissima*	900-1200	5-8	CA	
"	*Suaeda confusa*	800	6-8	C, EA	
"	*Suaeda eltonica*	900	5-7	CA	VU
"	*Suaeda prostrata* subsp. *prostrata*	1	6-9	N, WA	

Contd...

Table 21.1–Contd...

Families	*Taxa*	*Altitude (above msl)*	*Flowering Time (months)*	*Distribution*	*Threatened Categories*
Convolvulaceae	*Cressa cretica*	sea level	6-8	W, SA	
Fabaceae	*Alhagi mannifera*	0-2330	6-8	S, SEA	VU
"	*Alhagi pseudalhagi*	1-1200	6-8	*Turkey*	
"	**Astragalus karamasicus*	450-2060	6-7	E, C, S, WA	LR (lc)
"	*Astragalus shelkovnikovii*			EA	VU
"	*Lotus corniculatus* var. *tenuifolius*	1-2750	4-9	Turkey	
"	*Lotus strictus*	500-1300	6-8	"	
"	*Medicago minima* var. *minima*	1-1750	3-5	"	
"	*Melilotus messanensis*	1	2-4	WA	
"	*Prosopis farcta*	0-1450	6	S, EA	
"	*Sophora alopecuroides* var. *alopecuroides*	1-1750	4-7	Turkey	
"	**Sphaerophysa kotschyana*	1000	6	CA	LR (cd)
"	*Tetragonolobus maritimus*	800-1750	6-7	Turkey	
"	*Trifolium fragiferum* var. *fragiferum*	1-1350	4-8	NWA, S, CA	
"	*Trifolium repens* var. *repens*	500-2700	3-9	Turkey	
Iridaceae	**Gladiolus halophilus*	900-1200	6-7	CA	VU
Juncaceae	*Juncus acutus*	1-150	3-5	Turkey	
"	*Juncus articulatus*	1-3000	4-8	«	
"	*Juncus fontanesii* subsp. *pyramidatus*	1-1200	4-7	W, SA	
"	*Juncus gerardi* subsp. *gerardi*	1-2100	5-7	Turkey	
"	*Juncus gerardi* subsp. *libanoticus*	50-2500	5-8	NW, CA	
"	*Juncus heldreichianus* subsp. *heldreichianus*	1-1600	4-7	W, SA	
"	*Juncus littoralis*	*1-50*	*4-6*	Turkey	
"	*Juncus maritimus*	1-1050	5-7	W, Middle, SA	
"	*Juncus subnodulosus*	1-100	6-7	NWA	
"	*Juncus subulatus*	1	4-6	WA	
Lamiaceae	**Salvia halophila*	950-1000	8-10	CA	VU
Liliaceae	*Allium macrochaetum* subsp. *macrochaetum*	950-1650	6-8	"	
"	**Allium sieheanum*	900-1200	7-8	"	LR (lc)
"	**Asparagus lycaonicus*	1000	8	"	EN
"	*Asparagus persicus*	800-1700	5-8	Turkey	
"	**Bellevalia forniculata*	1800-2400	5	SEA	LR (lc)
"	*Linum seljukorum*	1100	6-8	CA	VU

Contd...

Table 21.1–Contd...

Families	*Taxa*	*Altitude (above msl)*	*Flowering Time (months)*	*Distribution*	*Threatened Categories*
Plantagina ceae	**Plantago anatolica*			EA	LR (cd)
"	*Plantago coronopus* subsp. *commutate*	0-800	2-7	N, W, S, E A	
"	*Plantago coronopus* subsp. *coronopus*	0-1750	4-11	N, W, S A	
"	*Plantago crassifolia*	1-900	5-10	W, Middle A	VU
"	*Plantago cretica*	0-600	4-5	W, S A	
"	*Plantago lagopus*	0-2000	4-8	NW, W, Middle, SA	
"	*Plantago lanceolata*			Turkey	
"	*Plantago maritime*	0-2400	5-8		
"	*Plantago scabra*	0-1250	5-11	Turkey	
Plumbagina-ceae	**Acanthalimon halophilum*	900-110	6	*C A*	VU
"	**Limonium anatolicum*	900-1000	6-9	"	VU
"	*Limonium angustifolium*	1	5-10	SA	
"	*Limonium bellidifolium*	1-1010	6-9	W, C A	
"	*Limonium echioides*	0-200	4-7	W, S A	
"	*Limonium gmelinii*	1-1450	5-10	Turkey	
"	**Limonium effusum*	0-750	7-8	W, SWA	VU
"	*Limonium graecum* var. *graecum*	sea level	5-7	WA	VU
"	**Limonium iconicum*	900-1040	6-9	CA	LR (lc)
"	**Limonium lilacinum*	900-1200	6-9	"	LR (lc)
"	*Limonium ocymifolium*	sea level	5-7	"	EN
"	*Limonium sieberi*	1-5	5-7	W, SW A	VU
"	*Limonium sinuatum*	1-100	5-7(10)	W,S A	
"	**Limonium tamaricoides*	-	-	C A	EN
"	*Limonium virgatum*	1-20	6-10	W, SA	
Poaceae	*Aeluropus littoralis*	1-1200	5-10	"	
"	*Arundo donax*			Thrace, W, S, NEA	
"	*Cynodon dactylon* var. *dactylon*	0-1830	4-9	W,NE, S A	
"	**Festuca rubra* ssp. *Pseudorivularis*	650-2000	6-7	Black Sea	CR
"	*Phragmites australis*	0-2400	8-10	Turkey	
"	**Puccinellia bulbosa* subsp. *caesarea*	1100-1200	7-8	C A	CR
"	*Puccinellia festuciformis*	sea level	5-6	N A	
"	*Sporobolus virginicus*	0-950	4-10	NW, S A	

Contd...

Table 21.1–Contd...

Families	Taxa	Altitude (above msl)	Flowering Time (months)	Distribution	Threatened Categories
Polygonaceae	*Rumex bucephalophorus*	0-150	2-5	N, W, S A	
Ranuncula-ceae	**Consolida glandulosa*	800-1600	6-9	C, E A	LR (nt)
Scrophularia-ceae	**Verbascum helianthemoides*	960-1200	6-8	C A	VU
"	**Verbascum pyroliforme*	1000	6-9	"	VU
Tamaricaceae	*Tamarix hampeana*	-	4	WA	
"	*Tamarix parviflora*	0-300	3-6	N, W, Middle A	
"	*Tamarix smyrnensis*	0-1000	4-8	N, Middle, W, S, SEA	
"	*Tamarix tetrandra*	0-1300	5	N, Middle,W, S A	
Zygophyllaceae	*Tribulus terrestris*	0-1200	6-9	Turkey	

*: Endemic; CR: Critically endangered; VU: Vulnerable; DD: Data deficient; EN: Endangered; LR: Lower risk; (cd): Conservation dependent; (lc): Least concern; (nt): Near threatened; A: Anatolia, E: East; W: West; N: North; S: South; C: Central; NW: Northwest; SE: Southeast; CA: Central anatolia; NE: Northeast).

Economical Prospects

More than 6 million ha of arable land is lost annually in industrialising countries where nearly one billion humans are facing food crisis. In view of this, guiding human activities towards an understanding of the changes in ecology and natural landscape will be highly fruitful. This will be possible only through interdisciplinary actions. One of the areas we need to ponder over deeply is management of saline-alkaline soils, since a great economic potential exists in the halophytic plants flourshing on these soils. For this purpose sound information is required on the plant diversity and vegetation cover of such degraded lands.

A large number of halophytes have been used as fodder, food, fiber, fuel, animal feed, medicine, or bread making, etheric oil and gum production by humans for 100's of years (Somers, 1982; Gallagher, 1985; Glenn *et al.*, 1991; Choukr-Allah, 1991; Khan and Duke, 2001). The data base of salt tolerant plants of the world compiled by Aronson (1984, 1989),Lieth (2002) contains more than 2500-3000 species in 550 genera and 117 families and focuses on economic uses as well as life form, plant type, geographical distribution, maximum reported salinity tolerance and photosynthetic pathways.

Salicornia species are used for vegetable oil production as well as forage crop in Kuwait, U.A.E and Saudi Arabia (Keen, 1986; Riley and Abdal, 1993). *Kochia, Salsola* and *Atriplex* together with *Salicornia* are good sources for leaf protein production (Dagar, 1995). *Salsola* is used for the production of bioactive compounds and the *Juncus* species for paper, pulp and fibre production. The *Atriplex* and *Spartina* species are used as an animal feed in Africa, Europe and the Americas (Ozturk *et al.*, 2006 b).

Suaeda has been used for exraction of soda for traditional production of soap in Iran whereas young leaves of *Suaeda aegyptiaca* are consumed by people. In some places the exract of *Salsola kali* is used for sterilization of raisins. The *Salicornia persica* is used as fodder particularly for goats (Akhani, 2006).

In Bahrain *Atriplex leucoclada, Cressa cretica, Cynodon dactylon, Heliotropium crispum, Limonium axillare, Phragmites australis, Plantago ovata, Prosopis farcta, Salsola villosa, Suaeda vermiculata* are used in the folk medicine whereas, *Alhagi graecorum* and *Chenopodium murale* are cosumed as food plants and used for medicinal purposes as well. The species like *Alhagi graecorum, Salsola baryosma, Aeluropus lagopoides, Aeluropus littoralis, Juncus rigidus, Tamarix spp.* are used as forage, fiber, oil, firewood, fuel and timber (Abbas, 2006).

According to Khan and Qaiser (2006) the species of *Arundo, Rumex, Suaeda, Alhagi, Atriplex, Chenopodium, Kochia, Prosopis, Salsola, Puccinellia, Sporobolus, Arthrocnemum, Salicornia, Tamarix, Plantago, Tribulus, Juncus, Imperata, Phragmites* and *Typha* are used in Pakistan for medicinal purposes, as food, forage and fodder, oil seed crops, fuel wood as well as for mats weaving, basket thatching and cordage.

Out of the species cited above the following taxa show distribution in Turkey as well; *Heliotropium europaeum* (Boraginaceae); *Artrocnemum fruticosum, Atriplex hastat*a *Atriplex lasiantha, Chenopodium album* subsp. *album, Chenopodium botrys, Chenopodium murale, Chenopodium opulifolium, Kochia prostrata, Salicornia europaea, Salsola kali, Salsola soda Salsola ruthenica, Suaeda prostrata* subsp. *prostrata* (Chenopodiaceae); *Cressa cretica* (Convolvulaceae); *Alhagi mannifera, Alhagi pseudalhagi, Prosopis farcta* (Fabaceae); *Juncus acutus, Juncus maritimus, Juncus subnodulosus, Juncus subulatus* (Juncaceae); *Plantago coronopus* subsp. *commutata, Plantago coronopus* subsp. *coronopus, Plantago crassifolia, Plantago cretica, Plantago lagopus, Plantago lanceolata, Plantago maritima, Plantago scabra* (Plantaginaceae); *Limonium angustifolium, Limonium bellidifolium, Limonium gmelinii, Limonium graecum* var. *graecum, Limonium sieberi, Limonium sinuatum, Limonium virgatum* (Plumbaginaceae); *Aeluropus littoralis, Cynodon dactlylon* var. *dactylon, Imperata cylindrica* subsp. *cylindrica, Phragmites australis, Puccinellia festuciformis, Sporobolus virginicus* (Poaceae), *Rumex bucephalophorus* (Polygonaceae); *Tamarix hampeana, Tamarix parviflora, Tamarix smyrnensis, Tamarix tetrandra* (Tamaricaceae); *Tribulus terrestris* (Zygophyllaceae). However, these are not evaluated to a large extent with a few exceptions. *Salicornia europaea* is used as salad by a large population in the Aegean region of Turkey. In some markets it was observed that *Arthrocnemum fruticosum* is sold in place of *Salicornia europaea,* so people consume the former in place of latter unknowingly, but up till now no adverse effect has been reported. *Limonum bellidifolium* is collected in spring season for indoor house decorations. *Halocnemum strobilaceum, Aeluropus littorali* and *Arthrocnemum spp.* are grazed by the cattle deliciously. The *Juncus* species are evaluated for fiber production, basket making, erosion control and lately for the wastewater treatment ponds. The halophytes can be used for biological desalination and reclamation of saline-alkaline habitats. The littoral halophytic taxa serve as a reclusive area for many animal species (Zeybek *et al.*, 1976). These plants can serve as indicators of salinity-sodicity, and areas with ground waters possessing high salinity could be used for halophytic forage crop production. Halophytes can serve in erosion control alongside the coastal zones due to their strong root systems.

During the last few decades this group of plants has attracted much attention in several parts of the world, in particular on biosaline agriculture (Ahmed and San Pietro, 1986; Khan and Weber, 2006; Ozturk et al.2006; Boer et al.2006). In fact there is a great potential in halophytic plant cover for consumption as well as amelioration of degraded lands in Turkey. Some work has already begun (Ozturk *et al.*, 1998; Akbas *et al.*, 1999; Guven *et al.*, 2001; Avci *et al.*, 2004; Dogan *et al.*, 2004, Gidirisli *et al.*, 2004), but many more studies and experiments need to be carried out.

References

Abbas, J.A. 2006. Economic halophytes of Bahrain, In: Khan et al.. (eds.), Sabkha Ecosystems Vol. II; West and Central Asia. Springer, The Netherlands, pp: 113-120.

Ahmad, R. and A. San Pietro. 1986. Prospects for Biosaline Research. Shamim Printing Press, Karachi, Pakistan, 587 pp.

Akbas, F., M. Ozturk, A. Guvensen and M. Gunduz. 1999. Ecology of the Halophytes of Soke Plateau. VII Kulturteknik Kongresi-Nevsehir, pp. 246-252, Ankara University Press.

Akhani, H. 2006. Biodiversity of halophytic and sabkha ecosystems in Iran, In: Khan *et al.* (eds.), Sabkha Ecosystems Vol. II; West and Central Asia. Springer, The Netherlands, pp:71-88.

Aronson, J.A. 1984. Economic halophytes- a global review. Plants for Arid Lands, Wickens G E (ed.), Allen and Unwin, London, pp.177-188.

Aronson, J.A. 1989. HALOPH; Salt Tolerant Plants for the World–A computerised global data base of halophytes with emphasis on their economic uses. Univ.of Arizona Press, Tucson, USA, pp. 77.

Avci, K., G. Kutbay and S. Gencoglu. 2004. Halophytic vegetation of Bafra plain. Directorate of Rural Services, Ankara, Turkey, Annual Report, 56 pp.

Birand, H. 1960. Erste Ergebnisse der Vegetations-Untersuchungen in der zentralanatolischen Steppe; 1- Hlophytengesellschaften des Tuz Gölü.- Bot. Jahrb. 79: 255-296.

Choukr-Allah, R. 1991. The use of halophytes for the agricultural development of south of Morocco. In: Plant Salinity Research New Challenges, Hassan II-C.H.A, pp. 377-387, Agadir, Morocco.

Dagar, J.C. 1995. Characteristics of halophytic vegetation in India. In: Biology of Salt Tolerant Plants, Khan and Ungar (Eds.), pp. 255-276, Book Crafters, Michigan, U.S.A.

DIE. 2000. Statistical Yearbook of Turkey. State Inst. of Statistics of Prime Ministry, Publ. 1720, Ankara, p.700.

Dogan, O., H. Ustun, N. Gurlesen, F. Tomul, R. Tipirdamaz and S. Sozeri. 2004. Ecology of halophytic vegetation in the Seyfe lake basin and determination of grassland and agricultural plants suitable for growth on arid soils. Directorate of Rural Services, Ankara, Turkey, Annual Report, 67 pp.

Ekim, T., M. Koyuncu, M. Vural, H. Duman, Z. Aytac and N. Adýguzel. 2000. Red data book of Turkish plants, Ankara, 246 pp.

Gallagher, J.L. 1985. Halophytic crops for cultivation at seawater salinity. Soil and Plant 89: 323-336.

Gehu, J.M. and T. Uslu. 1987. Donnés sur la Végétation Littorale de la Turquie du Nord-Ouest, Centre Régional de Phytosaciologie Heendries, F-59270 Bailleul, 75 pp.

Gidirisli, A., R. Cakýr and C. Yarcý. 2004. Halophytic vegetation of saline soils in Gelibolu-Gokbuet and Ipsala region. Directorate of Rural Services-Ankara, Turkey, Annual Report, 53 pp.

Glenn, E.P., J.W. O'leary, M.C. Watsoni, T.L. Thompson and R.O. Kuehl. 1991. *Salicornia bigelovii*: An oilseed halophyte for seawater irrigation. American Association for the Avdance of Science, Vol.251, 1065-67. pp.

Guven, E., M. Vural, I. Bahceci and B. Sonmez. 2001. Ecology of halophytic vegetation in inner Anatolia. Directorate of Rural Services, Ankara, Turkey, Annual Report, 264-280 pp.

Guvensen, A., I. Uysal, A. Celik and M. Ozturk. 1996. Environmental impacts on halophytic plant cover alongside the coastal area of Karabiga and Burhaniye. Symposium on Environmental Problems, Canakkale, pp.8-13.

Guvensen, A. and M. Ozturk. 2003. Halophytic Plant Diversity of South Aegean Coastal Zone in Turkey. Pakistan Journal of Botany 35: 853-864.

Kaynak, L., B. Imamgiller, N. Ersoy and K. Yazici. 2000. Salinity problems in agricultural soils. GAP nvironmental Congress, Sanliurfa, 555-561 pp.

Keen, M. 1986. A high grade fodder and seed crop which thrives on seawater. Arab World Agribusiness 2: 11-13.

Khan, M.A. and N.C. Duke. 2001. Halophytes–A resource for the future. Wetlands Ecology and Management 6: 455–456.

Khan, M.A. and M. Qaiser. 2006. Halophytes of Pakistan: characteristics, distribution, and potential economic usages. In: Khan *et al.* (eds.), Sabkha Ecosystems Vol. II; West and Central Asia. Springer, The Netherlands, pp: 129-154.

Khan, M.A., B. Boer, G.S. Kust and H.-J. Barth. (eds.) 2006. Task for Vegetation Science 42; Sabkha Ecosystems Vol.II; West and Central Asia. Springer, The Netherlands, 259 p.

Lieth, H., U. Herpin, B. Boer and A. Kotb. (eds.), 2002. Quest 2002 for Mena Conference Abstract Volume, Doha, Qatar, 27-30, January, 143 p.

Lieth, H. and M. Mochtchenko. (eds.), 2003. Task for Vegetation Science 38; Cash Crop Halophytes:Recent Studies, Kluwer Academic Publ., Dordrecht, The Netherlands, 250 p.

Ozturk, M., H. Ozcelik, L. Behcet, A. Guvensen and F. Ozdemýr. 1995. Halophytic flora of Van Lake basin -Turkey. In: Biology of Salt Tolerant Plants, Khan and Ungar (Eds.), pp. 306-315, Book Crafters, Michigan, U.S.A.

Ozturk, M., A. Guvensen, F. Akbas and Y. Fýrattekin. 1998. Relations between edaphic conditions and halophyte occurance in Soke plain, Turkey. In: Plant Life in South-West and Central Asia, pp. 139-148, Vth International Symposium-Tashkent, Uzbekistan.

Ozturk, M. and A. Guvensen. 2002. Studies on the Halophytic Vegetation of Saline-Alkaline Habitats in West Anatolia-Turkey. Proceeding of the International Symposium on Optimum Resources Utilization in Salt-Effects Ecosystems in Arid and Semiarid Regions, A. M. Hegazi *et al*. (eds.), pp. 241-244, DRC, Cairo, Egypt.

Ozturk, M., A. Guvensen, C. Gork and G. Gork. 2006a. An overview of coastal zone plant diversity and management strategies in the mediterranean region of Turkey. In: Biosaline Agriculture and Salinity Tolerance in Plants, Ozturk *et al*. (Eds.), pp. 89-100, Birkhauser Verlag (Springer Science), Basel.

Ozturk, M., A. Guvensen and G. Gork. 2006b. Halophytes of Turkey. Sabkha Ecosystems 3, Boer and Khan (Eds.), Kluwer-Springer Verlag, Berlin.

Ozturk, M., Y. Waisel, M. A. Khan and G. Gork. (eds.) 2006. Biosaline Agriculture and Salinity Tolerance in Plants. Birkhauser Verlag, Basel, Switzerland, 205 p.

Riley, J.J. and M. Abdal. 1993. Preliminary evaluation *Salicornia* production and utilization in Kuwait. Kluwer Academic Publishers 2: 319-329.

Somers, G.F. 1982. Food and Economic Plants: General Review. In: A. San Pietro (ed.), Biosaline Research, pp. 127-148, Phnelum Press, New York-USA.

Yurdakulol, E. and T. Ercoskun. 1990. Ecological and syntaxonomical studies on the vegetation of arid lands in central Anatolia. Turkey Journal of Botany 14: 109-123.

Zeybek, N., K.H. Sheikh and M.A. Ozturk. 1976. Coastal Halophytes of the Aegean Region. CENTO Symposium, pp. 148-153, Adana, Turkey.

22

Forage Production and Management in Dryland Areas of Uganda

E.N. Sabiiti*, S.K. Mugasi and S. Katuromunda
Faculty of Agriculture, Makerere University, P.O. Box 7062, Kampala, Uganda
**E-mail: esabiiti@agric.mak.ac.ug*

ABSTRACT

The dryland areas in Uganda support a diversity of plant forms that are adapted to either water stress or saline conditions and the plants are grazed by both livestock and wild life. Significant work has been done mainly on plants/forages under water stress conditions but there is hardly any research on forages under irrigation using saline waters because of limited areas with acidic conditions. There are several salty lakes in the rift valley that are surrounded by salt tolerant/resistant vegetation that supports grazing by animals but there appears to be reduced plant productivity. Equally the forages under dry/water stress conditions are overgrazed by livestock and this is affecting their productivity. This paper focuses on forage production in drylands used for livestock grazing and stresses the fact that new areas of research should be explored to increase on forage production under both rainfed conditions and irrigation to sustain livestock production without destroying the natural resource base.

Introduction

Uganda covers a total land area of 241,548 km^2, of which 36 330 km^2 is covered by water. More than 180,000 km^2 is available for both cultivation and pasture (Table 22.1). Pastures and grazing land are estimated at over 160,000 km^2, half of which (84000 km^2) is rangeland (Dryland) that is mainly used for grazing livestock. The major livestock species grazed in the drylands are cattle, sheep and goats. From an economic point of view, cattle are the most important livestock with significant contributions to milk and meat production. Majority of the cattle population in Uganda are found in the drylands and are owned by pastoralists. The drylands constitute the "cattle corridor" which stretches from Karamoja region in the north-eastern, through the central to the south-western region of the country (Figure 22.1). Almost all the meat and about 85 per cent of the milk marketed in Uganda are produced by indigenous cattle in this corridor. The corridor lies on a plateau at an average altitude of 1160 m above sea level, and the climate is sub-humid. The average daily temperature is 27 °C and annual rainfall ranges between 500 and 1000 mm (Payne, 1990; Pratt and Gwyne, 1977). The seasonal rainfall combined with moderately high temperature, grazing and periodic fires have led to the development of typical vegetation dominated by open grasslands interspersed by Acacia trees. The grasses form the main forage resource for the cattle. Throughout the cattle corridor, permanent water sources are limited. Seasonal rainfall is the principal source of water for the cattle, humans and the grazing resources. Rainwater harvesting is done using valley tanks and valley dams and is the most reliable means of ensuring sufficient water supply throughout

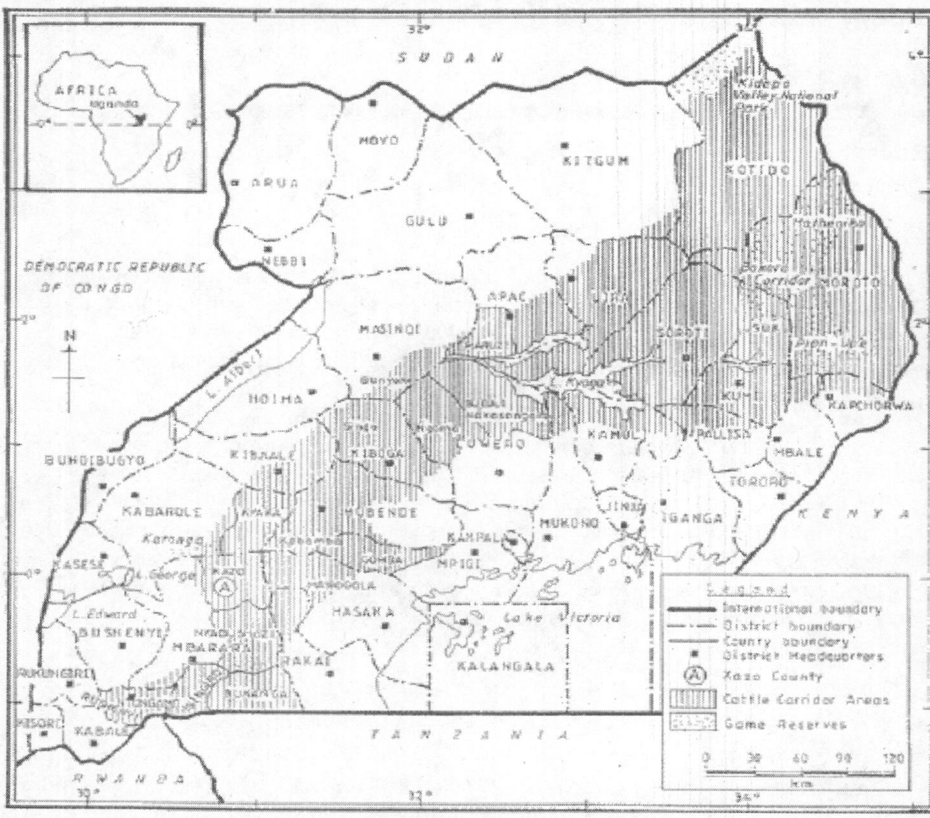

Figure 22.1: Map of Uganda Showing the "Cattle Corridor"

the year. However, there is no irrigation carried out for forage production. The forages have over centuries adapted themselves to water stress during the dry seasons. There are also limited areas in the drylands with forages which grow under severe salt conditions but are grazed by livestock and wild life.

Table 22.1: Current and Potential Grazing Areas of Uganda

Land Classification	*Square Kilometres*	*Per cent*
Grassland	51,118.6	21.16
Farmland	83,931.0	34.75
Woodland	40,277.7	16.67
Bush	14,198.6	5.88
Total potential grazing land	189,525.9	78.46
Total land area	241,548.0	100

Brief History of Pasture/Forage Research and Development

Forage research and development in Uganda has been going on since 1906 and considerable work on forage production has been achieved (Sabiiti and Mugerwa, 1990; Sabiiti and Teka, 2004) but all this work and up-to-date has been on forage production under rainfed conditions in drylands.

This paper will focus on forage production under dryland conditions with limited water conditions because of limited research on irrigation using saline waters has not been done and we do not have it as a serious problem.

Evolution of Agro-Pastoralism in the Drylands/Rangelands

In the traditional pastoral set up, land was used primarily for cattle keeping, and milk and meat constituted the staple food. Other food items would only be obtained by exchanging them with livestock products, particularly meat, milk and ghee. There has, however, been a gradual transformation of the pastoral economy into an agro-pastoral economy. Land is now used for both livestock rearing and crop production (Mugasi *et al.*, 2000). The major reason for cultivation becoming incorporated in the pastoral economy is the emergence of the market for milk. Because of the emergence of milk collecting centres in the remote areas, farmers now sell most of the milk which traditionally is the staple food. Furthermore, herd size per household is dwindling due to increased cattle sales. Thus crops have to be grown to supplement the dwindling livestock products.

Status of Forage Resources in the Drylands

Uganda's ruminant livestock in the drylands depend entirely on natural pastures as source of feed. During the wet seasons, natural pastures produce plenty of fresh forage. But their productivity declines sharply during the dry seasons when forage becomes coarse with low feeding value that cannot meet the nutritional requirements of livestock. The nutritional problem is worsened by drought and overgrazing due to overstocking, both of which have led to rangeland degradation. The major indicator of rangeland degradation is the encroachment by non-desirable shrubs, mainly *Acacia hockii*, *Lantana camara* and *Solanum incanum* (Sabiiti and Teka, 1999). Shrub/bush encroachment is affecting livestock nutrition by suppressing the growth of palatable pasture (grasses and legumes) and keeping their development in check. Shrubs are heavy users of water and minerals, both so essential to the proper development of the pasture. Naturally, cattle tend to exert more grazing pressure on the palatable pasture species while shrubs grow undisturbed. This selective grazing by cattle, coupled with competition for water and minerals by the shrubs, suppresses pasture growth and hence leads to reduced herbage dry matter yield. Studies conducted by Sabiiti and Wein (1991) and Mugasi *et al*. (2000) revealed increased bush encroachment into the rangelands and subsequent decline in herbage productivity. According to Mugasi *et al*. (2000), herbage dry matter yield and crude protein content of herbage were highest on cleared farms (Table 22.2).

Table 22.2: Average Herbage Dry Matter Yield and Chemical Composition of Pasture on Two Types of Drylands

Parameter Measured	*Cleared Farms*	*Bushy Farms*
Dry matter content (per cent)	27.49	30.13
Dry matter (kg ha^{-1})	2,040	906
Crude protein (per cent)	9.15	8.92
Crude protein (kg ha^{-1})	182	86
Neutral detergent fibre (per cent)	62.50	63.00
Legume component (per cent)	7.35	2.27

Effect of Declining Herbage Yields and Quality of Livestock Production

The low quality and irregular seasonal forage supply from native pastures are the major factors contributing to the low productivity of ruminant livestock (Haque and Tothill, 1988). Mugasi *et al*. (2000) observed that cattle grazed on cleared farms had a better body condition and gave more milk than those on bushy farms. In addition, the heifers grazing on cleared farms reached puberty earlier and hence calved earlier than heifers on bushy farms. An economic analysis on the effect of shrub encroachment revealed that there were economic losses incurred by farmers due to reduced levels of animal productivity (Table 22.3). According to the results in Table 22.3, the average gross income per cow per annum for cleared farms was higher compared to that for bushy farms. This was attributed to the higher income of cleared farms resulting from higher animal production in terms of milk yield, body condition, and reproductive performance.

Forage Resources Found in Uganda's Drylands

Uganda appears to be a home of important genera of grasses (Table 22.4) all of which are important for grazing by livestock and wildlife. Also in these grasslands there is a mixture of legumes that improve the quantity and quality of the pastures. Within the dominant genera there are important grass species as shown in Table 22.5. These grasses have medium to high feeding value since they are all palatable to cattle to varying degrees (Heady and Heady, 1982). There is limited research data on the majority of these natural grasses and legumes.

Table 22.3: Average Annual Gross Margin per Cow of Farms Under Two Types of Range Condition

Item *Income*	*Cleared Farms (Ushs)*	*Bushy Farms (Ushs)*
Milk and Cattle sales	45,393	23,351
Total Gross Income	45,393	23,351
Variable Costs		
Shrub Control	15,196	5,545
Cattle Drugs	8,463	5,969
Fencing and Fence Repairs	5,007	4,268
Veterinary Consultancy	729	348
Farm Labour	7,764	3,921
Calf bosses	750	1,125
Total Variable Costs	37,909	20,546
Gross Margins	7,484	2,805
TR/TVC Ratio	1.2	1.1

Forage Production in Dryland Under Natural Conditions

The dominant grass forages in Uganda are largely natural as described above but there are also introduced legumes which have performed well in terms of persistence and productivity as in Table 22.6. Chlroris gayana is an exotic grass species that is well adapted to dryland conditions. Most of the research on forage production has been under natural rainfed conditions largely because the soils are not acidic. Table 22.7 shows the productivity of a grass/legume mixture and this is a general trend of forage production in the country. It is important to relate forage production and animal production and Table 22.8 shows different levels of animal productivity (milk) from different forage mixtures under dryland farming. Irrigation could increase productivity of both forage and animals but this needs research in future.

There is no data on forage production with saline water because the dominant soils in Uganda are not acidic. The only serious problem is from areas in the rift valley around crater lakes with salt/salt waters where the vegetation is under high salt concentration. Some of the surrounding areas are grazed and there are also areas which have been desiccated by salt conditions.

Table 22.4: Major Forage Grass and Legume Genera and Species Found in Uganda

Forage Grasses		Forage Legumes and Browses	
Genera	No. of Species	Genera	No. of Species
Andropogon	10	Acacia	17
Aristida	7	Aeschynomene	7
Axonopus	1	Alysicarpus	5
Bothriochloa	4	Antopetitia	1
Brachiaria	21	Argyrolobium	5
Cenchrus	1	Calopogonium	1
Chloris	7	Canavalia	2
Cynodon	3	Centrosema	1
Dichanthium	2	Clitoria	2
Digitaria	18	Crotalaria	54
Echinochloa	7	Desmodium	12
Eleusine	3	Eriosema	20
Enteropogon	2	Erythrina	8
Eragrotis	38	Glycine	5
Eriochloa	3	Indigofera	42
Heteropogon	1	Lablab	1
Hyparrhenia	24	Latononis	1
Melinis	4	Lotus	4
Paspalum	6	Milletia	2
Panicum	29	Phaseolus	2
Pennisetum	17	Rhynchosia	16
Rottboellia	1	Sesbania	6
Rhynchelytrum	2	Stylosanthes	1
Setaria	21	Tephrosia	17
Sorghum	8	Teramnus	6
Urochloa	2	Trifolium	15
Themeda	1	Vigna	24
		Zornia	4

There is hardly any research going on in such areas and yet some of the plants appear to be disappearing. There is thus a need to embark on research to address such a problem in Uganda through collaboration with experts attending this conference.

Table 22.5: Important Forage Grasses in Uganda

Andropogoneae	
Andropogon gayanus	*Sorghum bicolor*
Dichanthium annulatum	*Sorghum halepense*
Hemarthria altissima	*Sorghumn drummondii (S. Sudanese)*
Hyparrhenia rufa	*Tripsacum andersonii (T. laxum)*
Lasiurus hirsutus	*Zea mays*
Sorghum almum	*Zea mexicana*
Paniceae	
Axonopus compressus	*Panicum antidotale*
Brachiaria arrecta	*Panicum coloratum*
Brachiaria brizantha	*Panicum maximum*
Brachiaria decumbense	*Paspalum dilatatum*
Brachiaria mutica	*Paspalum plicatulum*
Brachiaria ruziziensis	*Pennisetum clandestinum*
Cenchrus ciliaris	*Pennisetum pedicellatum*
Cenchrus setigerus	*Pennisetum polystachion*
Digitaria decumbens	*Pennisetum purpureum*
Digitaria pentzii	*Urochloa musambicensis*
Echinochloa polystachya	*Setaria sphacelata*
Melinis minutiflora	*Digitaria milanjiana*
Chloridoideae	
Chloris gayana	*Cynodon nlemfuensis*
Cynodon aethiopicus	*Eragrostis curvula*
Cynodon dactylon	*Eragrostis tef*

Source: Clayton, 1983.

Table 22.6: Recommended High Quality Forage Species for Drylands of Uganda

Forge legumes	*Forage grasses*
Thatching grass (*Hyparrhenia rufa*) (*H. filipendula*)	Siratro (*Macroptilium atropurpureum*)
	Green leaf desmodium (*D. intortum*)
Rhodes grass (*Chloris gayana*)	Silver leaf desmodium (*D. uncinatum*)
Guinea grass (*Panicum maximum*)	Glycine (*Neonotonia wightii*)
Congo signal grass (*Brachiaria spp*)	Stylo (*Stylosanthes guianensis*)
Buffel grass (*Cenchrus ciliaris*)	Centro (*Centrosema pubescens*)
Star grass (*Cynodon dactylon*)	
Nandi setaria (*Setaria anceps*)	

Table 22.7: Mean Dry Matter Yield and Botanical Composition of the Natural Grass Pastures Over-Sown with *D. uncinatum* at Makerere University Farm, Kabanyolo, Uganda

Number of Cuts	*Pasture Component*	*Grass Pasture + Legume*		*Grass Pasture Without Legumes*	
		DM (kg ha^{-1})	*BC (%)*	*DM (kg ha^{-1})*	*BC (%)*
1st	Legume	6,350	50.6	-	-
	Grasses	5,350	42.6	2,462	46.4
	Weeds	846	7.8	2,845	53.6
	Mean total	12,546	100	5,307	100
2nd	Legume	7,642	52.5	-	-
	Grasses	6,544	45	3,114	50.5
	Weeds	362	2.5	3,048	49.5
	Mean total	16,385	100	13,555	100
3rd	Legume	7,150	43.6	1,200	8.9
	Grasses	8,235	50	4,680	34.5
	Weeds	1,000	6.4	7,675	56.6
	Mean total	16,385	100	13,555	100
4th	Legume	9,965	50.6	1,543	9.3
	Grasses	8,974	45.6	6,022	36.4
	Weeds	760	3.8	8,971	54.3
	Mean total	19,699	100	16,536	100
5th	Legume	9,540	49.1	1,360	9.4
	Grasses	9,242	47.6	4,883	33.8
	Weeds	650	3.3	8,211	56.8
	Mean total	19,432	100	14,454	100

Source: Sabiiti *et al.* (1989)

Strategies for Improving Forage Production and Livestock in Drylands

The drylands in Uganda are very critical for both the livelihoods of the people and also the general economy of the country and so they need to be managed sustainably. The following strategies have been suggested:

1. There is need to widen research scope in new areas that have not been tackled but appear to be a threat to forage production in drylands, such as research in irrigation or salt tolerance of forage species.
2. There is need to educate / sensitize pastoralists about the stocking strategies that can prevent overstocking, which is the cause of overgrazing. Overgrazing promotes deterioration of pastures and this is widespread in the drylands (Sabiiti and Teka, 1999). Grassland management through

administrative measures to enforce carrying capacities can be tried. Figure 22.4 shows the approaches (strategies) to the balancing of forage supply and livestock forage requirements as described by Behnke (1992).

Table 22.8: Milk Production per Cow on Tropical Grass and Legume Pasture Mixtures Forage Production under Irrigation in Drylands

Species	*Mean Yield (kg/cow/day)*	*Breed*	*Management*
Pennisetum clandestinum Neonotonia wightii	12.2	Guernsey and Jersey	Continuously grazed for 10 months
Pennisetum clandestinum	9.8	Guernsey and Jersey	Continuously grazed N-fertilised pasture
Pennisetum clandestinum Neonotonia wightii	8.2	Grade Jersey	Indoor feeding for 9 weeks
Pennisetum clandestinum	16.5	Friesian	Short-term trial grazing 3-4 weeks regrowths fertilised with N
Pennisetum clandestinum Subterranean clover	7.8	Guernsey	Full lactation for 2 years–some conservation 3 and 5 weeks. Regrowths for periods of 8 weeks ad lib grazing.
Setaria sphacelata and *Chloris gayana*	7.0	Jersey	
Dolichos purpureus	10.5	Jersey	Ad lib grazing
Panicum maximum/ Neonotonia wightii	10.3-14	Friesian	Mean yield for 300 day lactation of 15 cows. Some conservation.
Setaria sphacelata/ Desmodium sp.	9.0	Jersey	Full lactation–some concentrates
Phaseolus atropurpureus and *Desmodium intortum*	7.7	Jersey	Sort-term trial under lenient grazing. Cows in various stages of lactation.
Digitaria decumbens	9.0	Jersey	

Source: Stobbs, 1971

Conclusions and Recommendations

Uganda's drylands are important for sustaining the livelihoods of people as well as contribution to the general economy of the Country. The existing information is from research on forage production under rain fed conditions and it is important that research be widened to include new areas facing reduced forage productivity due to high salt concentrations. There is need to engage in collaborative research programmes with experts in irrigation in dry areas using saline waters or those with experience in forage production of halophytic plants.

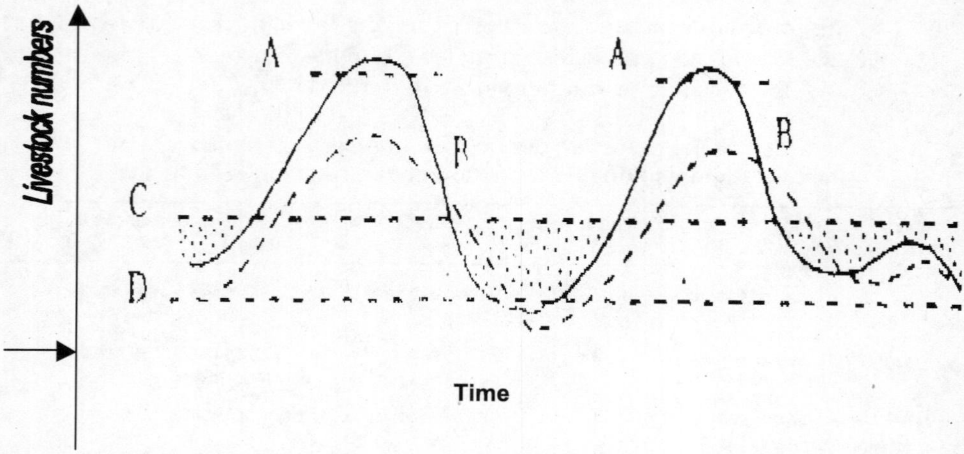

Figure 22.4: Stocking Strategies for Coping with Irregular Seasonal Forage Supply

Key:
Stocking strategies:
A = Migratory or transhumant strategy
B = Tracking strategy
C = Agro-pastoral or mixed crop-livestock strategy
D = Extremely light stocking strategy

Dotted areas represent periods of supplementary feeding;
Solid line represents ecological carrying capacity;
Dashed line represents livestock numbers under stocking strategies A to D

Acknowledgement

The senior author wishes to acknowledge Uganda National Academy of Sciences and the Uganda National Council of Science and Technology for nominating him and NAM S&T Centre for sponsoring him to attend the International Workshop on Crop and forage production using saline waters in dry areas.

References

Haque, I. and J.C. Tothill. 1988. Forages and pastures in mixed farming systems of sub-Saharan Africa. In: Land management and management of acid soils in Africa. Proceedings of the 2nd regional workshop on Land Management and Management of Acid Soils in Africa, 9-16 April 1987. Lusaka, Zambia. 107-131 pp.

Heady, H.F. and E.B. Heady. 1982. Range and wildlife management in the tropics. Intermediate Tropical Agriculture Series. Longman, London, New York, 140 pp.

Mugasi, S.K., E.N. Sabiiti and B. Tayebwa. 2000. The economic implications of bush encroachment on livestock farming in rangelands of Uganda. African Journal Range and Forage Science 17: 64-69.

Payne, W.J.A. 1990. An introduction to animal husbandry in the tropics. 4th Edition. Tropical Agriculture Series, Longman, U.K. 881 pp.

Pratt, D.J. and M.D. Gwyne. 1977. Rangeland management and ecology in East Africa. London, Hodder and Stoughton.

Sabiiti, E.N. and T. Teka. 2004. Dryland Husbandry in Uganda: Community participation in Development.OSSREA, Ethiopia. 98 pp.

Sabiiti, E.N. and T. Teka. 1999. Research and pastoral management in Uganda. In: Mugasi, S.K. E.N. Sabiiti, B. Tayebwa and T. Teka (Eds.), Dryland Husbandry in Uganda. Research Report. DHP Publication No. 5; December 1999. Organization for Social Research in Eastern and Southern Africa (OSSREA), Addis Ababa, Ethiopia, 1-26 pp.

Sabiiti, E.N. and R.W. Wein. 1991. Effects of fire intensity and browsing by goats on the dynamics of Acacia/encroachment in rangelands of Uganda. Proceedings of the IVth International Rangeland Congress, 22-26 April 1991, Montpellier, France. 2: 860-862.

Sabiiti, E.N. and J.S. Mugerwa. 1990. Forage research and development for livestock production in Uganda. In: B.H. Dzowela, A.N. Said Asrat Wendem-Angenehu and J.A. Kategile (Eds.), Utilization of Research results on Forage and agricultural By-Product Materials as Animal Feed resources in Africa. International Livestock Centre for Africa, Ethiopia. 186-206 pp.

Sabiiti, E.N., J.S. Mugerwa and F.B. Bareeba. 1989. Forage legume germplasm evaluation in crop/livestock farming systems in Uganda. In: H.B. Dzowela (Ed.), The Pasture Network for Eastern and Southern Africa (PANESA) Newsletter. 11: 7-11.

Stobbs, T.H. 1971. Production and composition of milk from cows grazing Siratro (Phaseolus atropurpureus) and green leaf desmodium (Desmodium intortum). Australian Journal of Experimental Agriculture and Animal Husbandry 11: 286.

23

Indigenous Knowledge and Agriculture Development in Semi-arid Southeastern Indonesia

Benyamin Lakitan
Ministry of Science and Technology, Republic of Indonesia,
Jalan M.H. Thamrin 8, Jakarta 10340, Indonesia
E-mail: blakitan@ristek.go.id, blakitan@yahoo.com.sg

ABSTRACT

Semiarid Southeastern Indonesia has not been seriously considered as food production centre in the past, due to its unfavourable climate and availability of more fertile agricultural lands in other part of the country; however, at present, more efforts are dedicated to manage these sub-optimal lands for food production in order to keep up with continuous increase in domestic food demand. Dry season in this semiarid area (less than 150 mm annual rainfall) lasts for about 7-8 months per year, and 3-4 months of the dry season are almost totally without rainfall. Even though crop productivity is still considerably low, local farmers have practiced agriculture for centuries in this dry area, coping with the erratic climatic condition. This long experience has resulted in some indigenous knowledge (by local farmers) that we can use it as a reference in developing new technologies for improving crop productivity and sustaining food production system under this climatically stressful environment. Technologies developed should cover areas of crop breeding for high-yielding and drought tolerant varieties, seed production system, improved cultural practices including ecologically-sound water and land management, and post harvest handlings. Modern drought tolerant varieties for major food crops, *i.e.* upland rice, maize, and peanut, have been developed and some have been adopted by local farmers. Integrated farming system (crop-ruminant) has been practiced and some new models for the integrated system are being evaluated.

Introduction

Southeastern part of Indonesian archipelago is climatically different than most of the country's territory, well known as a humid tropical country. Due to its location in rain shadow of Australian continent, much of southeastern Indonesia is characterized by a semi-arid tropical environment (Borrell and Van Cooten, 1999). There are three provinces in this dry area, *i.e.* Bali, West Nusa Tenggara, and East Nusa Tenggara. The latter is the driest province.

Islands in East Nusa Tenggara Province are drier where total rain days range from 9 to 18 annually (Silitonga, 1997). Major islands in this province are Flores, Timor, Sumba, and Alor. A period of nine months without rain is not uncommon in Sumba Island (Nitis, 1999). In general, dry season in this province lasts for 7 months, from May to November.

In the past, efforts to optimize use of this dry area for agriculture had been limited because more arable lands were available. However, at present, most of fertile lands have been used for agricultural activities, except for forest reserve. In fact, this arable land has been continuously converted for non-agricultural purposes. It was estimated that the conversion rate of agricultural lands (rice fields and farms) into other uses in Java Island–the most densely populated island in Indonesia- at about 47.5 hectares annually (Verburg *et al.*, 1999). Nationwide, the conversion rate is estimated at more than 110 hectares per year.

Decrease in total acreage of fertile lands and continuous increase in food demand due to population growth have forced the nation to seriously consider extending agricultural activities to suboptimal lands, including land in semi-arid region of East Nusa Tenggara Province.

There are two options in coping with such unfavourable conditions for crop production: altering soil characteristics to make it more suitable for farming and/or improving the adaptability of crops to these unfavourable environments. Providing water for agriculture use in this dry area is a major obstacle, since water sources are very limited. Use of saline water has not been considered due to techno-economical reasons but may become significant in near future.

Climatic Variability in Indonesia

Climate in Indonesian archipelago is characterized by monsoon because it is located in between two major continents of Asia and Australia, and between two major oceans of Pacific and Indonesian oceans. Most of the areas have a 6-month period of relatively higher rainfall (called wet or rainy season) and another 6-month period of less rainfall (called dry season). Dry season basically a period of the year with less rainfall and in most cases, it is not completely dry (Figure 23.1).

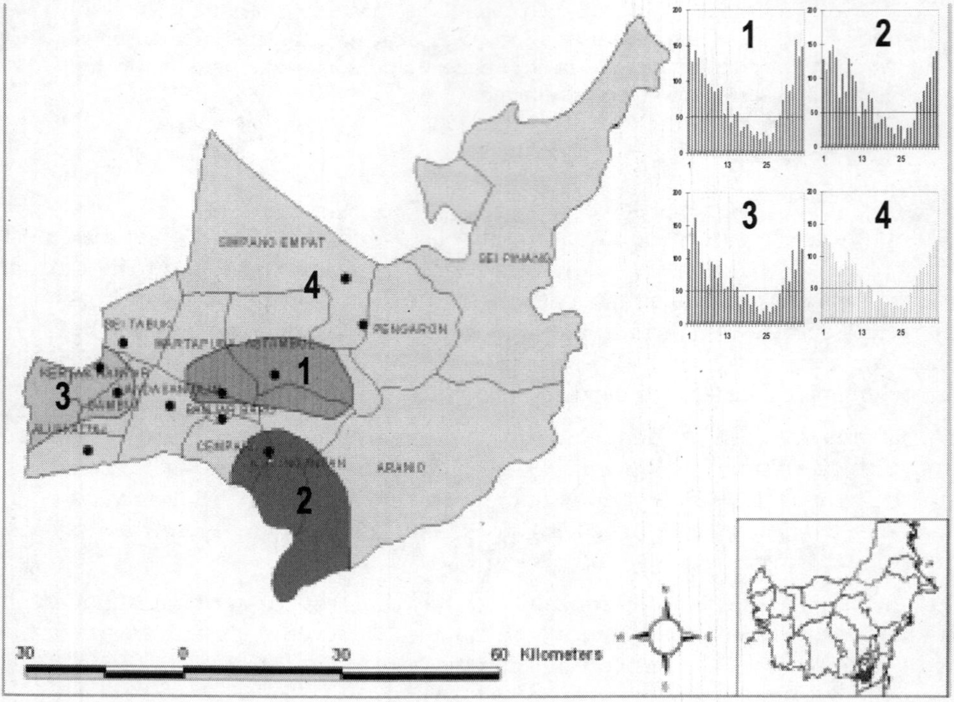

Figure 23.1: Rainfall Pattern in Banjar Regency, South Kalimantan Horizontal solid lines in each bar graph indicate threshold line between wet and dry months (at 150 mm).

Rainfall pattern as shown in Figure 23.1 represents most area of Indonesia, including central and southern part of Sumatera; most area of Java, Kalimantan (previously known as Borneo), Sulawesi (Celebes, except northern part), and Papua. Higher annual rainfall is commonly reported at mountainous area of Papua.

Northern Sumatera and Northern Sulawesi show a slightly different rainfall pattern, with a more well-distributed rainfall throughout the year (Figure 23.2). This rainfall pattern frequently classified as bi-modal rainfall pattern, since it shows two peaks of rainfall during the year. However, even though the pattern is different with the more distinct wet and dry season, total annual rainfall is not significantly different.

Much less or even lack of rainfall during the dry season is recorded in the southeastern part of Indonesian archipelago, including Bali, islands of West Nusa Tenggara and islands of East Nusa Tenggara (Figure 23.3). In semi-arid southeastern Indonesia, more than 90 per cent of rain falls in a distinct wet season between November and April (Borrell and Van Cooten, 1999).

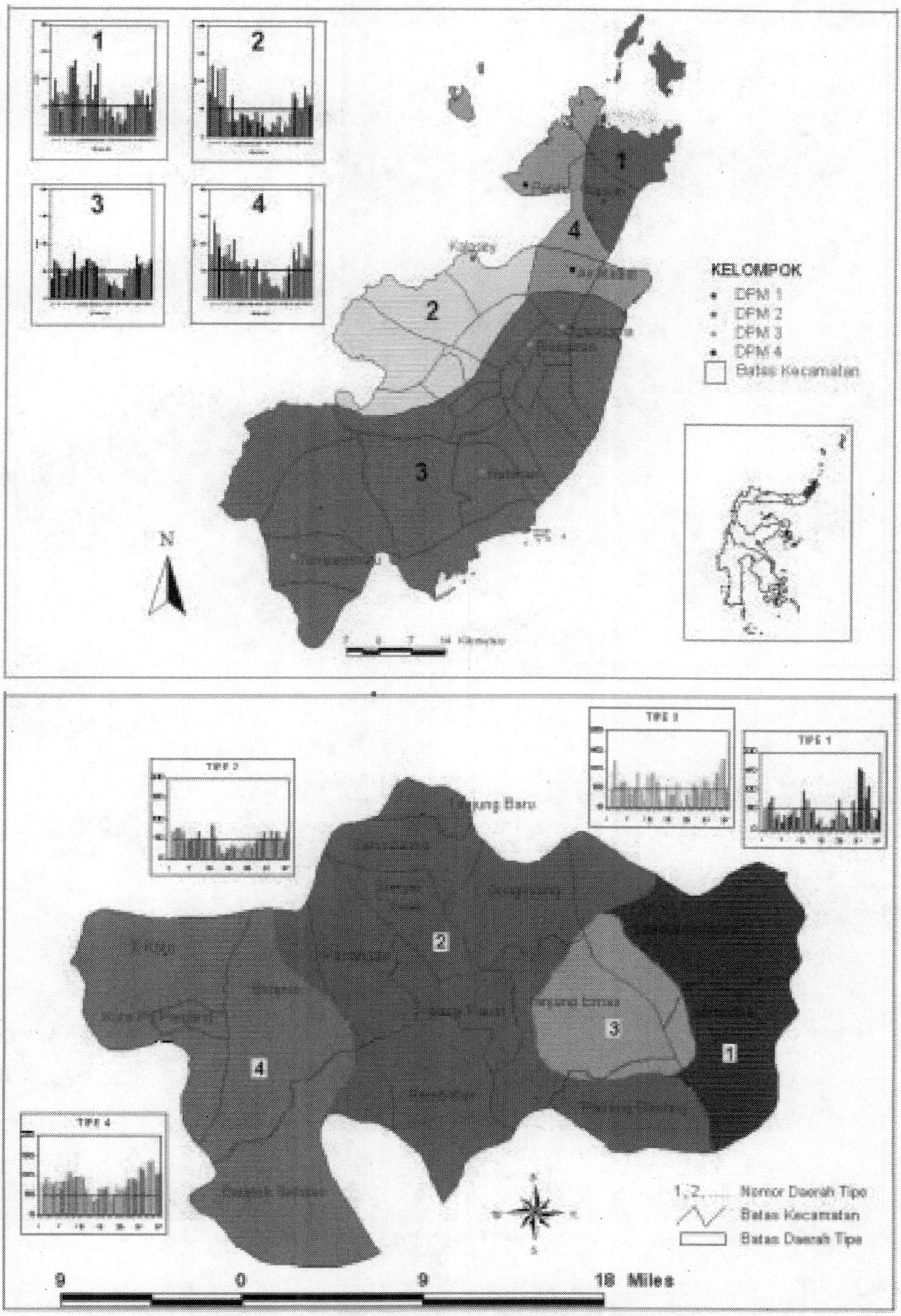

Figure 23.2: Rainfall Pattern in Minahasa Regency, North Sulawesi (Top) This pattern similar to rainfall pattern in North Sumatera (Bottom), with an even distribution of rainfall all year around

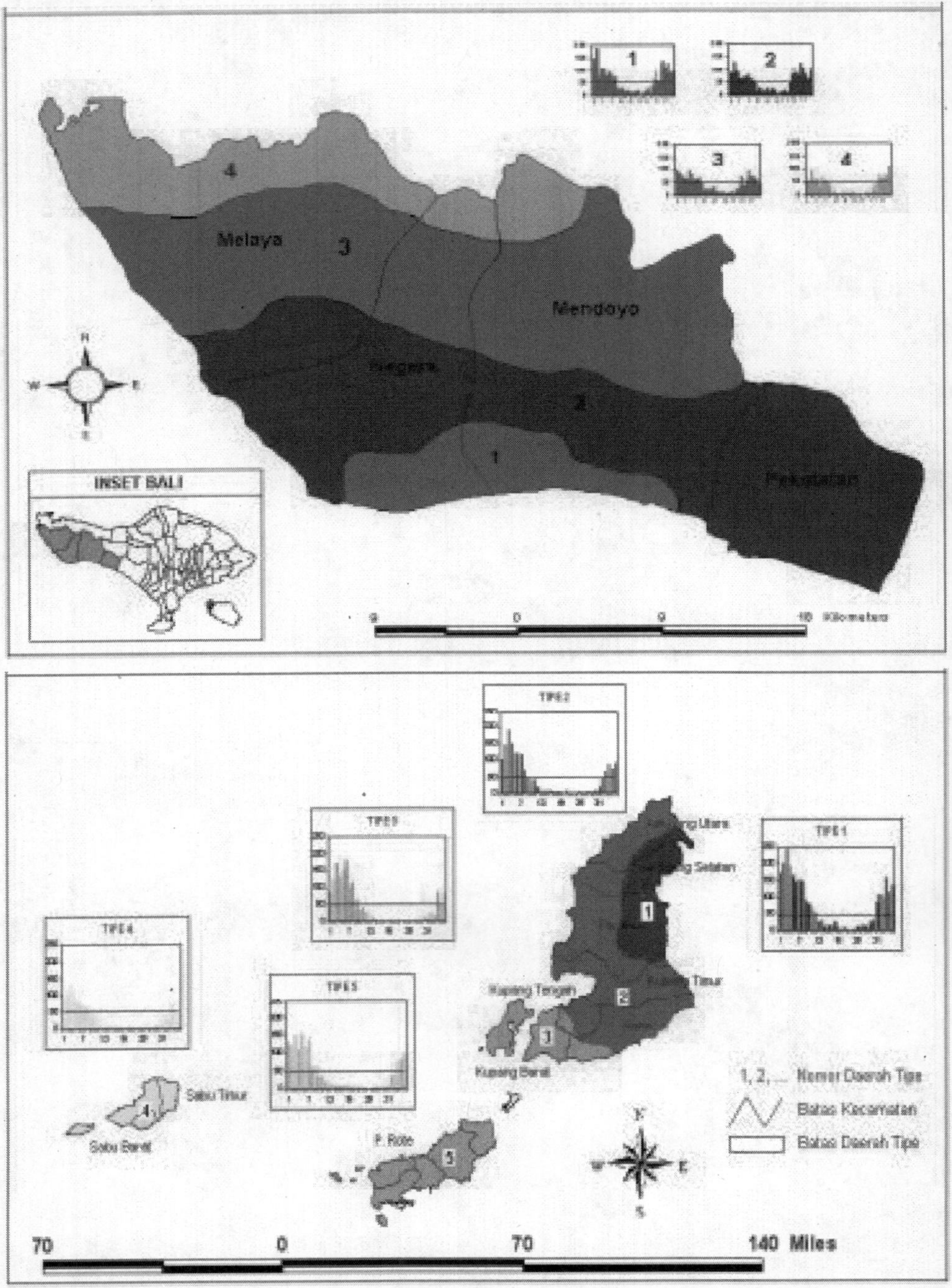

Figure 23.3: Rainfall Pattern in Jembrana Regency, Bali (Top) and Kupang Regency, East Nusa Tenggara (Bottom)

Note that there are almost no rainfalls during period of about 5 to 7 months.

Different total annual rainfall was reported between highland and lowland areas in East Nusa Tenggara. In Timor Island, mountain areas received significantly more rain of more than 1500 mm annually, while coastal areas received less than 800 mm annually and only occur in less than 3 month period; however, mountain areas are more vulnerable to heavy winds (Kieft, 2001).

In addition to very low total annual rainfall, year-to-year rainfall variability is also high. This variability and unpredictability of rain pattern, when it starts to rain and how much rain received every year frequently result in serious yield loss. Late arrival of wet season resulted in estimated 34 per cent decline in maize production (Barnett *et al.*, 2003).

There are also some dry spots in central and eastern Java. Rainfall pattern in Gunung Kidul Regency, situated at southern coast of central part of Java Island, is similar to semiarid area of southeastern Indonesia. This is an enclave of dry area in relatively high annual rainfall region.

Indigenous Knowledge of Local Farmer in Semiarid Southeastern Indonesia

Local farmers have cultivated this semiarid tropic area for a long period of time. They know this agroecosystem better than anyone else. Therefore, it is logical that they already know how to cope with this stressful environment. There are many examples of traditional knowledge, indigenous technologies and local innovations that can be used effectively as a foundation for improved farming systems (Salinger *et al.*, 2005).

Some indigenous knowledge of local farmers in semiarid tropic of Southeastern Indonesia that we should appreciate include: (Barnett *et al.*, 2003) selecting adaptable crops and/or varieties for dry agro-ecosystem, (Borrell and Van Cooten, 1998) Scheduling time of planting to match rainfall pattern in each location; (Borrell *et al.*, 1998) development of cropping system as a strategy to avoid total loss under unpredictable rainfall pattern, and (Fox, 1995) integrating crops and animal husbandry in farming system.

The farmers of Timor islands have their own unique varieties of major staple crops such as maize, sorghum, foxtail millet, upland rice, and cassava. The varieties have been developed and maintained over time, in order to cope with a climate that is characterized by erratic rainfall (Kieft, 2001). Corn is the most widely-grown by local farmers. Sorghum is commonly grown during dry period. Upland rice is also a very important crop due to its function not only as source of carbohydrate but also its use in socio-cultural ceremonies.

Instead of maize, dryland rice is used in rituals, both agricultural and social. A specific rice variety, "Ane Meal" is used for ritual purposes during ceremonies, because it is linked to nobility and purity (Kieft, 2001)

In 1997, cultivated local rice germplasms were collected from East Nusa Tenggara and drought tolerant cultivars were identified. The drought tolerant cultivars were locally named as Mama Wangi Kero and Mama Babe (Silitonga, 1997). There is some evidence that the traditional genotype filled grain better when water was limiting during grain growth by restricting vegetative production and enabling the crop to finish its life cycle (Borrell *et al.*, 1998)

Local farmers have developed a set of strategies to select seeds, maintain a seed stock, and anticipate climate changes. These strategies, which are holistic in nature and include both physical and spiritual indicators, are widely practiced and are of significant relevance to agricultural development. Crop and variety selection methods are used not only for indigenous land races, but also for newly introduced hybrid varieties (Kieft, 2001). High yielding varieties were introduced in the late 70's and early 80's.

Time of planting is critical in a short growing season area, such as area with short rainy season of East Nusa Tenggara. Local farmers usually predict time of planting by observing specific animal behaviour or other natural phenomena. Early-sown crops (rice) may escape end-of-season drought (Borrell *et al.*, 1998; Borrell and Van Cooten, 1999).

One of the characteristics of agriculture on Timor is its mixed cropping system. Although farmers grow maize as a major staple, a wide variety of other crops are also grown to maintain food security (Fox, 1995). To spread risk, farmers usually grow different varieties in different locations. Most farmers grow crops on 2-4 different sites, preferably located on different soil types (Keift, 2001).

Integrated farming system has been implemented in traditional agriculture in many parts of the globe, including in semiarid southeastern Indonesia. Due to its dry agroecosystem, local farmers commonly integrate their food crop production with forestry and animal husbandry. Alley cropping-growing food crops in between rows of frequently-pruned leguminous shrubs, used for green manure or livestock feed-has been practiced in dryland farming in Nusa Tenggara (Nitis, 1999).

Relay-cropping involves upland rice and legume is also commonly practiced by local farmers. This system has been developed mainly to maintain productivity and to reduce soil erosion and degradation. Legume in this cropping system is primarily used for forage or green manure. Feed shortage during dry season is not uncommon in Nusa Tenggara. Fodder shrubs and trees have been used to overcome such feed shortage (Nitis, 1999).

Government Intervention

At present, crop productivity in East Nusa Tenggara is lower than national average. For instance, yields of rice, corn, and soybean in East Nusa Tenggara are 2.8 ton ha^{-1}, 2.4 ton ha^{-1}, and 1 ton ha^{-1}, respectively, lower than national averages of 4.6 ton ha^{-1}, 3.5 ton ha^{-1}, and 1.3 ton ha^{-1}, respectively (Indonesian Central Bureau for Statistics, 2006).

To improve crop productivity in dry areas, Indonesian Agency for Agricultural Research and Development had developed and released several droughts tolerant cultivars of food crops, including corn, upland rice, and peanut. In general, however, drought tolerant cultivars yield less than recommended high yielding varieties. For instance, yield potential of drought tolerant upland rice cultivars (Batugigi and Danau

Gaung) varies only around 3 to 3.4 tons ha^{-1}. Drought tolerant peanut cultivars only yield about 2 tons per hectare.

Not all of introduced cultivars are adopted by local farmers. Some cultivars were grown only for few years and no longer used by local farmers due to their susceptibility to certain pest in the field or during storage.

Integrated farming system has been promoted by government in order to increase land productivity and/or to reduce land degradation. Three strata forage system has been introduced. This system consists of planting grass and legumes (first stratum), shrubs (second stratum), and fodder trees (third stratum) surrounds food crops. The system consists of 0.25 ha (50 m x 50 m) land, centre part (core, 0.16 ha, 40 m x 40 m) is planted with local food crops, outside 5 m wide belt (total area is 0.09 ha) is planted with fodder grass and/or legume, four-side borders are planted with 2,000 shrubs and 42 fodder trees (Nitis, 1999). The Government of Indonesia has established nine Forage and Fodder Development Centres, including in West Nusa Tenggara and East Nusa Tenggara.

Strategies for Improving Agriculture Productivity in Semi-arid

Majority of population in Indonesia consume rice as their staple food. Therefore, population growth in Indonesia, along with in other rice consumer countries, will directly correlate with increase in global rice demand. Khush (1997) predicted that rice production must increase by 60 per cent by 2025 to feed the additional rice consumers.

In 2004, Indonesia regained its status as self-sufficient country for rice, however, since then the country is not able to fulfil its domestic demand in rice. In 2005, Indonesia imported about 100,000 tons rice. Since rice can be grown under many different growing conditions such as irrigated, rain-fed lowland, rain-fed upland, and flood-prone ecosystems, therefore, it is also possible to develop rice production system in dry areas.

Rice selection and adaptation to diverse environments has resulted in numerous cultivars. It is estimated that about 120,000 varieties of rice exist in the world (Khush, 1997). Rice varieties differ greatly in their ability to tolerate an aerobic soil and moisture deficit. The greatest ability to grow and produce some grain with chronic moderate water deficit is found in *japonica* varieties. Notable levels of drought tolerance are also observed in the early-maturing varieties (Lafitte *et al.*, 2004).

Mechanism of drought resistance is probably due to the reduced density of stomata in the leaves. Lower stomatal density reduces water loss due to transpiration, thus maintains plant water content for a longer period. This is a drought avoidance mechanism (Trijatmiko, 2005). Total root biomass and root distribution pattern have also been reported to associate with ability to tolerate drought stress. Larger root biomass increase crop ability to absorb more water from soil.

One accession of *Oryza longistaminata* had greater stomatal conductance and maintained leaf elongation under stress better than most other genotypes. Several upland rice cultivars from *japonica* subspecies also maintained greater conductance and leaf expansion than an *indica* cultivar. In addition, *Oryza longistaminata* accessions and some *japonica* cultivars show either an increase in total root biomass or an increase in the proportion of root biomass in deeper soil levels when water deficit was imposed (Liu *et al.*, 2004).

Continuously improved knowledge on characteristics of drought tolerant varieties combined with abundant varieties of cultivated rice and identified wild species of rice which can be used as gene source, should lead us to well-focused breeding program in developing upland rice varieties which adapt and yield optimally in semiarid agroecosystems. There are twenty-one wild species of genus *Oryza* and two cultivated species: *Oryza sativa*, the Asian cultivated rice is grown all over the world and *Oryza glaberrima*, the African cultivated rice, grown in West Africa. Of the two cultivated rice species, there are about 120,000 known varieties (Khush, 1997).

Our knowledge of how rice genes respond to stress is increasing daily. Our understanding of how these changes translate into plant growth and crop level differences in performance under stress lags behind, and will require a focused effort at synthesis in order to convert the exciting results of genomics into tools and guidelines that plant breeders can use (Lafitte *et al.*, 2004)

Even though upland rice is a significant crop, agriculture development in semiarid Southeastern Indonesia will not be focused solely on rice. Other food crops and other food commodities will also be considered, including maize, cassava, goat, and cattle.

Roadmap of Agriculture Development for Semiarid Southeastern Indonesia

Agricultural development in semiarid region of Indonesia will integrate indigenous knowledge of local farmers who are rich in experiences in dealing with this unfavourable climatic conditions and modern technology which strongly supported by recent progress in scientific researches (Figure 23.4).

Long history of field trials by local farmers has ended up with some selected crops or cultivars which have been proven to adapt well under low rainfall conditions or even under drought stress environment. In addition, these crops have been socially accepted by local consumers in term of taste and other socio-cultural reasons. However, almost all of these crops produce low yields.

Lower yields produced by local farmers are associated with less intensive cultural practices or low technological inputs due to limited resources of local farmers. Development of appropriate technology suitable for each local condition is more likely able to increase crop production. In semiarid agroecosystem, the most significant limiting factor is water availability. Therefore, developing irrigation system to provide sufficient water during crops development will significantly increase yield if other factors are also managed.

Genes controlling superior traits on yield and drought-tolerant from local cultivars and introduced high yielding and/or drought tolerant varieties should be tailored to produce new varieties which have ability to withstand drought stress environment and to produce high yield. Conventional breeding program supported

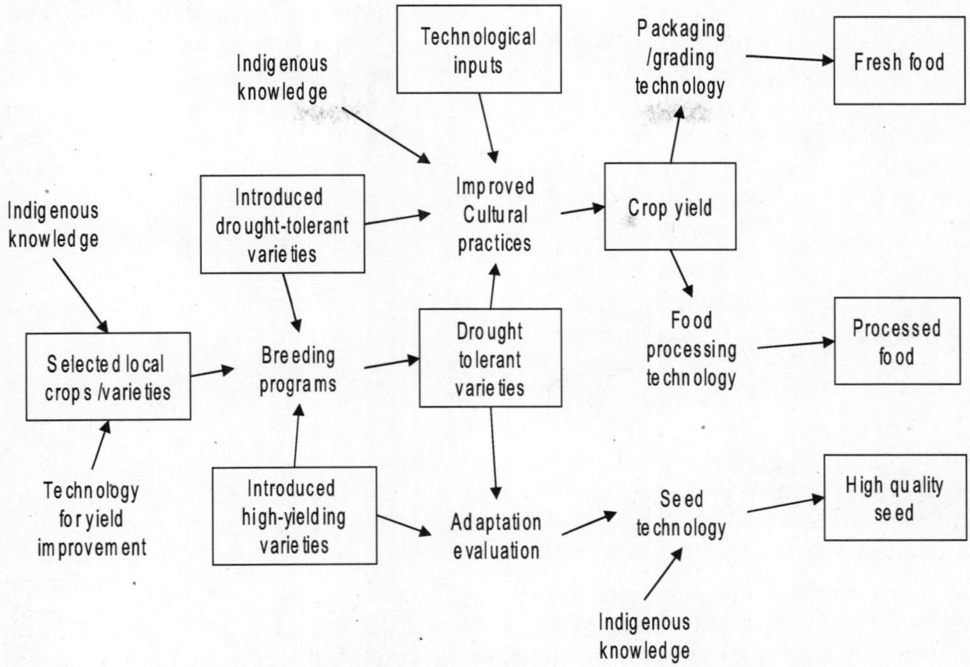

Figure 23.4: Roadmap for Agriculture Development in Semiarid Southeastern Indonesia

by biotechnological applications will be the key success factor for agricultural development in this semiarid agroecosystem.

Field trials to evaluate adaptability of newly developed varieties and all introduced varieties should be conducted prior to use by local farmers in their crop production activities. Site selection for the trials should be based on its soil and climatic conditions that represents majority of agricultural land in the area. Technology applied during trials should not be too far ahead of that applied by local farmers in order to ensure its adoption by the farmers.

Improvement of cultural practices for both newly developed and introduced varieties should consider agricultural knowledge and skill of local farmers to adopt introduced technology, financial capacity to purchase technological inputs, and socio-cultural aspects of local communities. Indigenous knowledge of local farmers may be used as primary reference to start with, and improve from there.

Intermediary indicator of success in this agricultural development scenario is an increase in yield of crops grown by local farmers, not yield gained at controlled trial plots. Yield gap almost always exists between trial plots with average yield of crop managed by local farmers, in some cases, the gap is relatively huge. It is best if the gap is minimal.

Post harvest technology is required not only at time when total food produced exceeds total consumption but also needed every time after harvest, since crops harvested are not consumed at once and the crops are perishable and/or vulnerable to pest if not handled properly. At present, crop losses during and after harvest are reported still at significant rates. Post harvest technology developed should cover technology of washing, waxing, grading, and packaging for fresh food products; also preserving, processing and packaging for processed food products.

High quality seed is one of pre-requisites for achieving higher yield and good quality crops. Seeds of recommended varieties of each major crop should be available to farmers at time of planting and at affordable prices. Therefore, establishment of seed industry should be encouraged.

Conclusions

Local farmers in semiarid area of Southeastern Indonesia have been practicing food production system based on their indigenous knowledge after century-long experience coping with the stressful environment. The knowledge includes screening for tolerant crops and/or varieties, selecting vigorous seeds, mixed cropping systems, and integrated farming suitable for dry climate.

Government interventions on improving food production capacity in this semiarid environment include introducing high-yielding and/or drought tolerant varieties of some major food crops and introducing improved mixed/integrated farming systems.

Local farmers screened the introduced crops/varieties with their best practices but with minimum additional inputs (fertilizers, pesticides, irrigation). As results, some varieties performed close to expected level and are adopted by local farmers, in addition to local varieties that have been used and proven tolerant to dry climatic condition.

Integrating crop and ruminant in food production system is commonly practiced by local farmers. Mixed cropping consists of many food crops and fodder grass, shrub, and trees are also commonly practiced in this semiarid area. However, the cropping pattern seems to be random. There is an opportunity to improve the cropping pattern to increase land use efficiency and productivity. Some improved cropping pattern has been evaluated at experimental plots.

Acknowledgements

I would like to thank Prof. Carunia M. Firdausy, Ministry of Science and Technology, Republic of Indonesia, for facilitating my participation in this workshop; Dr. Andi Eka Sakya, Indonesian Geophysics and Meteorology Agency, for providing all rain pattern maps presented in this paper.

References

Barnett, J., S. Dessai and R. Jones. 2003. Climate change in Timor Leste: science, impacts, policy, and planning. http://www.geography.unimelb.edu.au (accessed on 6 April 2006)

Borrell, A.K. and D.E. Van Cooten. 1999. Enhancing food security in semi-arid eastern Indonesia through permanent raised-bed cropping: a review. Australian Journal of Experimental Agriculture 39: 1035-1046.

Borrell, A.K., R.M. Kelly and D.E. Van Cooten. 1998. Improving management of rice in semi-arid eastern Indonesia: response to irrigation, plant type, and nitrogen. Australian Journal of Experimental Agriculture 38: 261-271.

Fox, J.J. 1995. Social history and government policy of agricultural development in Eastern Indonesia. Paper presented at the International Conference on Agricultural Development in Semi-arid Areas of East Nusa Tenggara, Kupang, 10-16 December 1995.

Indonesian Agency for Agricultural Research and Development. 2005. Varietas Unggulan. http://www.deptan.go.id (accessed 6 April 2006)

Khush, G.S. 1997. Origin, dispersal, cultivation, and variation of rice. Plant Molecular Biology 35: 25-34.

Kieft, J. 2001. Indigenous variety development in food crops strategies on Timor: their relevance for in situ biodiversity conservation and food security. Indigenous Knowledge and Development Monitor 9: 8-13.

Lafitte, H.R., A. Ismail and J. Bennett. 2004. Abiotic stress tolerance in rice for Asia: progress and the future. In: New Directions for a Diverse Planet. Proceedings of the 4th International Crop Sciences Congress, 26 September–1 October 2004, Brisbane, Australia.

Liu, L., R. Lafitte and D. Guan. 2004. Wild Oryza species as potential sources of drought-adaptive traits. Euphytica 138: 149-161.

Nitis, I.M. 1999. Country Pasture/Forage Resource Profile: Indonesia. http://www.fao.org (accessed 6 April 2006)

Salinger, M.J., M.V.K. Sivakumar and R. Motha. 2005. Reducing vulnerability of agriculture and forestry to climate variability and change: workshop summary and recommendations. Climate Change 70: 341-362.

Silitonga, T.S. 1997. Exploration and collection for cultivated rice in East Nusa Tenggara and Moluccas, Indonesia. A Report on CRIFC-IRRI Cooperation. Central Research Institute for Food Crop, Indonesia.

Trijatmiko, K.R. 2005. Comparative analysis of drought resistance genes in Arabidopsis and rice. PhD thesis, Wagenigen University, The Netherland.

Verburg, P.H., T. Veldcamp and J. Bouma. 1999. Land use change under conditions of high population pressure: the case of Java. Global Environmental Change 9: 303-312.

24

Saline Water Use in Agriculture: Highlights of Indian Research

R.K. Trivedy

Department of Environmental Sciences, University of Pune, Pune – 411 007, India

E-mail: rktrivedy@unipune.ernet.in, rktem@pn3.vsnl.net.in

ABSTRACT

India with a geographical area of 2.5 per cent of the world supports 15 per cent of its population. The total available agricultural land in India is about 145 million ha, but half of the land surface is arid or semiarid with seasonal and erratic rainfall. Consequently a huge quantity of saline ground and surface water is used for irrigation in India. Irrigation by sewage (usually untreated or partially treated) and treated or untreated industrial waste is rampant. While in some areas where rainfall is adequate the salt build-up has not taken place due to extensive seasonal leaching. However extensive soil salinity build-up is reported in many other areas use of industrial waste has degraded vast tracts of land. Irrigation by saline water has also affected the crops yields. Present paper highlights quality of irrigation water available in India, effect of saline water irrigation and soil properties and impact on crop yield. Use of sewage and industrial wastes in irrigation in India is also briefly touched upon. A vast number of studies carried out in India are reviewed in the paper to have an overall view of the situation for better management of saline water and its judicious use in agriculture.

Introduction

India with a geographical area of 2.5 per cent of the world supports 15 per cent of its population. India receives an annual average rainfall equivalent to about 4000 billion cubic meter (BCM). This only source of water is unevenly distributed both spatially and temporally. Most of the rainfall is confined to the monsoon season, from June to September and levels of precipitation vary from 100 mm a year in western Rajasthan to over 9000 mm a year in Meghalaya. Thus while India is considered rich in terms of annual rainfall and total water resources, its uneven geographical distribution causes severe regional and temporal changes (MOEF2001). Of the 4000 BCM of water from precipitation, the mean flow in its rivers is about 1900 BCM. The utilizable groundwater resources have been assessed at 395.6 BCM. Out of which 325.6 BCM is available for irrigation. A rapid rise in India's population in the last five decades has put a great pressure on agriculture. The net sown area has increased from 119 million ha in 1951 to 142 million ha in 1997. High cropping intensity has also resulted in an increased demand for water resources. The total available agricultural land in India is about 145 million ha. Half of the land surface is arid or semiarid with seasonal and erratic rainfall. Consequently a huge quantity of saline ground and surface water is used for irrigation in India. Irrigation by sewage (usually untreated or partially treated) and treated or untreated industrial waste is rampant. While in some areas where rainfall is adequate the salt build-up has not taken place due to extensive seasonal leaching, the soils have not shown much degradation. However, extensive soil salinity build-up is reported in many other areas. Use of industrial wastes has degraded vast tracts of land. Irrigation by saline water has also affected the crop's yields.

Present paper reports highlights on quality of irrigation water available in India, effect of saline water irrigation and soil properties and impact on crop yield. Use of sewage and industrial wastes in irrigation in India is also briefly touched upon. A vast number of studies carried out in India are reviewed in the paper to have an overall view of the situation for better management of saline water and its judicious use in agriculture.

Characteristic of Irrigation Water in India

A huge diversity exists in the characteristics of water available for irrigation in India, depending on the source. Table 24.1 gives the general chemical characteristics of different sources of water. Tables 24.2 to 24.11 describe quality of water available at various locations in India. It is clear from the Tables that water quality varies drastically in different areas, and highly saline water is available for irrigation in a number of areas. Table 24.7 provides the estimates of the area underlain with saline groundwater. Rajasthan, Gujarat and Haryana account for maximum percentage of area with saline water (Sharma, 2000). Table 24.8 gives per cent distribution of groundwater's of Rajasthan in different EC classes. It is apparent that the problem is very serious in the arid areas where about 90 per cent waters come under high salinity class. A study by Minhas *et al.* (1988) has shown that (Table 24.9) most of the water except in Ananthpur, Prakasham and Nellore district are good quality. However about 16 per cent waters were found to be highly alkaline. A study by Paliwal and Sharma (1993) indicates that in 6 districts of Rajasthan, groundwater salinity is high in more than 50 per cent of places. About 9 districts have saline area less than 10 per cent (Table 24.10). Table 24.11 shows the characteristics of groundwater at 5 locations in Haryana, where the water quality was found to be very low and the levels of heavy metals were also significantly high.

Table 24.1: General Characteristics of Water Available for Irrigation

Origin of Water	*Properties of Water*					
	Silt	*Chemistry*	*pH*	*Organic Matter*	*Seed Spores/ Bacteria*	*Fertility*
Rain	Tr	Fresh	6-7	Tr	tr	Poor
Springs	Tr	Fresh alkaline saline	6-9	Tr	tr	Poor to rich
Ground Water	Tr	Fresh alkaline saline	7-9	Tr	tr	Poor to Rich
Rivers	2-5 gl^{-1}	Fresh alkaline	7-9	Medium	Medium	Rich
Lakes	Little	Fresh alkaline saline	7-9	Little	Much	Rich
Drainage	Little	Brackish saline	7-8	Little	Much	Rich
Sewage Water	Little	Fresh slightly brackish	7-8	Much	Very much	Very much

Source: Kovda *et al.* (1973); tr: Traces.

Table 24.2a: Average Chemical Composition of Water Calcutta, India in 1967

	Constituents											
	HCO_3	*Cl*	*SO_4*	*NO_3*	*Fe*	*B*	*Ca*	*Mg*	*Na*	*K*	*SiO_2*	*pH*
Maximum	31.95	10.92	7.72	0.58	0.04	0.06	11.20	1.95	3.26	1.96	4.00	6.80
Minimum	tr	0.50	0.73	0.10	0.01	0.002	0.30	tr	0.32	tr	0.90	5.50
Average	7.5	2.87	3.73	0.26	0.016	0.025	3.83	0.58	1.17	0.43	2.23	6.13

Table 24.2b: Average Chemical Composition of Water Lucknow, India in 1980

*Constituents (*EC and ions expressed as μs cm^{-1} and mg L^{-1}, respectively)*										
pH	*EC*	*HCO_3*	*Cl*	*NO_3*	*Ca*	*Mg*	*Na*	*K*	*SiO_2*	*Ag*
4.65	7.20	24.9	8.83	1.14	2.12	1.70	0.59	0.32	0.24	0.41
Ca	*Zn*	*Co*	*Mo*	*Cd*	*Cs*	*Sr*	*Li*	*Rb*	*Fe*	*Mn*
3.08	9.74	0.60	1.10	0.19	0.41	3.63	0.96	0.61	26.5	2.45

Major Conclusions Drawn by Indian Studies on Effect of Saline Water Irrigation on Crops

1. Table 24.16 gives critical limits of EC_{iw} for crop tolerance (Gupta and Gupta, 2003)
2. Highly tolerant forest species to sodicity were found to be *Prosopis juliflora, Acacia nilotica, Eucalyptus hybrida, Zizyphus sp, Capparis decidua* to sodicity and *Avicennia marina, Prosopis juliflora, Rhizophora sp, Causuarina equisetifolia* to salinity (Yadav, 1989)

Table 24.2c: Average Chemical Composition of Water Canal Water in (UP), India

Properties	*Location of Lower Ganga*		
	Etah (9)	*Farukabad (14)*	*Kanpur(23)*
pH	8.2	7.9	8.5
EC (dsm^{-1})	0.39	0.29	0.44
SAR	1.4	0.4	2.5
RSC (meq L^{-1})	0	0	1.0
Ions : meq L^{-1}			
Ca	1.78	1.31	1.16
Mg	0.44	0.82	0.60
Na	1.54	0.41	2.35
CO_3	0.13	0.10	0.34
HCO_3	2.03	1.81	2.74
Cl	0.24	0.15	0.26
SO_4	1.35	0.32	0.32
NO_3	–	–	–

Source: Pandey and Sahi (1980).

Table 24.3: Salinity Levels dS m^{-1} of Some Rivers During Low Flow at Various Points on their Course in India

Location	*Salinity (dsm^{-1})*	*Location*	*Salinity (dsm^{-1})*
Ganga		**Yamuna**	
Haridwar	0.20	Tejewala	0.31
Kanpur	0.38	Delhi	0.49
Allahabad	0.80	Agra	0.75
Narmada		**Sutlaj**	
Garudshewar	0.34	Slapper Bridge	0.30
Ankleshwar	1.02	Nangal	1.10

Source: Handa (1972).

Table 24.4: Water Quality of River Ganga

Parameter	*pH*	*Temp.*	*Turbidity*	*Conductivity*	*Alkalinity*	*Chloride*	*DO*	*BOD*	*COD*	*TKN*
Value	8.48–8.44	15.5–16.5	70–470 NTO	504–522	–	0.62–0.64	7.20–6.92	0.49–1.38	–	0.01–0.01
Parameter	Nitrite + Nitrate	Total hardness	Calcium	Mg	SO^{4--}	Na^{+}	TDS		Phosphates	boron
Value	0.11–0.04	196–200	1.71–1.82	1.01–1.82	0.70–0.84	1.15–1.29	409–640		0.05–0.08	0.18–0.23

Source: CWC (1987) all values except pH are in mg L^{-1}.

Table 24.5: Chemical Composition of the Ganga River at Dakshineshwar Affected during the Bore Tide (May–1969)

EC (dsm^{-1})	pH	Ions							
		Ca	Mg	Na	KH	CO_3	Cl	SO_4	NO_3
8.6	7.8	108	199	1466	65	279	2227	375	0.35

Source: Handa (1972).

Table 24.6: Quality of Surface and Drainage Waters at Two Locations in India

State	Location	EC (dsm^{-1})		
		Irrigation Water	Drainage Water	
			Surface	Subsurface
Haryana	Sampla	0.5–0.7	0.6–0.8	8.0–32.0
Kerala	Kottanand Kari land	0.3–2.3	0.2–1.9	2.6–3.9

Source: Gupta (2000).

Table 24.7 Estimate of the Area (mha) Underlain with Saline Ground Water

Sl.No.	State	Total Area of the State	Area Underlain with Saline Ground Water
1.	Rajasthan	34.2	14.10 (41.22)
2.	Gujarat	19.6	2.43 (12.39)
3.	Haryana	4.4	1.14 (25.90)
4.	Karnataka	19.1	0.88 (4.60)
5.	Tamil Nadu	13.0	0.33 (2.53)
6.	Punjab	5.0	0.30 (6.00)
7.	U.P.	2.9	0.13 (4.48)
8.	Delhi	0.14	0.01 (7.14)
	Total	125.0	19.3 (15.44)

Figures in parenthesis express percentage. *Source*: Sharma (2000).

Table 24.8: Percent Distribution of Ground Water of Rajasthan in Different EC Classes

EC Ranges	Arid Districts	Semi Arid	Humid Districts
(dsm^{-1})	(2817)*	(4000)	(2614)
<0.75	10	23	41
0.75–2.25	29	48	49
2.25–5.00	27	19	9
5.00–10.00	21	8	2
10.0–15.0	9	2	-
>15.0	5	-	-

Source: Gupta (1979), * No of Samples

Table 24.9: Percent Distribution of Ground Waters in Various Categories in Some Districts of Andhra Pradesh, India

Districts	*No. of Samples*	*Good*	*Marginally Saline*	*Saline*	*High SAR Saline*	*Marginally Alkaline*	*Alkaline*	*Highly Alkaline*
Ananthapur	684	44.7	1.2	0.1	0.3	21.6	25.7	6.3
Prakasam	166	37.3	10.2	3.0	5.4	–	44.0	–
Guntur	837	64.2	14.1	3.5	0.7	1.3	14.8	1.4
East Godavari	140	75.0	7.9	2.1	0.7	–	14.3	–
West Godavari	825	83.0	8.4	0.9	–	7.5	0.1	0.1
Krishna	501	66.0	12.0	2.0	1.8	13.6	1.2	3.4
Khammam	544	79.4	1.3	–	0.2	11.0	6.6	1.5
Kurnool	621	66.7	5.8	0.8	1.3	14.3	6.6	4.5
Nellore	927	36.8	5.7	0.4	3.3	21.0	20.1	12.7
Total	5245	61.3	7.2	1.2	1.3	12.1	12.6	4.3

Source: Minhas *et al.* (1988).

Table 24.10: Area Distribution of Ground Water Salinity in Rajasthan (1998)

Sl.No.	*Districts*	*Total Area (mha)*	*Potential Zone Area (%)*	*Saline Area EC > 8 (dsm^{-1}) (%)*
		Saline Area > 50 per cent		
1.	Barmer	2.84	44.20	54.4
2.	Bikaner	2.72	40.00	59.9
3.	Churu	1.68	46.91	53.0
4.	Ganganagar	1.16	13.3	86.6
5.	Hanumangarh	0.95	13.3	86.6
6.	Jaisalmer	3.84	24.33	95.0
		Saline Area 10 to 30 per cent		
7.	Bharatpur	0.51	67.70	26.5
8.	Jalore	1.06	77.33	19.0
9.	Bholpur	0.30	68.12	-
10.	Jodhpur	2.28	74.63	24.6
11.	Nagpur	1.77	79.98	7.6
12.	Pali	1.23	59.58	25.8
		Saline Area < 10 per cent		
13.	Ajmer	0.85	38.04	-
14.	Alwar	0.87	78.28	4.31
15.	Dausa	0.34	90.22	-

Contd...

Table 24.10–Contd...

Sl.No.	Districts	Total Area (mha)	Potential Zone Area (%)	Saline Area EC > 8 (dsm^{-1}) (%)
16.	Jaipur	1.10	90.35	3.1
17.	Jhunjhuna	0.59	98.96	2.0
18.	Sawai Madhopur	0.50	88.76	-
19.	Karauli	0.50	77.45	-
20.	Sikar	0.78	92.12	1.2
21.	Tonk	0.72	90.63	4.1

Source: Paliwal and Sharma (1993).

Table 24.11: Range and Depth of Occurrence of Minimum and Maximum Values of Different Chemical Parameters including Heavy Metals (mg L^{-1}) in Aquifer Substratum of Five Sites in Gohana, Haryana, in India

Sl.No.	Soil Property	Site 1 (Bidal)	Site 2 (Katwal)	Site 3 (Moi)	Site 4 (Barota)	Site 5 (Lath)
1.	pH	8.2–9.0	8.3–8.7	8.1–9.2	8.0–9.2	8.4–9.2
2.	EC(dsm^{-1})	0.28–1.88	0.5–4.0	0.21–1.41	0.29–0.67	0.7–2.94
3.	SAR	1.2–4.5	0.5–3.9	2.9–13.2	2.3–7.6	2.3–16.6
4.	As	14.9–89.0	21.1–146.2	43.1–181.1	21.1–94.6	40.3–116.4
5.	Cr	0.2–4.7	1.8–10.8	0.73–9.1	1.48–5.12	3.4–13.2
6.	Cd	0.002–0.062	0.002–0.122	0.05–0.20	0.014–0.12	0.09–1.12
7.	Ni	0.12–0.99	0.004–0.74	0.02–0.85	Nil	0.09–0.47
8.	Pb	0.17–1.54	0.9–2.7	0.2–12.6	0.8–3.24	1.5–3.34
9.	Zn (mg L^{-1})	0.32–1.99	0.03–0.71	0.27–0.51	0.24–0.54	0.18–0.44
10.	Cu (mg L^{-1})	0.15–1.37	0.17–1.48	0.18–0.84	0.13–0.78	0.14–1.69

Source: Kamra *et al.* (2000).

Table 24.12: Relative Plant Growth as Affected by Chlorides and Sulphates

ECe (dsm^{-1})	Relative Plant Growth (per cent)	
	Chlorides	Sulphates
4	6	100
4–8	6–12	80–90
8–12	12–20	50–70
12–20	20–40	20–50
20	40	0

Source: Gupta and Gupta (2003)

Table 24.13: Effect of Irrigation with Saline Water (tcc = 112 meq l^{-1} sar = 8) having 5 mg/ca Ratio as 5.4 on Properties of Irrigated Soil (A: After; B: Before)

Depth of Soil (cm)	*TCC (meq L^{-1})*		*SAR*		*ESP*	
	A	*B*	*A*	*B*	*A*	*B*
0–20	16	131	3	9	8	4
20–40	11	72	4	7	6	18
40–60	11	54	3	5	13	22
60–100	49	64	5	8	14	27
100–150	91	98	8	9	24	40
150–200	60	97	6	8	29	42

Source: Gupta and Gupta (2003).

Table 24.14: Wheat Yield as Affected by Irrigation and Groundwater Salinity with Shallow Water Table (around 1m)

1990–1991				*1991–1992*			
Water Expense (cm)	*G.W. Salinity (dSm^{-1})*		*Mean Yield (tha^{-1})*	*Water Expense (cm)*	*G.W. Salinity (dSm^{-1})*		*Mean Yield (tha^{-1})*
	3.0	*5.5*			*3.0*	*5.5*	
13.5	4.5	2.5	3.5	12.5	4.6	3.7	4.1
19.0	4.8	4.0	4.4	18.0	5.1	5.4	5.2
24.5	5.0	3.7	4.8	24.0	5.9	5.2	5.5
30.0	5.3	4.2	4.6	30.0	5.7	5.4	5.5
35.5	4.8	4.4	4.3	35.0	5.8	5.3	5.5
Mean	4.9	3.8	-	-	Mean	5.4	5.0

Source: Rao *et al.* (1992)

3. The factors which influence the salt tolerance of crops are plant variety, growth stage, soil fertility and climate and irrigation practices.
4. Quality of various crops was also found to be affected by saline water irrigation like decrease in total sugar of the beet. A 0.5 per cent chloride was found to impact saltish taste to jaggery prepared from sugarcane (Krishan, 1949)
5. Quality of Tobacco was found to decrease in saline irrigation conditions.

Major Conclusions Drawn by Indian Studies on Effect of Saline Water Irrigation on Soils

1. Every irrigation adds some amounts of salts to the soil. If a good irrigation water is assumed to contain 1000 mgL^{-1} (EC 1.5 dSm^{-1}) of soluble salts and a total depth of irrigation requirement per year of about 60 cm, the amount of salt added to the soil annually would work out to be about 6 t ha^{-1} (Gupta and Gupta, 2003)

2. Soil salinity may or may not increase considerably after one season of irrigation with saline water, the actual increase will depend on the initial salinity of the soil, salinity of the irrigation water, physical properties of the soil, growing season and rainfall during the growing period.
3. Irrigation with waters having RSBC as high as 10 meqL^{-1} could be practical if SAR is <10 for raising wheat in semiarid regions. However, if winter rainfall is also appreciable (>150 mm) the permissible SAR could be relaxed up to 20. The relative toxicity of anions was observed to be $CO_3 > Cl > HCO_3$ (Gupta, 1980, 1983)
4. At a given salinity, increase in SAR of water tends to cause increase in ECC, SAR and pH of the irrigated soils, decrease in hydraulic conductivity and reduction in crop yields (Anonymous, 1986)
5. Salt build-up is due to salinity of water than with increasing SAR and Mg/Ca ratio. (Kumar *et al.*, 1986)
6. Significant effects on microbial characteristics of soils were also reported following saline water irrigation. Initially the fungal counts increased but declined at higher levels. Amount of microbial biomass, ninhydrin–N and dehydrogenase activity reduced with increasing RSC of water (Gupta and Gupta, 2003)
7. More studies are needed on microbial characteristics of soils under saline water irrigation characteristics.

Table 24.15: Crop Varieties Tolerant to Saline Water Irrigation

Crop	*Location*	*Varieties*
Wheat	Agra	WH 383, PBW 134, WL 2328, Raj 2560, HD 2374
	Dharwar	Vijaya, DWr 244–20, DWr 306–7–6, M.P. 157, CSW 540–2
	Jobner	K 65, Raj 2936, Raj 2917
Barley	Agra	T-59, Prakash, RH–30
Triticale	Agra	DT 6–960, DT 6–957, DT 6–962, DT 6–976, DT 6–1007, DT 6 - 1184
Mustard	Agra	DYS–1, Puja Kalyani
Cotton	Bapatla	Sangam
Green gram	Bapatla	LGG–127
Black gram	Bapatla	LBG–17
Chillies	Bapatla	X–235 and CA.960
Sunflower	Dharwad	Ramson record, RC.101495, EC.68413
Maize	Dharwad	Deccan 103, Vijaya
Sufflower	Dharwad	19–185 and A–1

Source: Anonymous (1987).

Table 24.16: Critical Limits of EC_{iw} for Crop Tolerance

Crop	*Soil Texture*	*Location*	*Critical Limits of Ec_{iw} for Yield Reduction of*		
			10 per cent	*25 per cent*	*50 per cent*
Wheat	Sandy loam	Agra	9.0	12.9	17.6
	Loamy sand	Jodhpur	7.3	10.3	13.6
	Sandy loam	Karnal	8.5	11.0	14.2
	Sandy loam	Hisar	7.4	10.2	13.1
	Black clay	Dharwad	2.7	7.4	18.0
	Black clay	Indore	3.8	8.8	–
	Sand (dune)	Karnal	12.5	16.0	–
Barley	Sandy loam	Agra	7.6	16.2	–
Rice	Clay	Baptala (S)	1.8	4.0	7.5
		Baptala (W)	1.5	2.6	4.7
	Black clay	Dharwad	3.8	6.5	9.9
Maize	Sandy loam	Agra	2.7	5.5	10.3
	Black clay	Dharwad	1.2	3.1	9.5
Sorghum	Sandy loam	Agra	8.3	12.8	17.8
	Black Clay	Dharwad	2.4	6.5	13.4
Pearl millet	Sandy loam	Agra	7.3	11.5	16.0
Finger Millet	Sand	Baptala	1.6	2.4	8.0
Mustard	Sandy loam	Agra	7.9	10.9	14.6
Sunflower	Sandy loam	Agra	6.0	8.5	11.6
Safflower	Sandy loam	Agra	6.2	11.4	–
	Black clay	Dharwad	4.9	12.2	24.7
Ground nut	Sand	Baptala	1.7	3.1	5.4
Sesamum	Sandy loam	Pali	5.6	8.5	12.6
Cotton	Black clay	Dharwad	3.3	7.7	15.1
Pigeonpea	Sandy loam	Agra	1.8	3.3	6.2
Cow pea	Black clay	Dharwad	2.0	4.6	8.8
Cluster bean	Sandy loam	Jobner	7.3	9.31	2.3
Onion	Sand	Baptala	2.5	5.4	10.2
	Sandy loam	Agra	1.7	2.8	5.1
Lady's finger	Sandy loam	Agra	3.4	7.2	13.5
Fenugreek	Sandy loam	Jobner	6.7	8.4	11.8

Source: Gupta and Gupta (2003).

Table 24.17: Effect of Irrigation with Normal, Saline and Alkaline–Sodic Waters on Soil Properties

Irrigation Water			*Irrigated Soil*		
EC (dSm^{-1})	*RSC (meq L^{-1})*	*SAR*	*EC*	*pH*	*SAR*
		Normal Water			
0.4	0.6	1.0	2.7	8.4	7.5
		Saline Water			
5	0	12.8	26.3	8.0	49
10	0	18.1	38.5	8.0	72
		Alkaline Sodic Water			
2	5	9.5	9.3	9.2	54
2	10	15.6	8.6	9.4	60
2	15	23.2	7.4	9.8	71

Source: Gupta and Abichandani (1973).

Effect of Management Practices on Salt Build-up

1. Surface irrigation methods have overall low water application efficiency and thus aggravate problems of salinity and alkalinity.
2. Sprinkler irrigation was found to be less damaging. According to Agarwal and Khanna (1983) the salt accumulation in sprinkler irrigated plots was 30 to 40 per cent less as compared to surface irrigation.
3. Agarwal and Khanna (1983) reported that the water use efficiency was higher for drip irrigation as compared to surface irrigation both for fresh and saline water.
4. Sharma and Tyagi (2000) has reported beneficial effects of pre-sowing irrigation with good quality water followed by mixed water over all the irrigation with mixed water throughout the crop growth. The yields at the same average salinity were higher in each case when pre-sowing irrigation was given with good quality water.
5. Fallowing of the lands for one or more seasons in arid and semi-arid regions was found to be a beneficial practice.
6. A number of studies in India have proved that increasing the seed rate by 25 per cent over and above the seed rate recommended for normal lands and normal irrigation waters. For transplanted crops, this effect can be achieved by decreasing the plant to plant or row to row distance. (Gupta and Gupta, 2003).
7. A number of studies have proved that dressing of farmyard manure (FYM) and fertilizers was beneficial in case of wheat and pearl millet grown under saline water irrigation. Application of manures enhanced nutrient uptake by crop root (Maliwal and Paliwal, 1971).

8. Application of Zinc at the rate of 20 kg ha^{-1} $ZnSO_4$ has been reported to increase the yield when irrigated with Saline/Sodic water in Rajasthan (Lal *et al.*, 1980).
9. Application of biofertilizers like blue-green algae was found to be highly beneficial. (Anonymous, 1986).
10. Gypsum application has been found to be very effective in removing the toxicity of boron, lithium and fluorine which are commonly associated with saline water. (Gupta, 1979, 1980).

Sewage Farming Practices in India: An Overview

There are over 220 sewage farms in India. The sewage farms range in size from 40 to 1500 ha. The sewage applied is from raw to secondarily treated. The sewage quantity used is from 2 to 66 million gallons/day; dosages of sewage given are from 45 to 900 m^3/ha/day and the major crops cultivated are paddy, wheat, maize, vegetables, Jowar (*Sorghum vulgare*) etc.

Surprisingly irrigation with sewage was not found to create significant soil related problems. There was a considerable improvement when the land is irrigated with wastewater in comparison to fresh water irrigation. This seems to be due to increase in organic carbon content of the soil. Data on soil total nitrogen and total phosphate show that in general there was an increase in these plant nutrients thereby resulting in improvement in soil fertility. The cations K, Na, Ca and Mg both in total and exchangeable forms increased under sewage irrigation.

Table 24.18: Chemical Characteristics of Domestic Wastewater in India

Parameters	*Name of the Town/City*		
	Rohtak	*Faridabad*	*Nagpur*
pH	7.1–7.6	7.1–8.0	7.2–7.8
EC, μS/cm	1.30–1.70	1.40–1.83	675–984
Na, Meq/L	5.20–7.20	7.20–8.10	4.22–6.08
Ca, Meq/L	2.60–3.40	3.80–4.50	1.64–2.72
Mg, meq/L	4.30–6.40	4.50–5.80	0.82–1.33
Cl, meq/L	6.85–9.20	5.00–7.20	1.84–3.10
HCO_3, meq/L	5.90–6.40	6.80–8.90	4.92–6.45
SO_4, meq/L	1.40–2.10	1.04–1.25	0.21–0.34
SAR	2.85–3.32	3.54–3.58	7.54–8.43
BOD, mg/L	105–170	85–130	340–409
N, mg/L	42.5–67.3	36.2–73.8	54.9–68.3
P_2O_5, mg/L	18.6–024.5	14.60–21.4	18.30–23.1
K_2O, mg/L	32.2–49.6	28.0–43.2	37.5–44.2

Extensive guidelines have been developed in this area (Kale, 2003). It is usually recommended in India (may not be followed though) that some kind of pre-treatment is essential before sewage application on land. Table 24.18 gives chemical characteristics of domestic wastewater in India for wastewater irrigation parameters.

High Rate Transpiration Systems

In India high rate transpiration systems have been developed. In such systems plants render the system to consume more water. The plants like Bamboo (*Dendrocalamus tinctus*), Neem (*Azadirachta indica*), Shishum (*Dalbergia sisoo*), Acacia (*Acacia mangium*), Gvavelliol (*Gravelliae pterodifolia*) and Eucalyptus (*Eucalyptus* hybrid) are capable to transpire water equivalent to 5 to 13 times of potential evapotranspiration. Since all the wastewater is consumed, groundwater pollution problems are minimal. The nutrients present in the plants are also used by the plants and partly retained in the soil system.

Table 24.19: Design Details of HRTs for Some Plants is Given Below (Juwarkar *et al.*, 1997)

Type of Plants		*Ridges*		*Furrows*		*Plant to Plants Distance on Ridges*
		Width (m)	*Height (m)*	*Width (m)*	*Depth (m)*	*(m)*
Bamboo	*Dendrocalamum strictus*	1.5	0.5	3	0.5	3
	Bambusa arundincea	2	0.5	4	0.5	2
Neem	*Azadiractha indica*	0.8	0.5	1.6	0.5	2
Shishum	*Delbergia sissoo*	0.8	0.5	1.6	0.5	2
Acacia	*Acacia mangium*	0.8	0.5	1.6	0.5	2
Eucalyptus	*Eucalyptus* hybrid	0.8	0.5	1.6	0.5	2
	Eucalyptus robusta	1	0.5	2	0.5	1.5

Conclusion and Recommendation

The present study is not a status report of the country but only highlights some of the important issues related to use of saline water in agriculture. India can be taken as a very important test case where all kinds of waters have been used in agriculture for a very long time. A large number of data are also available in India on water quality of diverse sources. Besides, there have been extensive studies on saline water irrigation with reference to crop tolerance, effects on crops and on soils. Another angle is added to the whole scenario by extensive use of sewage and industrial wastes in agriculture. I thus recommend:

1. Water quality data of irrigation waters generated in the last 40 years should be meticulously collected and analyzed.
2. Extensive survey (probably in the form of an All India coordinated project) on different crops grown in different regions of India under saline water irrigation is crucial.

3. Scientific studies on the effects of use of saline water on crops encompassing all aspects of crop productivity and physiology are collected and collated.
4. Effect of prolonged saline water irrigation on soil quality and build-up of salinity and alkalinity is investigated and the available information in this area is compiled and collated.
5. Sewage farming is very common in India, however, there have been only few studies related to effect of sewage farming on crops, soils, heavy metal built-up, disease transmission and groundwater pollution. Sewage farming if scientifically done may prove to be a proper solution to scarcity of water. More studies are urgently needed in this area.
6. It is presumed that about 5000 medium and major industries in India use the untreated/ semi- treated or treated wastewater for irrigation. However, very few studies are available on short and long term effects of such wastewater usage.
7. Information on best management practices in saline water use are collected, collated and transmitted to actual users.

References

Agarwal, M.C. and S.S. Khanna. 1983. Efficient Soil and Water Management in Haryana. CCSHAU, Hisar Bulletin, 118p.

Anonymous. 1986. Co-ordinated Project on management of Salt Affected Salts and Use of Saline Water in Agriculture. Ann. Rep. RBC College, Agra, U.P., India.

CWC. 1987. Water quality studies, Ganga system status report (1978-85), Central Water Commission, New Delhi, p.116.

Kumar, V.D., D. Kumar and C.P.S. Chauhan. 1986. Effect of saline waters with varying levels of SAR and Ca:Mg ratio on growth and yield of wheat and soil properties. Ann. Arid Zone 25, p.73-80.

Gupta, I.C. 1979. Note on the effect of leaching and gypsum on the detoxification of Boron in a saline-sodic soil. Curr. Agric., 3,pp 231-34.

Gupta, I.C. 1980. Effect of irrigation with high sodium waters on soil properties and the growth of wheat. Intern. Symp. Salt. Affected soils, Karnal, pp. 287-288.

Gupta, I.C. 1983. Quality of irrigation water and the concept of residual sodium carbonate. Cum. Agric. 7,pp. 150-155

Gupta, S.K. and I.C. Gupta, 1987. Management of saline soils and waters. Oxford and IBH Publishing Co. New Delhi.,p339.

Gupta, I.C. and S.K. Gupta. 2003. Use of saline water in Agriculture (A study of Arid and semi-arid zones of India) 3rd edition., Scientific Publishers, Jodhpur

Gupta,I.C. and C.T. Abhichandani. 1973. Seasonal variations in the salt composition of some saline water irrigated soils of western Rajasthan II: Effect of fallowing and irrigation. Bhartiya Krishi Anusandhan Patrika, 1, 53-56.

Handa, B.K. 1972. Geochemistry of the Ganga River water. Indian Geohydrology.8, pp.71-78

Juwarkar, A., P.R. Thawle, H.P. Jambulkar and Asha Juwarkar. 1987. Management of wastewater through crop irrigation. Ecotechnology for Pollution Control and Environmental Management, Enviro Media, India.p.25-48.

Kale, C.K. 2004. Wastewater farming practices in India including design of wastewater farms. Recent advances in wastewater Engineering, Book Enclave, Jaipur, India, p.333-381

Kamra, S.K., O.P. Singh and A.D. Mongia. 2000. Proc. National Conference Salinity Management in Agriculture. CSSRI, Karnal.

Krishan, P.G. 1949. Investigations on the alkalinity and saltiness in sugar made from Coimbatore sugar-cane varieties. Indian J. Agric. Sci. 19, pp. 163-179.

Kumar, V.D. Kumar and C.P.S. Chauhan. 1986. Effect of saline water with varying levels of SAR and Ca:Mg ratios on growth and yield of wheat and soil properties. Ann. Arid Zone 25, pp73-80

Lal, P., M. Singh and F. Lal. 1980. Response of wheat to micro-nutrient fertilization under different qualities of irrigation water. Curr. Agric. 4, p.67-71

Maliwal, G.V. and K.V. Paliwal. 1971. Effect of fertilizers and chemical composition of wheat irrigated with saline water. Agrochimica, 16, p. 450-458.

Pandey, S. and B.P. Sahi. 1980. Quality of irrigation waters in the Gandak command of North Bihar, Madras Agri.J., pp.724-728.

Paliwal, K.V. and P.D. Sharma. 1993. Irrigation water quality and crop management in the Union territory of Delhi. Res. Bull9, IARI, New Delhi, pp. 166.

Sharma, D.P. and N.K. Tyagi. 2000. Disposal of drainage water, recycling and reuse. Proc. 8th, Int. Drainage workshop, New Delhi, Vol.III, p.199-219.

Yadhav, J.S.P. 1989. Problems and potentials of reforestation of salt effected soils of India. Field Document No.14, RWEDP, Asia, FAO, Bangalore, P-56

25

Salinity and Use of Saline Waters for Irrigation in Crops and Forages in Cuba

Aurelio Álvarez Menédez
Pastures and Forages Research Institute, Ministry of Agriculture, Cuba
E-mail: auralva@cima-minag.cu

ABSTRACT

In Cuba the soils with primary salinity, only occupy restricted areas, often near sea coast. Secondary salinization is the main cause of salinity in the the majority of Cuban soils. Human activities such as deforestation of hilly lands and use of saline water for irrigation are the main reason for salinisation. Different organic and inorganic industrial residues were studied combined with irrigation with good results, to amend soils with high Sodium Absorbtion Ratio (SAR) and Electrical Conductivity (EC), such as ethanol distillation residues, solid and liquid residues from sugar cane production and acid and alkaline residues from the nickel processing industry. These result of several greenhouse and field trials in dry regions of Cuba are reported. These experiments were aimed to evaluate saline soil tolerance of introduce and naturalized crops and forage species. The results showed that the numbers of plants adapted to saline soil conditions are very narrow. Some conclusions about the salinity problems of soil and the use of saline water for crops irrigation were formulated.

Introduction

One of overall problems for mankind in the future will be strongly related to soil and water salinity and methods to use these sources. Soils with natural problems of salinity are located in areas with negative balance of annual rainfall to evapotranspiration. In arid and semiarid regions this characteristic is predominant, even in areas with better average annual rainfall, as dry tropic and dry temperate climates, the presence of longer dry periods could enhance the process of salinization.

Besides, factors like: adverse climate conditions, saline ground waters, lowland close to sea coasts, littoral marsh and lagoons and in areas next to saline mines and domes must be considered favouring occurrence of salinization.

The above mentioned factors imply that the natural salinization is widely distributed throughout the world and its magnitude is increasing depending on climatic changes, soil erosion and material redistribution related to geomorphologic process, as well as the underground and superficial hydrologic cycles.

On the other hand, the salinization is also a consequence of the human society development, due to non scientific soil practices and waters use in agricultural production, known as atrophic or secondary salinization. In the last century, human activity contributed to break the dynamic balance between production and degradation of natural resources. This situation introduced several environmental changes which threaten the availability of non-renewable resources and the acceleration of the exhaustion of renewable resources.

The Salinization in Cuba

The natural sources of salts in Cuba are:

1. Marine aerosols carried by atmospheric circulation of predominant winds (East-Southeastern and southeastern trade winds) and accompanying North winds cold wave.
2. Old marine sediments in North and South of the Cuban archipelago constituted by gypsum and sandy saline rocks.
3. Accumulation of continental sediments carried by rivers from the mountains to plane areas during the Quaternary era.
4. Sea intrusion in calcic underground aquifers below sea level.

There are limited areas of soils with primary salinization in Cuba and which are associated to littoral marsh and lagoons. The natural saline soils are a consequence of extremely high tides that frequently occur during hurricane periods. Therefore, other causes are the turf deposits and calcareous sediments, currently found in the littoral saline marshes.

In some small valleys in the Southeastern part of Cuba, at Guantánamo province, where the annual rainfall average is less than 700 mm, it is possible to find primary sodic soils (solonetz), even in zones where saline waters or saline sediments rocks emerged to the upper layers. In this place it is possible to find soils with saline-alkaline characteristics (solod) in B-horizon. These soils were naturally formed from intense rainfall leaching of old salts at the beginning of Holocene period.

Most of the soils affected by salinization in Cuba were originated from secondary salinity. Main causes of secondary soil salinization are:

1. Deforestation of hilly lands with the consequent change in the water regime, rainfall occurrence and intensification of erosion. The rivers began to sediment large quantities of clay and sandy particles in the river's mouth. Thus, flooding of lowlands was more frequent and underground water level increased. Wherever those underground saline water areas were near to surface, salinization of soils occurs.
2. Loss of vegetation in lowlands and marshes in coastlines: The endemic species of scrubs and trees were cut, such as red mangle (*Rhizophora mangle*) to tan hides and *Conocarpus erecta* to make charcoal. As a consequence of overuse, the native vegetal barrier disappeared, the coast became unprotected and consequently the sea entered more easily.
3. Use of saline waters for irrigation from calcic underground aquifers: Agricultural areas where the main source of underground waters had this characteristic were overexploited with crops with high water requirements.
4. Elevation of underground saline water levels: The excessive irrigation in arable lands, together with inadequate soil management practices (lack of shallow and deep drainage, organic matter application, and excessive chemical fertilization) had provoked damages to vertisol and gley (clay) soils.

Presently, more than a million hectares of saline soils are reported in Cuba and the same amount of current agricultural lands could become saline in the future (Ortega *et al.*, 1986). Most of these soils are located at the South and North of Central and Eastern regions of the country, in areas with an annual rainfall average between 400-700 mm and a negative rainfall/evaporation balance.

The main improvement practices applied to saline soils for increasing crop and forage production have been: monitoring the water quality for irrigation and the salt and water balance in soils, use of efficient irrigation and drainage systems, application of chemical and organic amendments and use of salt tolerant plant species which will be discussed in the next sections.

Chemical and Organic Amendments of Soils

The application of amendments to saline soils is a management practice, which guarantees maintenance of exchangeable sodium percentage (ESP) and pH at acceptable levels for cultivated plants. The purpose of this practice is to remove, totally or partially, the excess of salts in root zones through overdoses of local irrigation or by soil leaching by rainfall or artificial irrigation (Cabrera, 1992). Saikian (1991) indicated that during soil leaching, there are changes in electrical conductivity (EC) as well as in quantitative relations between liquid and solid phases of soil, as a result of variations in chemical composition of soil solution, which depend on concentration of solution itself and size of soil particles.

The use of amendments mainly is based on addition of calcium-rich substances to increase its levels in soil solution. It brings the possibility of replacing exchangable sodium by calcium and avoids the sodium passage to soil absorbent complex (Otero, 1990, 1993).

The chemical amendments for saline soils are grouped in three categories:

1. Calcium soluble salts
2. Low solubility calcium salts
3. Other Acid forming substances

Otero (1993) reported the positive effect of mud residues from acetylene factories on saline soils without carbonates and pH lower than 7.5 at the Western provinces of Cuba.

The solid alkaline residues (gypsum pulp) of nickel processing industry were studied by Hernández *et al.* (1992) for soils improvement at Eastern provinces of Cuba. These researchers reported reductions of water requirements for soil leaching, sodium and magnesium substitution by calcium in soil solution and decreasing in bicarbonates formation during soil leaching.

Moreover, the liquid acid residues of nickel processing industry were tested as an amendment for saline soils (Otero and Ortega, 1986). These authors reported the positive effect of sulfuric acid (H_2SO_4) to improve soil permeability, reduction of sodium exchange percentage and increase of growth and biomass yield of Rhodes grass (*Chloris gayana*).

The objective of amendment application in saline soils is to replace Na^+ in the exchange complex in the soil. With the translocation of sodium to soil solution, is increasing its concentration and will also induce the formation of new sodic salts. These salts must be leached out from soil. If leaching did not happen, the new salts induced for the amendment, could affect the plants by increasing the osmotic pressure of soil solution and its inherent toxicity for the plants.

The organic methods for saline soil amendments have been considered to increase the efficiency of the other corrective practices (chemical techniques, salt leaching). Nevertheless, the experimental results have shown that their utilization could increase the effectiveness of salt leaching. It is possible to obtain acceptable crop yields in soils with high sodium content by organic amendments in the cases of unavailability of chemical products (Gálvez *et al.*, 1996).

Otero (1990) studied the effect of different waste sources as organic amendments combined with saline water irrigation of rhodes grass (EC = 2.6 dSm^{-1}) on chemical soil properties and forage biomass yields. His results have been shown in Tables 25.1 and 25.2.

The Sugar cane filter cake mud (SCFCM) treatment was able to control the calcium quantities existing in the soil absorbent complex, when saline water was used for irrigation. The substitution of sodium by calcium and magnesium in the soil complex seems to be a consequence of the high calcium content in the sugar cane filter cake mud. This control contributed to reduce the negative irrigation effect of saline water. The ethanol distillation residues (EDR) increased the total electrolyte and potassium concentration. The soil potassium increase should be related to the chemical EDR composition and represent a negative aspect, considering the dispersion properties of this ion.

Table 25.1: Variations of Some Chemical Characteristics of Saline Soil with the Application of Sugar Cane Filter Cake Mud (SCFCM) and Ethanol Distillation Residues (EDR) Combined with Saline Water Irrigation in a Short Term Greenhouse Pot Experiment (Otero, 1990)

Chemical Indicators	*Initial Soil Indicators*	*Irrigation and Amendment Doses, t.ha^{-1}*			
		0	*60*	*120*	*180*
		Sugar Cane Filter Cake Mud			
Ca^{2+}/BEC (Bases exchange capacity)	74.0	58.8	59.0	60.5	62.5
Mg^{2+}/BCE	13.1	32.2	31.2	30.1	28.4
Na^{+}/BCE	1.5	2.2	1.9	1.8	1.3
K^{+}/BCE	11.5	6.8	7.7	7.6	7.7
Organic matter (%)	3.5	2.9	3.4	4.1	4.1
pH	7.5	7.6	7.5	7.4	7.4
Chemical Indicators	*Initial Soil Indicators*	*Irrigation and Amendment Doses, m^3.ha^{-1}*			
		0	*750*	*1500*	*2250*
		Ethanol distillation residues			
Ca^{2+}/BCE	74.0	63.5	50.9	40.2	38.1
Mg^{2+}/BCE	13.1	23.9	23.3	22.8	28.2
Na^{+}/BCE	1.5	3.7	3.3	4.2	3.2
K^{+}/BCE	11.5	8.9	22.5	32.7	30.6
Organic matter (%)	3.5	2.7	3.9	3.9	4.05
pH	7.5	7.4	7.3	7.3	7.3

Biomass yield increased with the application of both sources of organic amendments. This result was related to Na^{+} and Mg^{2+} reduction in the soil solution. Similar results were obtained by Valdés (1990) in *Sorghum bicolor* grain harvested in greenhouse pot experiments, with the application of 250, 500 and 750 m^3.ha^{-1} of ethanol distillation residues and biomass yield of sugar cane (*Saccharum officinarum*) with four sugar cane filter cake mud doses (20, 35, 50 and 70 t.ha^{-1}) in the field conditions.

Table 25.2: Biomass Yield (g. pot^{-1}) of Rhodes Grass with the Application of Sugar Cane Filter Cake Mud (SCFCM) and Ethanol Distillation Residues (EDR) Combined with Saline Water Irrigation in Short Term Greenhouse Pot Experiment

Treatments	*Biomass Yield at Cutting*			
	Number of Cuttings			
	I	*II*	*III*	*IV*
Sugar cane filter cake mud				
Irrigation (control)	7.1	10.3	9.6	5.2
Irrigation + SCFCM 60, t.ha^{-1}	7.3	11.3	9.8	5.5
Irrigation + SCFCM 120, t.ha^{-1}	8.3	11.9	10.4	6.2
Irrigation + SCFCM 180, t.ha^{-1}	7.4	11.7	9.1	6.0
Ethanol distillation residues				
Irrigation (control)	6.4	11.1	8.6	2.3
Irrigation + EDR 750, m^3.ha^{-1}	7.0	11.1	10.8	5.9
Irrigation + EDR 1500, m^3.ha^{-1}	6.4	12.1	10.4	6.4
Irrigation + EDR 2250, m^3.ha^{-1}	6.0	14.4	11.9	5.1

In rice (*Oryza sativa*) commercial production areas in Granma province, three different sources and dosages of organic matter amendment in saline clay soils were compared (Tables 25.3 and 25.4) (Otero, 1990).

Table 25.3: Primary Characterization of Saline Clay Soils for Rice Production in Granma Province (Otero, 1990)

Soil Depth (cm)	*% of BEC*				*Organic Matter (%)*	*EC (dSm^{-1})*	*pH*
	Ca^{2+}	*Mg^{2+}*	*Na^+*	*K^+*			
0-20	47.5	36.6	15.1	0.76	0.78	1.56	7.9
20-40	59.2	20.0	18.1	0.70	0.70	2.52	7.3
40-60	63.5	17.0	18.9	0.59	0.54	3.12	7.3
60-80	61.3	17.7	20.4	0.59	0.65	2.89	7.4
80-100	54.7	23.6	21.1	0.59	0.66	2.82	7.4

Grain yields of all organic matter application levels from rice hulls (RH) and the combination of ethanol distillation residues plus sugar cane filter cake mud (EDR + SCFCM) were higher than the control. For sugar cane filter cake mud (SCFCM) alone, the 30 t.ha^{-1} variant of organic matter application was higher to the control treatment. The application of any of the organic matter levels had a beneficial effect on the reduction of alkalinity, substitution of Na^+ for Ca^{2+} in the soil solution and increased the organic matter content of soil.

Table 25.4: Rice Yield Under Commercial Conditions with Different Sources and Dosages of Organic Matter Applications in a Saline Clay Soil in the Province of Granma (Otero, 1990)

Treatments	*Seasonal Grain Yield, t.ha⁻¹*	
	Wet and Warm Season	*Dry and Cold Season*
Control	4.4	4.2
Sugar cane filter cake mud (SCFCM)		
SCFCM, 15 $t.ha^{-1}$	2.6	2.6
SCFCM, 30 $t.ha^{-1}$	4.5	4.8
SCFCM, 45 $t.ha^{-1}$	3.9	3.7
Rice hulls (RH)		
RH, 5 $t.ha^{-1}$	6.7	6.5
RH 10, $t.ha^{-1}$	5.3	5.2
RH 15, $t.ha^{-1}$	4.3	4.9
Ethanol distillation residues + Sugar cane filter cake mud (EDR + SCFCM)		
EDR 125 $m3.ha^{-1}$ + SCFCM, 5 $t.ha^{-1}$	5.7	5.1
EDR 250 $m3.ha^{-1}$ + SCFCM, 10 $t.ha^{-1}$	5.4	4.5
EDR 375 $m3.ha^{-1}$ + SCFCM, 15 $t.ha^{-1}$	4.7	4.4

Monitoring the Irrigation with Saline Waters

The use of low quality waters for irrigation of crop and forage, including sea water, is an interesting possibility. Saline waters for irrigation will depends on the type of crops or forages, the soil characteristics, climatic conditions and irrigation and drainage methods. Not always the chemical quality is the most important, but rather its agronomic quality (Orellana, 2002).

Effect of salinization and irrigation water salinity on the yield of vegetable crop showed a variable effect (Table 25.5). The predominant ions were between Na^{+}/Ca^{2+} and Mg^{2+}/Ca^{2+}.

The control of salt concentration in the soils and irrigation waters is very important because its osmotic effect on plant nutrition by decreasing total water potential in the soil. In this situation the root absorption decrease as well as shoot transpiration rate and water and nutrient flows in the plant (Monedero *et al.*, 2005).

Bat *et al*. (1991) reported that the interaction between salinity and fertility in the soil is strongly affected by use of poor quality waters for irrigation. Casanova and Savón (1995) observed a linear relation between soil salinity and vegetable crop yields.

In Matanzas province to the Eastern part of Cuba, Duarte and Pérez (2005) during 15 years of monitoring the effects on soil and fresh biomass yield of sugar cane plantations (*Sacharum officinarum*) irrigated with waste waters from sugar cane factory. The soils are mainly Inseptisol, with pH of 7.0-8.6. Changes in cations, organic matter, EC and pH shown in Table 25.6.

Table 25.5: Characterization of Soil and Irrigation Water to Protect Crop for Horticulture Production, Located in Different Western and Southwestern Municipalities of the Havana Province (Adapted from Monedero *et al.*, 2005)

Municipality	*EC in the Soil, dSm^{-1}*		*SAR*		*EC in the Irrigation Water, dSm^{-1}*	*Vegetable Crop Yield Reduction (%)*
	0-20 cm	*20-40 cm*	*0-20 cm*	*20-40 cm*		
Güira de Melena # 1	22.9	14.8	2.0	1.6	1.4	50-75
Güira de Melena # 2	3.5	2.2	0.3	0.2	0.8	10-25
Güira de Melena # 3	7.6	6.0	0.7	0.4	0.4	>25
Quivicán	4.6	1.3	0.4	0.5	0.6	10-25
San Antonio de los Baños	4.2	2.8	0.4	0.3	0.7	10-25
Alquizar	4.3	1.9	0.6	0.5	1.4	25-50

Table 25.6: Initial and Final Soil Chemical Analysis After 15 Years of Irrigation with Sugar Factory Waste Waters in the Province of Matanzas

Year	*cmol (+).Kg^{-1}*				*Organic Matter (%)*	*pH*	*EC (dSm^{-1})*
	Ca^{2+}	*Mg^{2+}*	*Na^{+}*	*K^{+}*			
1988	48.0	11.5	1.0	1.4	3.2	7.0	0.38
2003	46.9	6.6	0.8	0.8	3.8	7.7	0.42
Variation	-1.1	-4.9	-0.2	-0.6	+0.6	+0.7	
% of variation	2.3	42.6	20.0	42.8	15.8	9.1	

There was no significant change in soil chemical properties (Table 25.1) or any other environmental damages from irrigation with sugar factor wastewaters (Figure 25.1).

Fresh yield of sugar cane was increased in irrigated areas with waste waters (Figure 25.2). Moreover persistence sugar cane of shoot increased in plantation irrigated with waste waters, compared with those plantations irrigated with natural water Duarte and Pérez (2005).

In areas near the coast and in keys of Cuban archipelago, related with increasing demand of fresh vegetable for tourist business, some efforts were made to develop horticulture based on organic substrates and irrigation with saline and waste waters (Orellana *et al.*, 2002).

Orellana *et al.* (2002) pointed out that any of the preliminary experiences in horticulture using saline waters for irrigation were closely related to adequate management of waters, organic substrates and crops, particularly the physical and chemical conditions of organic substrates. Irrigation water quality was shown in

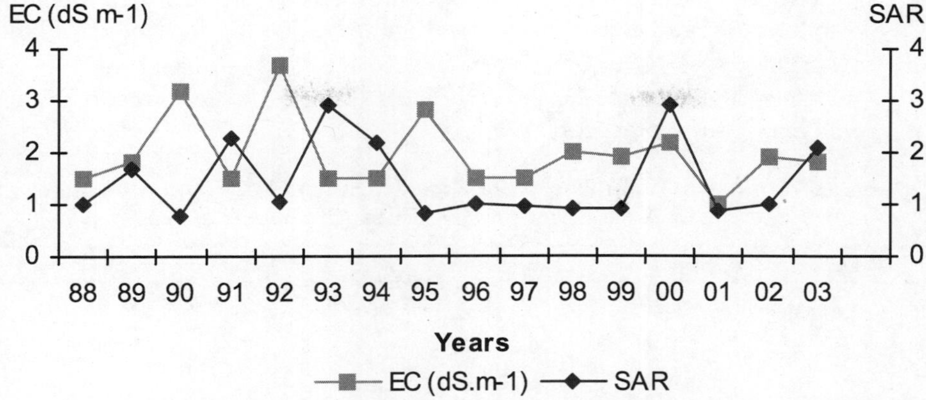

Figure 25.1: Annual Quality Average of Factory Waste Waters from Sugar Cane Production During 15 Years in the Province of Matanzas (Duarte and Pérez, 2005)

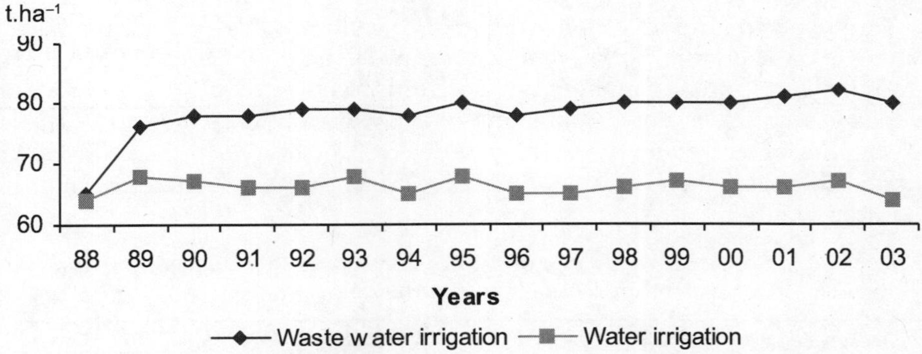

Figure 25.2: Fresh Biomass Yield of Sugar Cane During 15 Years Irrigated with Industrial Waste Water from Sugar Cane Factory (Duarte and Pérez, 2005)

Table 25.7. Following are the recommendations on the methodology to produce vegetable under Cuban littoral climatic and soil conditions:

1. Systematic application of organic matter on nursery in a proportion higher to 50 per cent of total substrate.
2. To produce seedlings for transplanting by controlled methods.
3. To apply microbiological strains for improving plant nutrition and attenuation of salinity problems (free and associated bacteria for nitrogen fixation, phosphorus mobilization in the soil and fungus such as Azotobacter, Azospirillum, Phosphor bacterias, Mycorrhiza fungi).
4. To apply the saline water directly to the soil and not under crop foliage.
5. To use of planting and harvesting calendar, in order to take advantage of littoral rainfall.

6. Continue maintenance of humidity in root area to safeguard the water flow to the plant and avoid the evaporation through mulch and other organic covers.
7. To keep all time a strict technological discipline and efficiency and efficacy in agricultural practices.

Table 25.7: Irrigation Water Quality for Organic Orchard in Urban Horticulture Production in Some Coast Areas of Cuba (Orellana *et al.*, 2002)

Locality	*EC (dSm⁻¹)*	*SAR*	*HCO³⁻*	*Na⁺*	*Cl⁻*
Guanabo beach	2.90	5.62	4.51	14.0	20.0
Cojimar town	2.40	5.03	6.19	12.0	14.5
Celimar town	2.01	6.06	4.20	12.0	12.2
Cojimar river	6.30	14.60	5.61	41.0	46.0
Cayo Largo (key)	4.73	13.85	3.90	32.4	36.4

Three water pits for irrigation in organic horticulture parcels placed in different localities of Camagüey province, were selected by Hernández *et al*. (2005) to conduct a study on water quality monitoring. The results showed high concentration of bicarbonates, chlorines, and sodium ions. In two water pits (El Imán and La Rubia) the increasing of water salinity and EC were detected. In the other pit (Los Olivos) the water used for irrigation had high content of bicarbonates, followed for chlorides and nitrates ions.

Table 25.8: Some Characteristic of Waters and Substrates Utilized for Urban Horticulture Production (Hernández *et al.*, 2005)

Vegetables	*Irrigation Water EC, dSm⁻¹*	*Organic Substrate*				*Organic Matter (%)*
		pH		*EC, dSm⁻¹*		
		Control	*Irrigated*	*Control*	*Irrigated*	
Radish	2.48	7.3	7.4	0.71	0.96	13.7
Flowers	2.44	7.2	7.3	0.90	1.12	12.2
Onion	1.55	7.5	7.6	0.92	0.94	16.5

The inadequate quality for vegetable production of water in monitored pits is consequence of saline ground waters. This result is in agreement with those reported by Borroto and Gandarilla (1997) about the saline conditions of some water pits in the same city.

Hernández *et al*. (2005) conclude that besides the quality of water is not optimal for vegetable irrigations, in correspondence with Cuban Regulations of Water Quality for Agricultural Use (Ministerio de la Agricultura, 1990), the pragmatic necessity to continue using these waters for irrigation in urban horticulture, will need to consider some corrective actions:

1. To search for salt tolerant species

2. Drainage methods improvement and nursery flattening
3. Over irrigation and mixed saline water with less or not saline ones before irrigation
4. To select calcium rich organic substrates and/or enriching with some source of calcium

Table 25.9: Crop Yields and Leaf Composition of Three Vegetables Irrigated with Saline Water (Hernández *et al.*, 2005)

Vegetable	*Average Yield ($Kg.m^{-2}$)*	*Leaf Composition*			
		Ca^{2+}(%)	*Mg^{2+}(%)*	*Na^{+}(%)*	*NO^{-3} (ppm)*
Radish	23.7	2.68	0.73	1.33	10456
Flowers	103.5	0.94	0.63	1.67	3400
Onion	27.6	1.19	0.40	0.21	5174

In Cuba there are little commercial availability of species with confirmed tolerance to saline soils and irrigation with saline waters.

Borroto (1990) in several experiments carried out to determine plant adaptation and productivity in different saline stress conditions, tested more than 52 species, varieties and clones of root, tubers, vegetables, cereals and pastures. The authors reported the reduction of yields at EC = 2 dSm^{-1} and only few salt tolerant plants reached some production at 5 dSm^{-1}.

Cabrera *et al.* (2005) in a controlled experiment under laboratory conditions evaluated the germination and relative growth of root and shoot biomass in seven varieties of tropical wheat (*Triticum aestivum*). At minor values of electrical conductivity (3 and 6 dSm^{-1}) the morphological aspects of all varieties had similar response in comparison to the control. However from EC = 9 dSm^{-1} the height of shoot reduced compare to the control (Figure 25.3).

The germination rates of wheat varieties and control was superior to 50 per cent which show the varietal resistance to saline stress at moderate to high saline water values (Figure 25.4) (EC 9 dSm^{-1})

Borroto (2006) studied the reduction of forage yields of some pastures used for livestock feeding in the region of San Antonio del Sur, province of Guantanamo in soils between low to moderate salinity (Table 25.10).

For 20 years the researchers of Institute of Pastures and Forages have been working to evaluate a large collection of introduced and native grasses and legumes for animal feeding. In these studies we are testing more than 500 species of pastures and forages in different conditions of soils, clime and management (77.3 per cent of total agroecosystems for livestock production). During the first version of regionalization of pastures and forages in Cuba (published at 1985), which recommended 20 grasses and 3 legumes as pastures and forages for 5 large grouping categories of soils and considering four main soil limiting factors. In the second version (published at 1990) it was considered 27 grasses and 16 legumes, adapted to 22 types of soils and eight limiting factors, included soil salinity which were shown in Table 25.11 and 25.12.

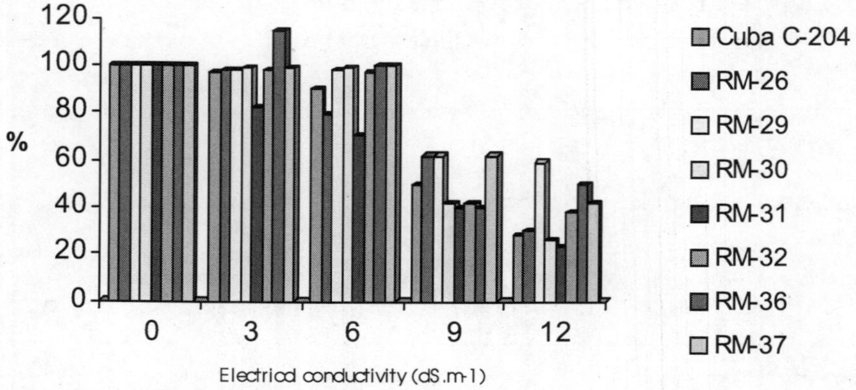

Figure 25.3: Relative Percentage of Plant Height Respect to Control in Seven Varieties of Tropical Wheat (Cabrera *et al.*, 2005)

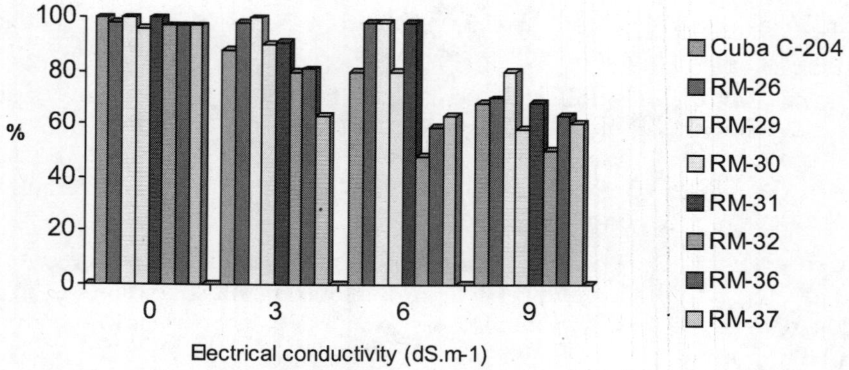

Figure 25.4: Relative Percentage of Germination in Seven Varieties of Tropical Wheat (Cabrera *et al.*, 2005)

Table 25.10: Determination of Forage Yield Reduction (%) in Six Roughages Sowed in Saline Soil in the Region of San Antonio del Sur. (Adapted from Borroto, 2006)

Forage	*Dry Matter Yield*(t.ha^{-1})*		*Difference (%)*
	Optimal soil	*Saline soil*	
Cenchrus ciliaris (Buffel grass)	11.0	9.0	18.2
Cynodon nlemfuensis (Star grass)	13.5	11.1	17.7
Chloris gayana (Rhodes grass)	10.0	8.3	17.0
Pennisetum purpureum (King grass)	15.7	11.9	38.0
Panicum maximum (Guinea grass)	14.3	9.1	36.4
Leucaena leucocephala (Ipip-Ipil)	7.2	6.6	8.3
Saccharum officinarum (Sugar cane)	24.0	10.4	56.6

* Without irrigation and chemical fertilization.

Table 25.11: Adapted Legumes to Different Levels of Salinity (Paretas, 1990)

Species	*Varieties*	*Electric Conductivity, dSm⁻¹*		
		<2.4	*2.4-4.8*	*4.8-6.0*
Clitoria ternatea		X	X	---
Lablab purpureus (Cowpea)		X	---	---
Leucaena leucocephala	Ipil-Ipil	X	---	---
Macroptilium atropurpureum	Siratro	X	---	---
Medicago sativa (Lucerne)	Gilboa	X	X	---
Neonotonia wightii		X	X	---
Teramnus labialis		X	---	---

Table 25.12: Adapted Grasses to Different Levels of Salinity (Paretas, 1990)

Species	*Varieties*	*Electric Conductivity, dSm⁻¹*		
		<2.4	*2.4-4.8*	*4.8-6.0*
Brachiaria purpuracens	Aguada	X	---	---
Cenchrus ciliaris	Biloela	X	X	---
	Formidable	X	X	---
	Verde-azul	X	X	---
Chloris gayana	Callide	X	X	---
Cynodon dactylon	Coat cross # 1	X	---	---
	67	X	---	---
	68	X	---	---
	de Costa	X	X	---
Cynodon nlemfuensis	Jamaicano	X	X	---
	Panameño	X	X	---
	Tocumen	X	---	---
Digitaria decumbens	Pangola	X	---	---
Panicum maximum	Común	X	---	---
	Likoni	X	---	---
	Uganda	X	---	---
Pennisetum purpureum	King grass	X	---	---

In general there a wider range of species (grasses and legumes) adaptation to low salinity levels in the soil, only when salt content in soils is increased, the availability for species able to surviving and produce in this stressing situation, is reduced considerably.

Conclusions

1. The most of saline soils in Cuba is due to secondary salinization, through the negative action of humankind along of the history. However there are a considerable resurging of saline soils due to appearance of large drought periods and desertification conditions in some areas.
2. The importance of amendments for saline soils is a well known method to stop or at least, to retard the salinization. More emphasis will be placed to identify new sources of organic matter to amend of saline soils.
3. The hardware necessary to apply large quantities of organic amendments will be acquired and new techniques will be developed.
4. There are little and recent experiences in Cuba about saline water irrigation in dry areas. The Cuban regulations for water quality in agricultural use will be reviewed.
5. The number of adapted species of crop and forages to saline soils conditions is low. New research will be conducted to solve this situation.

Acknowledgements

The authors are indebted for the assistance in this work to:

1. Dr. Fernando Funes and Dra Marta Monzote, Lic. Dulce Maria Vento. Pastures and Forages Research Institute. Ministry of Agriculture of Cuba
2. Institute of Irrigation and Drainage, Ministry of Agriculture of Cuba.
3. Institute of Soils, Ministry of Agriculture of Cuba.
4. Mrs. Jorge Luis Fernández and Luis Gómez. International Cooperation Department. Ministry of Science, Technology and Environment of Cuba.
5. And last but not least to Prof. Arun P. Kulshreshtha, Director of The Centre for Science and Technology of the Non-Aligned and Other Developing Countries (NAM S&T Centre) for the support to attend to this international meeting.

Abbreviations

BEC: Bases Exchange Capacity. (per cent or cmol (+).Kg^{-1})

CE: Electrical Conductivity of soil saturation extract. (dSm^{-1})

SAR: Sodium Adsorption Rate in soil solution.

$$SAR = \frac{N^+}{\sqrt{(Ca^+ + Mg^{2+})/2}}$$

References

Bat, T. A. L., A. S. Feigenbaum and D. L. Sparks. 1991. Institute of soil and water, Agricultural Research Organization. Bet Bagan, Israel. Irrigation Science (Germany) 12: 27-35.

Borroto, M. 1996. Disponer de especies y variedades apropiadas para la explotación económica en suelos con diferentes tenores de salinidad. Informe final del proyecto de investigación. 004-14-32. Instituto de Suelos. Ministerio de la Agricultura. 88 pp.

Borroto, M. and L. Gandarilla. 1997. Estudio preliminar de sales en pozos de 3 Consejos Populares de la ciudad de Camagüey. Informe Técnico. Instituto de suelos. Ministerio de la Agicultura. 25 pp.

Borroto, M. 2006. Rehabilitación ecológica del valle de San Antonio del Sur. Zona costera árida Guantánamo-Maisi. Memorias. VI Congresote la Sociedad Cubana de las Ciencias del Suelo. 8-10 de marzo. Ciudad de La Habana, Cuba. pag. 1-14.

Cabrera, M., L. Gutíerrez, H. Penichet, E. Lago, M. Pérez and M. J. Mendoza. 2005. Comportamiento de variedades cubanas de trigo (Triticum aestivum) ante condiciones de estrés hídrico y salino. Memoria. Congreso Internacional de Riego y Drenaje. Octubre, 25-28. La Habana, Cuba. pag: 428–437.

Cabrera, R. 1992. Fundamentos de las medidas para el mejoramiento y recuperación de los suelos salinos del area cañera del valle de Guantánamo. Tesis en opción del grado de PhD en Ciencias Agrícolas. Ministerio del Azúcar. Instituto nacional de Investigaciones de la Caña de Azúcar. Estación Provincial de investigaciones de la Caña de Azúcar. Cuba 117 pp.

Casanova, A and J. R. Savón. 1995. Producción biointensiva de hortalizas. Agricultura Orgánica. Revista de la Asociación Cubana de Técnicos Agrícolas y Forestales (ACTAF). Año 1: 13-16.

Duarte, J. G and J. R. Pérez. 2005. Resultados del fertirriego con residuales líquidos en la Empresa Azucarera "Mario Muñoz": 15 años de experiencias. Memoria. Congreso Internacional de Riego y Drenaje. Octubre, 25-28. La Habana, Cuba. pag: 823–829.

Galvéz, V., O. Hernández, G. López, R. Orellana, L. Otero and Magali Valdés. 1996. Prácticas de manejo de suelos. In: Salinidad un nuevo concepto. Ed. A. Flores, V. Gálvez, O. Hernández, J. G. López, A. Obregón, R. Orellana, L. Otero y M. Valdés. Universidad Autónoma Metropolitana, Universidad de Colima, México y Centro de Información y Documentación Agropecuaria e Instituto de Suelos, Cuba. pag: 58-86.

Hernández, O., L. Otero and V. Galvéz. 1992. Efectividad de la papilla de yeso en el mejoramiento de suelos afectados por sales. Memorias. XXV Congreso de las Ciencias del Suelo. México. pag. 45

Hernández, O., M. González V. Gálvez, M. Borroto, T. de la Guardia. 2005. Efecto de las aguas de mala calidad en el riego de tres organopónicos de la ciudad de Camagüey. pag. Memoria. Congreso Internacional de Riego y Drenaje. Octubre, 25-28. La Habana, Cuba. 915-919.

Ministerio de la Agricultura. 1990. Norma cubana para el uso de la calidad del agua en la Agricultura.

Monedero, Milagros, Carmen Duarte, C. A. Alfonso, Maria Isabel Hernández, L. Gómez, Juana Maria Dantín and Elia Deisy Ramos. 2005. Evaluación de la salinidad de los suelos ferraliticos rojos bajo condiciones protegidas. Memoria. Congreso Internacional de Riego y Drenaje. Octubre, 25-28. La Habana, Cuba. pag: 980–985.

Orellana, Rosa, Melba Cabrera, A. Prats, J. M. Moreno, V. León and Caridad Pérez. 2002. Utilización de aguas salinas para la producción de hortalizas en zonas costeras. Agricultura Orgánica. Revista de la Asociación Cubana de Técnicos Agrícolas y Forestales (ACTAF). Año 8: 15-17

Ortega, S. F., F. Peña and N. Castillo. 1986. La salinidad de los suelos de Cuba. Aspectos económicos globales. Revista Ciencias de la Agricultura. 27:137-144.

Otero, L. and F. Ortega.1986. Posibilidad del uso de los efluentes ácidos de la Empresa "Comandante Pedro soto Alba" en el mejoramiento de los suelos salinos. Ciencias de la Agricultura. 27: 156-158.

Otero, Lazara. 1990. Efectividad de la utilización de materiales de origen mineral y orgánico como mejoradotes de suelos salinos y salinizados. Informe final del proyecto de investigación. 004-13-31. Instituto de Suelos. Ministerio de la Agricultura. 26 pp.

Otero, Lazara. 1993. Particularidades del intercambio catiónico de los suelos oscuros plásticos de la provincia de Granma, relacionado con su manejo. Tesis en opción del grado de PhD en Ciencias Agrícolas. Ministerio de la Agricultura. Instituto de suelos. 100 pp.

Paretas, J. J. (1990). Ecosistemas y regionalización de pastos en Cuba. Instituto de Investigaciones de Pastos y Forrajes. Ministerio de la Agricultura. 178 pp.

Saikian, S. V. 1991. Procesos de desalinización-desolonetización y métodos de su aceleración con el mejoramiento químico y el lavado de los solonchak-solonetz sódicos del valle de Ararat. Tesis en opción del grado de PhD en Ciencias Biológicas. Universidad de Moscú. 36 pp.

Valdés, M. 1990. Utilización del mosto de destilería y cachaza en el mejoramiento de los suelos salinos. Informe final del proyecto de investigación. Instituto de suelos. Ministerio de la Agricultura. 40 pp.

26

Salinity Development in the Dry Zone of Sri Lanka: A Review

M.M.M. Najim* and K.P.K. Jayakody
Department of Agricultural Engineering,
Faculty of Agriculture, University of Peradeniya, Peradeniya, Sri Lanka
**E-mail: bmarambe@pdn.ac.lk, nazimhn@yahoo.com*

ABSTRACT

Soil salinity development is identified as one of the problem in the Mahaweli development scheme and the dry parts of the agricultural areas in the country. The cause of salinity development in Sri Lanka is identified as the elevation of the groundwater tables and evaporation especially in the dry regions. In Nochiyagama block of the Mahaweli system H showed that 30 per cent of the lands are affected by salinity. In the areas where the salinity has developed more than 5 dS m^{-1} has shown reduced paddy yields. It was calculated that about 757 kg ha^{-1} of paddy yield is reduced when salinity is elevated to 5 dS m^{-1}. The salinity development has deteriorated the drinking water quality too. Few irrigation tanks in the Mahaweli system showed elevated levels of salinity, *i.e.* Alli Tank and Vakanneri Tank. Parakkrama Samudraya, Kanthali and Minneriya tanks also show occasional salinity levels. In the areas, the presence of salt deposits on the surface and also the decline in rice yields are reported by farmers. Certain localities have shown higher salinity levels in empirical studies too. Poor drainage, undulating topography, reuse of irrigation water, mono cropping for a longer period etc. are the major factors contributing to the salinity development in the area. Introduction of policies, research on salinity and related issues, maintenance of irrigation infrastructures with proper and adequate drainage, improving water productivity, crop rotation and good agronomic practices etc. are the management practices needed to curb the salinity development. Salt affected lands could be used for alternative crop productions such as forage production in order to improve livestock raring.

Introduction

Sri Lanka is a tropical country and agriculture is the main economic activity. Dry zone of Sri Lanka is the main agricultural area in the country where salinity has been associated with irrigated agriculture since its early days. Irrigation often aggravate the effects of salinity, which occurs due to many reasons such as natural weathering of saline parent material from sea water deposits or other sources etc.. Salinity has also been associated with the rise of groundwater tables resulting from excess irrigation and poor drainage in large-scale, perennial irrigation systems; resulting shallow water tables bring salts to the upper layers of the soil profile. Salinity can also be induced by the use of pumped groundwater. In the dry zone of the country, water drained from one agricultural area is used in some other areas. The drainage water is high with some salt concentrations, which could lead to salinity development. The evaporation of water due to dry weather also concentrates salts on the soil surface. In these cases, the physical process underlying salinization is the absence of a downward soil water flux of sufficient degree to leach the salts from the root zone. Salinity can also result from sea water intrusion into coastal areas where the water tables have been lowered by the mining of groundwater.

In the recent years, some farmers have been abandoning their rice fields in the dry zone irrigation schemes due to the incidences of salinity. Many organizations and individuals have carried out few research works on salinity and associated problems in the dry zone. Mahaweli system H that includes Kalaweva Basin was the first project under the Mahaweli development program. Under the System H, approximately 38533 ha of lands are being cultivated by about 31000 farm families. Paddy, chili, vegetables, onion and legumes are the major crops that are being cultivated in the area. During the main rainy season Maha, paddy is the main crop due to the ample availability of water. During the off-season Yala the paddy is only cultivated in the areas where sufficient water supply is secured. In the other areas, the major crops are chili, vegetables, onion and other field crops. In general, Mahaweli system H contributes 25 per cent of paddy production, 30 per cent of chili and 20 per cent of big onion production of the country. Salinity development in these lands will have tremendous effect on the agricultural productivity from this area. In this paper the Irrigation-induced salinization in irrigation systems in dry zone and salinity problems in the Mahaweli system H of Sri Lanka will be discussed.

Mahaweli System H

Land degradation due to salinity and water logging is primarily associated with coastal areas and irrigated lands in the dry zone of Sri Lanka, which covers half a million hectares (TAMS/USAID, 1980). The dry zone of Sri Lanka is the most important area as far as irrigation is concerned. In Sri Lanka, the dry zone occupies nearly two thirds of the land where people depend mainly on arable farming for food and income. The Mahaweli system H (Figure 26.1) is one of the major agricultural areas in the dry zone, which receives transbasin diversion water.

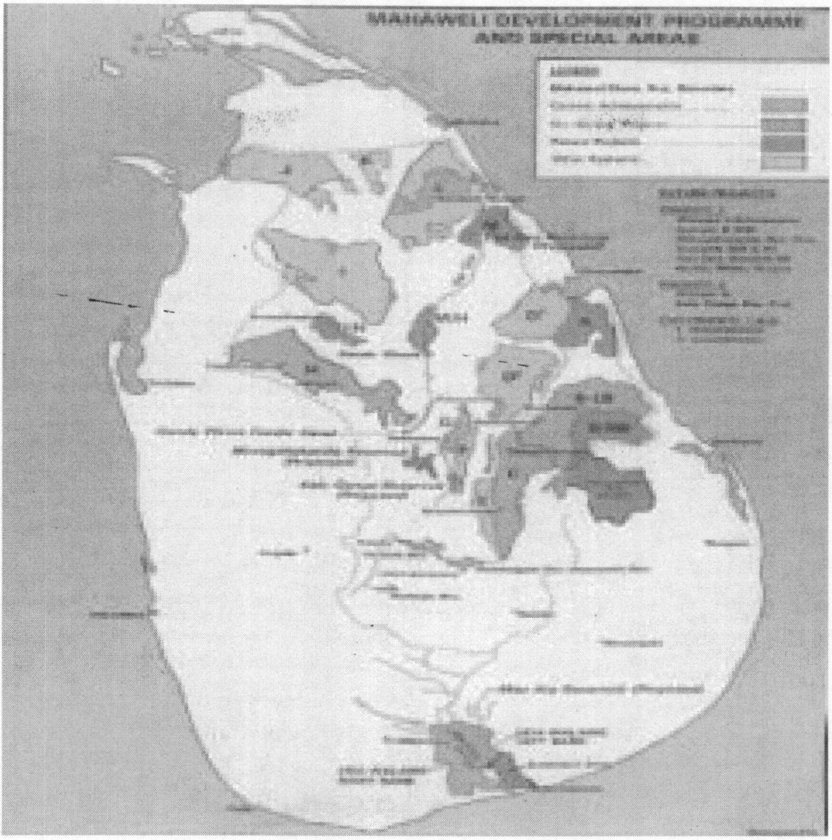

Figure 26.1: Mahaweli Development Schemes

Topography and Soil

The landscape of the Mahaweli system H is undulating with slopes ranging from 1 to 4 percent (De Silva, 1987). The approximate slope along the transect from top to bottom are 3-10 per cent on tales slope, around 2 per cent in the convex portion and less than 1 per cent in the concave part. Rock landscape and erosion remains are found only in the concave part. The landscape has been transformed into bounded paddies. The side slope of the lower order valleys lead to the development of two main land types. The upper part of the side slope of the land is wet only during the wet weather and leaving rest of the year dry (De Silva, 1987). The lower part, on the other hand, is wet for a longer period about six month or more. The poor drainage in these wet areas has accumulated salts leading to salinity development.

Geochemistry of Landscape

The two main land types that are found on the side slope of the lower order valleys of the inland dry zone have specific geochemical characteristic. The land located on the upper convex slope loses water, finer soil particles and soluble elements to the lower concave slope. Consequently soil in the convex part of the side slope becomes dry and well drained, coarse textured and relatively depleted of soluble elements. The soil in the concave part of the slope on the other hand becomes wet with a retarded drainage, finer textured and enriched with soluble elements (De Silva, 1987). This portion of the landscape is much prone to salinity development. Many ancient minor irrigation schemes, which are still operational, have experienced salinity patches within the command areas.

Irrigation Pattern

Irrigation water delivery in the area is a rotation system. Under this system water is delivered to each farmer in sufficient quantity for a fixed period of time under a prearranged schedule. In this system, the farmer does not have sufficient irrigation water when he wants. For other field crops, farmers use supplementary water from shallow agro wells. High application efficiencies have not been achieved in the system. Effective rainfall can be utilized fully and standing water is not maintained continuously in order to prevent deep percolation losses. Utilization of rainfall is beneficial for flushing the accumulated salts.

Salinity Status of Soil and Water in Mahaweli System H

Accelerated Mahaweli development scheme supplied water into the existing major and micro irrigation systems. These systems had reached a stable equilibrium over many centuries. Introduction of trans-boundary diversions into such systems have brought many changes to the water and salt balance in the area. Establishment of major irrigation schemes destroying forest cover in well drained land and supplying additional water to the whole landscape over a longer period of time has interrupted the established equilibrium for both salt and water in the region (Handawala, 1983). It was also shown that the field salinity observed in the opened lands in the Mahaweli system H can be attributed to the emergence of underground salt reserves. There was a gradual increase in the sodium absorption ratio (SAR) of irrigation water from 1978 to 1986 which indicates that there is a gradual build up of salts in soil solution in the Mahaweli system H (Pathmanathan, 1999). Handawala (1983) showed that after the Mahaweli system development, the drainage capacity of the natural stream and canals were badly declined, and that there were cases when farmers blocked drainage canals to obtain more water for irrigation. Such drainage blockages maintained and circulated released salts within the location for a longer period than necessary without being flushed out, thus causing salinity hazards.

In Nochiyagama block of the Mahaweli system H showed that 30 per cent of the lands are affected by salinity. In the areas where the salinity has developed more than 5 dSm^{-1} has shown reduced paddy yields. It was calculated that about 757 kg ha-1 of paddy yield is reduced when salinity is elevated to 5 dSm^{-1}. The salinity development has deteriorated the drinking water quality too. As a solution, proper drainage is proposed in this area (Pathmanathan, 1999).

Few irrigation tanks in the Mahaweli system showed elevated levels of salinity, *i.e.* Alli Tank and Vakanneri Tank. Parakkrama Samudraya, Kanthali and Minneriya tanks also show occasional salinity levels. In the areas, farmers report the presence of salt deposits on the surface and also the decline in rice yields. Certain localities have shown higher salinity levels in empirical studies too (Wijesekera, 1981).

Sikurajapathy *et al.* (1983) estimated that about 4.96 per cent of the paddy fields were affected by salinity in the Mahaweli system H. Researchers indicated that more lands are likely to become salt-affected if improved drainage facilities are not provided. Dhruwasangary (1983) assessed the effects of drainage on salinity level and the cost involved in the Mahaweli H area and showed that subsurface drainage will improve crop yield significantly. Gangodawila (1988) in his study indicated gradual emergence of salinity problems in the Mahaweli system H.

Cascades in Mahawilachiya, Nuwaragampalatha central, Medawachchiya and Kahatagasdigiliya in the Anuradapura district showed salinity developments. Many paddy fields are seriously affected by salinity and the cultivation is totally stopped. These lands are neglected and the only vegetation that can be seen is salt tolerant native grasses and creepers. These areas can be used economically for forage production where the farming community can integrate it with animal husbandry.

Preliminary soil surveys in the Mahaweli system H prior to Mahaweli diversion showed that 80 per cent of the soils in the area contained manganese that is less soluble. The manganese content increased down the slope. Iron and aluminum, which are the least soluble of the elements, are hardly mobilized and generally settled in the well-drained soil (De Silva, 1987). According to preliminary soil surveys in the Mahaweli H area prior to its diversion there were low or medium salinity of soil and water in the region. Only certain locations have shown high salinity levels (Wijesekera, 1981).

The salinity levels of the irrigation water in the Mahaweli system H are always low or medium. Electrical conductivity (EC) of the irrigation water tends to increase from wet weather to dry. A survey of water quality of minor tanks in the area shows that the EC level increases in between the months March to November, which is the drier period in the area (De Silva, 1987). This phenomenon is similar in the major tank systems. Another observation is that the EC of tank water increases down the longitudinal slope of the valley, which is given in Table 26.1. This is a very clear indication that the drainage and return flows from the head end to the tail end concentrates along the gradient.

Table 26.1: Salinity Variation in Tanks in Dry and Wet Seasons

Name of the Tank	*Electrical Conductivity dSm^{-1} (Dry Season)*	*Electrical Conductivity dSm^{-1} (Wet Season)*
Kalawewa	0.40	0.20
Kalankuttiya	0.45	0.25
Katiyawa	0.45	0.35
Rajangana	0.70	0.35

Source: Salinity Status of Inland Dry Zone of Sri Lanka, De Silva, 1987.

Nochchiyagama block in Mahaweli system H is located at the end of the right bank channel of the main irrigation tank Kalawewa. This area has a very high vulnerability to salinity compared to other blocks in the system H. Table 26.2 gives the salinity test results in Nochiyagama Block of Mahaweli system H.

Table 26.2: Salinity Test Results in Nochchiyagama Block

Salinity Range (dSm^{-1})	*Salinity Category*	*Total Number of Farmers*	*Farmers (%)*
0-2	Low	33	30
2-5	Medium	43	40
5-8	High	26	23
>8	Severe	8	7

Source: Economic and environment assessment in land degradation caused by soil salinity in Nochchiyagama Block of the Mahaweli system H, Pathmanathan, 1999.

The salinity levels in Table 26.2 shows that 30 per cent of the farmers are operating under salinity free conditions (2 dSm^{-1}) and about 40 per cent are operating under medium level of salinity. The severity of salinity varies from plot to plot and also within the plot. The pH values of soil in the block were in the range of 7-9 that indicates that there is no alkalinity and sodic problem. The high pH values were recorded in the more water logged areas. The pH values of the major irrigation schemes, which are located in the dry zone, were found to be around 8 (Amarasiri, 1975). It has been reported that the surface water logging in Nochchiyagama could have been caused by gradual increase in water table since the inception of Mahaweli project. This may be the reason for high salinity in Nochchiyagama block.

In the Mahaweli system H, salinity is common in the following sites (De Silva, 1987).

1. Point at which the slope changes from convex to concave or vise versa, where seepage flow tends to emerge out.
2. Poorly drained sites with inadequate supply of water to maintain a layer of saline water.
3. Old tank beds and other depressions with insufficient drainage.
4. Heaps and elevations located within the capillary rise reachable stagnant water pools.
5. Fields wetted by capillary rise from canal seepage.
6. None irrigated fields that are receiving seepage or capillary rise from adjacent irrigated fields.

Crop Yield and Salinity

Salinity build up directly influences the crop yields that could be found out by Cobb-Douglas form of production function. According to Pathmanathan (1999), salinity values up to the threshold level, paddy production will not be affected. Beyond the threshold limit salinity level shows negative effect on yield. About 61 per cent of the variation in paddy yield is determined by many factors including salinity (Pathmanathan, 1999). The expected negative production elasticity of soil salinity was the important determinant of yield, which dominated over fertilizer, capital and labor inputs and also soil salinity coefficient was very much significant both magnitude wise and statistical wise at 1 per cent (Pathmanathan, 1999). This indicates that a 1 per cent increment in soil electrical conductivity reduces the yield by 0.8 per cent in Nochchiyagama block in Mahaweli system H.

According to Thiruchelvam and Pathmarajah (2003) in salinity-affected areas, the soil salinity is the principal factor that determines rice production. In moderately saline areas, the yield loss ranged from 10 per cent -15 per cent ; in high and severe soil salinity areas, yield was reduced by about one third. Therefore, it is important to identify such areas in the irrigation projects and reclaim the soil from permanent damage.

Salinity has reduced rice productivity by 33 per cent in 7 per cent of the areas away and about 1 per cent in the areas close to the main reservoir in the Mahaweli system H. About 55 per cent of the productivity loss in medium and high salinity areas was due to soil salinity, indicating the need for technological and management interventions to prevent the soil salinity damages. Drainage improvements are economically feasible in medium and high salinity affected areas and is highly uneconomical in severe soil salinity areas (Thiruchelvam and Pathmarajah, 1999).

Drainage and Impact on Yield

According to Table 26.2, about 30 per cent of interviewed farmers are affected by high soil salinity but only few farmers on this group uses considerable amount of labour to improve the drainage facilities through the techniques of cleaning and deepening the drainage channels. Table 26.3 shows that nine farmers used less than five labourers for drainage improvement while majority of affected farmers (18) used 5-10 labourers for drainage improvement. Even though this number of labourers for drainage action is insufficient, it gave 5 per cent yield increase than those who used less than five labourers. But the seven farmers who used more than 10 labourers for drainage improvement received about 12.5 per cent yield increase than those who used less than 5 labourers. Thus cleaning and deepening of drainage channels helped to get an advanced yield increase of about 5-12.5 per cent (Pathmanathan, 1999). This reveals that more labourer investment for drainage improvement may receive more advantages to the farmer at the present level of technology available in the area. A few number of farmers are using the paddy straw as organic matters to improve the drainage, but the amount applied is much below the required rate which is essential to show any significant effect on salinity (Pathmanathan, 1999).

Salinity Management Techniques for the Dry Zone

Maintenance of irrigation infrastructures could play a vital role in salinity management. This leads to reduce seepage and unexpected rise of the water table. Conservation of water in the catchments and rain in irrigated areas also help to reduce the salinity development.

Table 26.3: Relationship Between Labourer Employment in Drainage Improvement and Paddy Yield

No Labor Used for drainage	*Number of Farms Affected by Salinity*	*Average Yield (kg ha^{-1})*
<5	9 (8 per cent)	3520
5-10	18 (16 per cent)	3082
>10	7 (6 per cent)	3931

Source: Economic and environmental assessment of land degradation- caused by soil salinity in Nochiyagama block Mahaweli H, Pathmanathan, 1999.

Proper drainage is a good management to control salinity development. Construction and maintenance of new and existing drainage facilities to remove the excess water quickly from the fields will be very important. Reuse of drainage water should only be done with great care and precaution.

Good agronomic practices such as introduction of different crops with crop rotations, use of less-water demanding crops, use of more drought and salt tolerant ones, irrigate according to crop water requirement preventing excess water in the field, reduce irrigated area by using high tech irrigation methods such as drips and sprinklers, on-farm watercourse improvement and precision land leveling and application of soil amendments to improve root zone environment. Utilization of the salt affected lands for forage production will also be a good solution where the forage production could enhance livestock raring.

Policies should be introduced for water and land management with special reference to salinity development. Farmers also should be involved in water resource management and maintenance. Introduction of management information systems on land and irrigation management would ensure high-quality salinity management in dry zone in the future. Research on salinity and water quality in terms of salinity development is not much reported from Sri Lanka. Therefore, promotion of research activities on salinity, drainage and water productivity should be introduced and funded by the policy makers in the agricultural sector in the country.

At present, there are few areas of uncertainty that are identified in salinity development in the dry zone of Sri Lanka. The two main concerns are the present inability to quantitatively predict the effects of salinity using different quality irrigation waters on crop yields, and the infiltration rate and hydraulic conductivity of the irrigated soils. Future research studies should be involved with studies of the effect of using saline water throughout the longer period of time. Therefore physical and chemical changes occur in root zone with saline water, leaching requirement with different type of commercially valuable crops under different types of salinity level and different types of soil, and crop yield responses under different conditions of salinity should be carefully studied. In the future these findings may facilitate to increase the land productivity with lesser quantity of water supplied as economic value for good quality water increases day by day.

Conclusion

Land degradation due to salinity and water logging is primarily associated with coastal areas and irrigated lands in the dry zone of Sri Lanka. Mahaweli system H is one of the prime areas in the country, which produces considerable amounts of paddy and other field crops. Salinity development in this area shows an increasing trend. Poor drainage, undulating topography, reuse of irrigation water, mono cropping for a longer period etc. are the major factors contributing to the salinity development in the area. Introduction of policies, research on salinity and related issues, maintenance of irrigation infrastructures with proper and adequate drainage, improving water productivity, crop rotation and good agronomic practices etc. are the management practices needed to curb the salinity development. Salt affected lands could be used for alternative crop productions such a forage production in order to improve livestock raring.

References

Amarasiri, S.L. 1975. Water quality of major irrigation tanks in Sri Lanka, Tropical Agriculturalist, Vol., 129.

De Silva, A.M.L.D. 1987. Salinity problem in the low lying area of Mahaweli irrigation development scheme (System H), Project Report, PGIA, University of Peradeniya, Sri Lanka.

Dhuruvasangary, S. 1983. Assessment of drainage effects in the Mahaweli area. M.Phil Thesis. PGIA, University of Peradeniya, Sri Lanka.

Gangodawila, C.D. 1988. Environmental degradation and socio- economic impact under major irrigation development programmes in Sri Lanka - A Review Proc. of 5 ISCO Vol. I, Bangkok, Thailand.

Handawala, J. 1983. Occurrence of saline and alkaline soils in dry zone of Sri Lanka, Department of Agriculture Report.

Pathmanathan, K. 1999. Economic and environmental assessment of land degradation caused by soil salinity in Nochchiyagama Block Mahaweli H area, M.Phil. Thesis Report, Postgraduate Institute of Agriculture, University of Peradeniya, Sri Lanka.

Sikurajapathy, K., S.N. Jayawardena and M.A. Roonge. 1983. Survey of suspected salt affected soil in paddy lands in system H. A Report, Mahaweli Authority.

TAMS/USAID. 1980. Environmental assessment of accelerated Mahaweli development programme. Vol.1. New York, USA.

Thiruchelvam, S. and S. Pathmarajah. 1999. Hydrologic and economic feasibility of reducing soil salinity problems in the Mahaweli river irrigation system H Area, Sri Lanka, Proceedings of 99 International Conferences on Agricultural Engineering Beijing, China.

Thiruchelvam, S. and S. Pathmarajah. 2003. An economic analysis of salinity problems in the Mahaweli river system H Irrigation Scheme in Sri Lanka, International Development Research Centre, Ottawa, Canada.

Wijesekeras, R. 1981. The Determination of salinity in Mahaweli area. B.Sc. Dissertation, Faculty of Agriculture, University of Peradeniya, Sri Lanka.

27

Some Forage Substitutes for the State of Qatar

Yassin M. Ibrahim
Department of Biological Sciences,
College of Arts and Science, University of Qatar, Doha- 2713, Qatar
E-mail: yassin50@lycos.com

ABSTRACT

Forage production is one of the main sectors of agriculture comprising 33.39 per cent of the total agricultural land in Qatar. Alfalfa (*Medicago sativa* L.) is the main forage crop in the country, which faces low productivity during summer. Rhodes grass (*Chloris gayana* L.), a forage exhibiting inter and intra-cultivars variability in response to salt stress, is now competing with alfalfa and is replacing it in many areas. Other forages include Sudan grass (*Sorghum bicolor* var. *sudanensis*), pearl millet (*Pennisetum americanum* L.), sorghum (*Sorghum bicolor* L.), etc. Due to small holdings and low production, the country is experiencing a severe shortage in forage production especially during summer. Huge forage amount is imported daily from Saudi Arabia and other countries, and the supply is not regular. Research attempts were made to find some alternatives to help fulfil the need. They include:

1. *Forage halophytes*: Halophytic species like saltbush (*Atriplex sp.*) and fodder beet (*Beta vulgaris* L.) were studied to determine the optimal methods for their establishment and suitability as forage crops. They have been found productive in providing forage yields comparable to those of Rhodes grass. Additionally, the crops can be grown in environments which are not suitable for conventional crops, particularly in the salt affected lands. The nutritive value was found better or similar to that of the known forages.

2. *Native perennial grasses*: Two native perennial grasses *viz. Panicum* (*Panicum turgidum* L.) and buffel grass (*Cenchrus ciliaris* L.) were compared with Rhodes grass. Quantity and quality of the grasses were comparable or sometimes better than Rhodes grass, and the two may successfully replace Rhodes grass.
3. *Annual forages*: Experiments were conducted on annual forage legumes (*Hyacinth bean, Lablab purpureous* L., guar (*cluster bean*), *Cyamopsis tetragonoloba* L., cowpea, *Vigna unguniculata* L., *pigeonpea, Cajanus cajan* L.) and annual forage grasses (*Sorghum spp.*). The results revealed that most of the plants investigated have the potential to grow under saline conditions and at times where the shortage is at maximum during summer.
4. *Soil-less forage production*: Recently, growing barley (*Hordeom vulgare* L.) as a soil-less forage in a green hut was started in one location north of Qatar. The production was promising and the experiment is under investigation.

Introduction

Drought and salinity in both dry land and irrigated agriculture settings are major factors in limiting agricultural productivity. Attempts to generate plant varieties with improved salinity or drought tolerance using selection based breeding strategies have proved largely unsuccessful. The dry climate of the state of Qatar limits agricultural production in certain areas. The main constraints are the drought and salinity in both water and soil. Rangelands are very weak to support the feed requirement of increasing livestock numbers, and the main demand is on irrigated forage production.

The high heat and salinity tolerance of certain types of shrubs could be important for rangeland rehabilitation and provide alternative fodder at hard times of the year especially at arid and semi-arid areas. Substantial progress has been made in evaluating *Atriplex* species for their potential use as crop plants (Ibrahim, 1998; Nadaf *et al.*, 1999; Le Houerou, 2000; Walker, 2003; Malcolm, 2004). Ibrahim *et al.* (2004) added that they will recycle some accumulated salt, redistribute some of it and provide a long sustainable profitable land use. *Atriplex nummularia* found to be the best fodder plant as it can grow to over four meters high in around one year and with a good re-growth if not heavily grazed (Ibrahim and Maroff, 2003).

Fodder beet, being one of the chenopodes, grow very well in saline soils. It offers the highest yield potential than any other arable fodder crop (Nadaf *et al.*, 1998). The roots have an excellent feed quality and they are very palatable to ruminant stock (Nadaf *et al.*, 1998; Niazi *et al.*, 2000). Native perennial forages fit in perfectly with adding value to their stubbles and dry pasture, direct or hand feeding and ameliorating the adverse effects of salinity, erosion, silt, salt, and nutrient flow (Malcolm, 2003). Buffel grass and *Panicum* were found to be useful warm-season perennials which are excellent in semi-arid areas and having good forage characteristics that compensate for the forage shortage season. Annual summer forage legumes have a good potential in Qatar as proved by previous investigations. Since, warm-season grasses have low quality, production of these legumes is often a direct complement as indicated by Loi *et al.* (2005).

Summer annual forage grasses are normally used as emergency forage as shown in previous investigations. Mclaughlin *et al.* (2004) reported that annual forage grasses are characterized by rapid growth, excellent drought and salinity resistance. Netondo *et al.* (2004) stated that members of sorghum genus have fine fibrous secondary roots and tillers, giving them drought tolerance characteristics. Forage barley is a class by itself due to the quality forage, rather than on tonnage alone. Huang *et al.* (2006) noted its suitability as green-feed, grains, part of the animal concentrates or as seedlings in a soil-less culture.

The traditional forages in the State of Qatar (Alfalfa and Rhodes grass) are experiencing many difficulties in production, persistency and management. This is accompanied by the deficit in forage production. Hence, the aim of this work is to find other substitutes and alternatives to fill the forage deficit gap in state of Qatar.

Materials and Methods

A set of experiments and trials throughout the past five years were conducted at different locations of Qatar. Field trials of planting *Atriplex species viz. Atriplex canescens, A. halimus, A. halimus halimus, A. lentiformis, A. leucolata, A. nummulria, A. torreyi, A. undulata* and fodder beet (*Beta vulgaris*) have been established to study their suitability as irrigated alternative forage crops. Another field experiment was conducted with native perennial forages to determine the optimal methods for their establishment and suitability as forage compared to Rhodes grass. *Panicum*, Buffel and Rhodes grasses were evaluated for productivity and quality.

Field and glass house experiments were conducted for annual forages to test their suitability as a forage crop and tolerance to salinity. Hyacinth bean, cluster bean, cowpea, pigeon pea were evaluated in the green house at different salt concentrations (2000, 4000, 6000, 8000 ppm NaCl) and at the field during summer. Annual forage grasses mainly sorghum and barley were also evaluated at both controlled and field environments. Sorghum was tested at 2000, 4000, 6000, 8000 and 10000 ppm NaCl while barley was tested at 3000, 6000, 9000 and 12000 ppm NaCl and $CaCl_2$. Recently, growing barley as a soil-less forage in a green hut was started in one location North of Qatar to test the productivity under controlled conditions.

Yield and yield components were measured at both field and laboratory while relative yield was calculated under both saline and non-saline conditions. Criteria for *Atriplex* evaluation was after Watson (1991) methods while forage quality measurements were following Walinga *et al.* (1989). The data was subjected to statistical analysis with the aid of MSTAT-C computer program.

Results and Discussions

Results of field trials have shown that halophytic species offer a variety of desirable characters as irrigated forage crops. As shown in Table 27.1, the green fodder yield of four wing saltbush in one cut was 28.93 ± 1.74 tons. The re-growth was harvested every 2-3 months, hence the trees produced 4-6 cuts in one year. From the above data, the plant yielded an average total fresh crop of 144.65 t/ha which is higher than the yield of Rhodes grass (124.65 t/ha).

Table 27.1: Green Fodder Yields (t/ha) of Saltbush (Fourwing) and Fodder Beet (Permano) as Compared to Rhodes Grass (Katambora)

	Saltbush	*Fodder-beet*	*Rhodes grass*
1. (a) Fodder yield/cut	28.93+ 1.74	–	–
(b) Number of cuts/year	4-6	–	–
2. (a) Fodder yield/year	–	142.22+18.70	–
(b) Fodder yield /year			
1-First season	–	110.88+6.99	–
11-Second season	–	97.65+16.80	–
Total of the year	–	208.53	–
(c) Fodder yield/year			
1-First harvest	–	56.88+9.40	–
11-Second harvest	–	151.02+18.68	–
Total of the year	–	207.90	–
3. (a) Fodder yield/cut	–	–	14.25+3.12
b) Number of cuts/year	–	–	8-10
Total average yield /year	144.65	186.22	124.65

Table 27.1 also revealed the yield of fodder beet. Since, the plant is harvested once, the yield was 142.22 ± 18.70 t/ha. Fodder beet, when planted twice a year in the same land yielded 110.88 ± 6.99 t/ha and 97.65 ± 16.80 t/ha, respectively. The third trial yielded 65.88 ± 9.40 t/ha for the leaves and 151.20 ±18.68 t/ha for the re-growth of the same plant (tops and roots). The total average yield of the three trials was 186.22 t/ha. This was far better than the yield of Rhodes grass (124.65 t /ha).

Studies on nutrient composition is shown in Table 27.2.The results revealed that the fresh material of saltbush contained 34.30 per cent dry matter as compared to 19.80 per cent in case of Rhodes grass which is a popular perennial forage in the area. On dry matter basis, the material had 9.40 per cent crude protein (CP), 24.00 per cent crude fiber (CF), 1.50 per cent ether extract (EE), 19.60 per cent ash and 45.50 per cent nitrogen free extract (NFE). These results were comparable to nutritive value of Rhodes grass (Table 27.2).

Investigations on the nutritive value of fodder beet (Table 27.2) indicated that the plants were nutritious and could be a good source of carbohydrates. Fodder beets were rich in starch (NFE = 62.40 per cent) and low in crude fibre (CF = 7.90 per cent) as compared to Rhodes grass (NFE = 42.80 per cent, CF = 34.00 per cent). Crude protein content (10.20 per cent) was higher than that of Rhodes grass (8.80 per cent). However, dry matter content was very low (12.80 per cent).

Table 27.2: Nutrient Composition of Saltbush (Fourwing) and Fodder Beet (Permano) as Compared to Rhodes Grass (Katanbora)

Content (% DM-basis)	*Fodder-Beet (Whole Plant)*	*Rhodes Grass (Shoot)*	*Saltbush (Bulk Sample)*
Dry Matter (%)	10.2	8.9	34.3
Crude Protein (%)	10.2	8.8	9.4
Crude Fiber (%)	7.9	34	24
Ether Extract (%)	1.3	1.8	1.5
Ash (%)	18.3	12.6	19.6
Nitrogen Free extract (%)	62.4	42.8	45.5

Table 27.3: Cumulative Dry Matter Yield (t/ha) and Mean Dry Matter per cut ($t\ ha^{-1}$) of Different Grasses

Grasses	*Dry matter yield*	*SE*
(a) Cumulative Yield ($t\ ha^{-1}$)		
Rhodes grass	25	0.9
Panicum grass	24.2	
Buffel grass	23.8	
(b) Yield/cut ($t\ ha^{-1}$)		
First cut	2.9	0.3NS
Second cut	3.6	
Third cut	3.6	
Fourth cut	3.3	
Fifth cut	2.7	
Sixth cut	3.3	
Seventh cut	2.9	
Eighth cut	2.6	

Table 27.4: Nutrient Composition of Different Grasses (per cent DM-basis)

Parameter	*Rhodes Grass*	*Panicum Grass*	*Buffel Grass*	*SE*
Dry matter (DM)	19.8	19.3	18.6	0.8NS
Organic matter (OM)	87.8	87.3	86.5	0.44*
Crude Protien (CP)	8.8	09.4	10.7	0.85NS
Crude fiber (CF)	34.0	35.0	32.3	1.00**
Ether extract (EE)	01.8	01.7	01.9	0.14NS
Ash	12.6	12.6	12.8	0.44NS
Nitrogen free extract (NFE)	42.8	41.3	41.7	0.90

NS: Not significant; *: Significant (significant at 5 per cent level);

** : Highly significant (significant atv 1 per cent level).

Table 27.3 revealed the cumulative dry matter yield and mean dry matter per cut of three perennial grasses. *Panicum* and Buffel grasses have a dry matter yield comparable to Rhodes grass (24.20, 23.80 and 25.00 t/ha. per year respectively). The nutrient composition in percent dry matter basis is shown in Table 27.4. The crude protein of both grasses was comparable to that of Rhodes grass. There was a significant difference in nitrogen free extract and organic matter between grasses while there was a highly significant difference in crude fibre. Other results discussed else where showed that chopped fodder beet produced more cow milk than whole fodder beet or Rhodes grass. Cowpea was performing better than other forage legumes up to 6000 ppm salt, while high concentrations of NaCl affected barley more than high concentrations of $CaCl_2$. American sorghum was found to tolerate salinity better than other sorghum genotypes. Barley, under soil-less controlled condition can produce fresh forage comparable to traditional production.

In conclusion, results revealed clearly that plants under study present a potential forage crops for irrigation with saline water. Forage halophytes could be grown in environment which is not suitable for agricultural crop production, particularly in the salt affected lands or with saline irrigation water. Quantity and quality of the native perennial grasses were comparable or sometimes better than Rhodes grass. Results on annual forages revealed that most of the crops investigated have the potential to grow under saline conditions and at times where the forage shortage is at maximum (during summer). Soil-less barley is under investigation and the initial results were promising. However, further investigations are required pertaining to the management for sustainable forage production through special programs and / or projects.

References

Annual book of Agricultural Statistics. 2003. Department of Agriculture and Water Research. Ministry of Municipal Affairs and Agriculture. Doha, Qatar.

Huang, Y., G. Zhang, F. Wu, J. Chen and M. Zhou. 2006. Differences in physiological traits among salt-stressed barley genotypes. Comm. In Soil Sci. and Plant. 37: 557–570.

Ibrahim, Y. M. 1998. Salt tolerance of *Atriplex* during germination and early growth. Agric. Sc. J. 3: 55–58.

Ibrahim, Y. M. and Maroff, A. 2003. *Atriplex species*: A promising forage bush tree in Qatar. J. of Municipal and Agric. 44–45.

Ibrahim, Y., M. A. Maroff, M. G. Wafi, H. Al-Mary and M. Al-Shahwani. 2004. Comparative studies of three Atriplex species under Qatar University farm conditions. 5th. Scientific Workshop of Agricultural and Water Research Department. Ministry of Municipal Affairs and Agriculture. 10–12 /5 /2004. Doha, Qatar. The Quarterly J. of Agric. And Water Res. 4: 18–28.

Le Houerou, H. N. 2000. Utilization of fodder trees and shrubs in the arid and semi-arid zones of West Asia and North Africa. Arid land res. Mang. J. 14: 135–139.

Loi, A., J. G. Howieson, B. J. Nutt and S. J. Carr. 2005. A second generation of annual pasture legumes and their potential for inclusion in Mediterranean–type farming systems. Aust. J. of Exp. Agric. 45: 289–299.

Malcolm, C. 2003. Draining away a resource. Saltlist. Saltlist@senet.com.au

Malcolm, C. 2004. MBLS to Tackle. Salinity. Saltlist@ozonnet19.ozemail.com.au.

McLaughlin, M. R., T. E. Fairbrother and O. E. Rowe. 2004. Forage yield and nutrient uptake of warm season annual grasses in a swine effluent spray field. Argon. J. 96: 1516–1522.

Nadaf, S. K., Y. M. Ibrahim, M. Akhtar, M. G. ElHag and A. H. Al-Lawati. 1998. Performance of fodder beet in Oman. Annals of Arid Zone J. 37:377–382.

Nadaf, S., S. A. Al-khamisi, M. G. Al-haj, A. H. Al-lawati and Y. M. Ibrahim. 1999. Productivity of Saltbush (*Atriplex*) species under saline soil and water conditions in Oman. Agric. Sic. J. 4: 41–46.

Netondo, G. H., J. C. Onyango and E. Beck. 2004. Sorghum and salinity: 1. Response of growth, water relations and ion accumulation to Nacl salinity.

Niazi, B. H., J. Rozema, R. A. Broekman and Salim. 2000. Dynamics of growth and water relations of fodder beet and sea beet in response to salinity. J. Argon. And Crop Sci. 183: 101–110

Walinga, l., W. Vanvark, V. J. Houba and Vanderlee. 1989. Soil and plant analysis. Wageningen Agricultural University. Wageningen, Netherland.

Walker, M. 2003. Cropping halophytes to remove salt. Saltlist. Saltlist@oznet19.ozemzil.com.au.

Watson, C. 1991. Field evaluation of *Atriplex species* using saline drainage water. Plantsalinity Res. Proc. New Challenges. Agadir, Morocco.